Ma Zongjin, Fu Zhengxiang, Zhang Yingzhen,
Wang Chengmin, Zhang Guomin, Liu Defu

Earthquake Prediction

Nine Major Earthquakes in China (1966—1976)

Seismological Press Beijing
Springer–Verlag Berlin Heidelberg New York
London Paris Tokyo

Ma Zongjin, Fu Zhengxiang, Zhang Yingzhen,
Wang Chengmin, Zhang Guomin, Liu Defu

Center for Analysis and Prediction
State Seismological Bureau
63 Fuxing Avenue, Beijing, China

Published by Seismological Press Beijing
Distribution rights throughout the world, excluding The People's Republic of China, granted to Springer–Verlag Berlin Heidelberg New York London Paris Tokyo

ISBN–13: 978–3–642–64769–7 e–ISBN–13: 978–3–642–61269–5
DOI: 10.1007/978–3–642–61269–5

Softcover reprint of the hardcover 1st edition 1990

2132 / 3140–543210
Seismological Press Book No. ISBN–13: 978–3–642–64769–7/ P.83

Authors' Remarks of English Edition

The Chinese edition of this book has been widely appreciated among the Chinese seismological and earth science communities, of which more than six thousand copies were sold out within six months, and has been awarded the first prize for the nationwide science and technology books by the Publishers Association of China. In order to promote the international exchanges and cooperations in earthquake studies, now the English edition of this book has been brought out.

The nine large earthquakes of magnitude over seven occurred in succession in the North China and Yunnan–Sichuan regions during the period from 1966 to 1976. The higher frequency, the greater magnitude and the more disastrous losses of these earthquakes are rarely seen in the history of the human world. During this period, the Chinese seismological workers carried out a great number of observational investigations and research work. Contained in this book are research results on the basic geological and geophysical data in various seismic regions, especially on the data of the geophysical, geochemical, geodetic, environmental and animal anomalies before and after these earthquakes. Based on the comparison and analyses, some theoretical aspects have been dealt with for the continental seismicity and earthquake prediction, with the idea and working procedure for the comprehensive step–by–step earthquake prediction proposed. Owing to the complexity and time–space limitations of earthquake prediction, what we have recommended and brought forth shoud stand the new tests afterwards. The accomplishment of this book is what the scientists ought to describe, ponder on and excite after the human beings suffering from hundreds of disasters.

Mr. Gu Ping from the Division of Foreign Affairs, State Seismological Bureau, has undertaken the heavy translation of the whole book into English. Dr. John Beavan from the Lamont–Doherty Geological Observatory, Columbia University, was kindly requested to revise and approve the English version of Chapters from one through nine, and Mrs. Qian Jingyang from the Geological Institute of State Seismological Bureau, was invited to check the English version of Chapters from ten through fourteen. Ms. Li Jing, from Center for Analysis and Prediction, State Seismological Bureau has kindly typed the complete English manuscript of the book, and Mr. Sun Tong, from the same Center, has drawn all the figures and tables included in the book. We appreciate very much the joint efforts made by the Seimological Press and the Springer for publishing this English edition. Mrs. Yao Jialiu from the Seismological Press and Mr. Engel from the Springer–Verlag have done the editorial work of this edition. we would like to take this opportunity to express our sincere thanks to the people mentioned above for their hard work which made the publishing of this edition possible.

Professor Ma Zongjin
Deputy Director
Center for Analysis and Prediction State Seismlogical Bureau

Preface

Floods, earthquakes, hurricanes and typhoons are the major natural disasters in the world, among which the least understood are seismogenic processes. With highly concentrated populations and rapidly developing economy, social requirements for earthquake prediction are becoming more and more pressing.

The seismic region in Mainland China is situated at the eastern edge of the Eurasian seismic system, and is the largest intracontinental region of shallow strong earthquakes in the world, as a result of the collision of the Indian Ocean plate and the underthrusting of the west Pacific Ocean plate. Since the beginning of this century,there have been on average two strong earthquakes of M over 7 every 3 years,of which at least one has caused catastrophic losses.From 1966 to 1976,China was struck by successive large earthquakes, of which nine with M over 7 (that is, the 1966 M7.2 Xingtai earthquake;the 1969 M7.4 Bohai earthquake; the 1970 M7.7 Tonghai earthquake;the 1973 M7.9 Luhuo earthquake; the 1974 M7.1 Yongshan earthquake; the 1975 M7.3 Haichang earthquake; the 1976 M7.4 Longling earthquake; the 1976 M7.8 Tangshan earthquake; the 1976 M7.2 Songpan earthquake) occurred in the densely populated areas east of longitude 98 ° E, causing very severe casualties and damage. According to statistics, about 300,000 people have been killed by these earthquakes. These seismic disasters aroused great concern in the Party, the government and the people, who required seismological workers to immediately carry out research on earthquake prediction.

At that time in China seismological sciences were rather backward, and the basis for earthquake prediction was even weaker, severe earthquakes however occurred in rapid succession. Such a situation impelled us to begin with field observations, to undertake research on an interdisciplinary basis, and to make concurrent observations by various means, so as to try out different techniques and methods for earthquake prediction, and put them into practice as soon as possible. However, there are evident problems with such an approach, as work was done in a hurry in many directions. During those years, emphasis was placed on descriptions of superficial phenomena of earthquake anomalies and the summarization of events associated with earthquakes, so very often various earthquake emergencies were dealt with on an empirical and statistical basis. Some theoretical hypotheses and explorations were made based on the features of continental earthquakes, but were not followed up by further profound investigations and basic research on seismology was almost neglected. Today, after a certain part of the road toward earthquake prediction has been covered in haste, in reviewing the previous tortuous and difficult history of earthquake prediction practice, people will remember not only the successful prediction of the Haicheng earthquake, but also keep in mind the grievous lessons of failure to predict the Tangshan earthquake. Therefore, in the days to come, it will be necessary to synthesize these experiences and lessons, to sort out large quantities of accumulated data, to analyze the problems existing in the present research work and to strengthen as soon as possible the corresponding areas of theoretical research, so as to improve the art of earthquake prediction. However, a very long period of painstaking work will be needed.

In order to promote such research, we have compiled this book in accordance with the principal accomplishments of research in recent years. We have tried to describe, as objectively as possible, this historic period in the development of earthquake prediction and have done our best to sort out and scientifically interpret some data of sufficient merit, have listed some precursory phenomena which are currently difficult to explain but may be of a certain value, and recommended some important subjects that are in need of further research. There have been various specific studies and final reports on the details of each of the nine large earthquakes. Therefore, in this book, besides depicting the main individual characteristics for each of the earthquakes, we have concentrated on analyzing and synthesizing their common features and have made preliminary discussions both on some theoretical problems of earthquake prediction and on the principles and methods of prediction work.

Altogether there are 14 chapters in the book. The first chapter reviews the history of development in earthquake prediction, engendered by the nine large earthquakes, and the presentation of various scientific problems and conceptions for earthquake prediction; the second through eighth chapters describe in proper order the basic parameters, the seismotectonic environments, seismicity patterns, geophysical field anomalies, crustal deformation, stress variations, anomalous changes of groundwater level, anomalous changes in the atmosphere, and so on, pertinent to the nine large earthquakes; the ninth chapter analyzes the anomalous indicators of a large earthquake within an extensive region; the tenth chapter deals with the synthetic conclusions and comparative studies of the time–space distributions of the above–mentioned anomalies; the eleventh chapter discusses the focal mechanisms of tectonic earthquakes; the twelfth and thirteenth chapters explore problems related to earthquake fields and seismic episodes, and the last one summarizes the fundamental understandings so far obtained for Chinese continental earthquakes, and the designs formulated for earthquake prediction research.

For earthquake prediction research during the past decade or so, in line with the instruction issued by the late Premier Zhou Enlai ("there have been numerous records of earthquake disasters preserved in ancient China, but the experiences are insufficient. It is hoped that you can summarize such experiences and will be able to solve this problem during this generation"), Chinese scientific workers have conducted a great number of studies of the nine severe earthquakes. The basic data for this book have been provided by the national seismological system and other departments concerned, and we have drawn on research results obtained both at home and abroad. The authors for each chapter are as follws: Professor Ma Zongjin for the first; Fu Zhengxiang for the second; Ma Zongjin for the third; Fu Zhengxiang for the fourth; Zhang Guomin for the fifth; Zhang Yingzhen for the sixth; Wang Chengmin for the seventh; Liu Defu for the eighth;

Wang Chengmin and Zhang Yingzhen for the ninth; Zhang Guomin and Fu Zhengxiang for the tenth; Fu Zhengxiang for the eleventh; Zhang Yingzhen for the twelfth; Ma Zongjin for the thirteenth; Ma Zongjin and Wang Zhengmin for the fourteenth. Ma Zongjin is the editor–in–chief for this book and Fu Zhengxiang and Zhang Yingzhen are managing editors responsible for editorial work.

Any criticism and corrections will be greatly appreciated, since there are certainly some shortcomings and mistakes which could hardly be avoided owing to the authors' limited knowledge.

Contents

CHAPTER 1

Introduction —— History of Earthquake Prediction Work and Subject Matters for Prediction Research

Since the birth of New China in 1949, because of the needs of large–scale construction work,forecasting of seismic intensity, that is, research on long–term earthquake prediction, has been carried out nation–wide, and some observations, investigations and studies on seismology and seismogeology have been conducted accordingly. When the national scientific program was in the making in 1956, Professor Fu Chengyi took part in the formulation of the research program for earthquake prediction. Since then, earthquake prediction, that is research on short–term prediction, has been taken up into the national program. However, no large earthquakes occurred in the densely populated areas of China until the Xingtai earthquake of 1966. Therefore, not much observational and research work had been performed on earthquake prediction until then. But since the Xingtai earthquake, prediction studies have been carried out on a vast scale. In the following, we will, in order of occurrence of nine large earthquakes, outline briefly the situation of each of them, and review the history of research work on earthquake prediction and some related matters in China. The principal subject matters for earthquake prediction studies are also summarized.

The Xingtai earthquake of March 1966 occurred in the densely populated North China Plain (central Hebei Province), causing some 8000 deaths and great economic losses. The problem of how to develop earthquake prediction research was raised immediately after that earthquake. In the light of "going down to the spot for investigation, and catching hold of the Xingtai earthquake firmly for research" as instructed by the late Premier Zhou, the State Science and Technology Commission rapidly organized a great number of professionals from Academia Sinica, the Ministry of Geology, the Ministry of Petroleum, the State Bureau of Surveying and Cartography and other related universities and colleges to visit the quake–stricken area for investigations, and to carry out field observations and prediction research as best they could. Various observations of anomalies on seismicity, gravity, geomagnetism, telluric current, crustal deformation, stress, underground water, atmospheric and animal behaviour were thus started in the epicentral region and its surrounding areas. Meanwhile, the seismogenic environments, like surface seismogeology, deep structure and astronomy were also studied and investigated. Earthquake prediction research and practice in China have entered thus a new stage whose working basis is the observation of numerous precursors. There was obvious foreshock activity in the Xingtai earthquake sequence, with the first strong one of M6.8 occurring 14 days before the mainshock of M7.2. The sequence contained 39 earthquakes with M over 5, with small and moderate events continuing to the time of writing. The spatial distributions of small and large shocks were also complicated. In the sequence, there were very often phenomena of "concentration–quiescence–mainshock" (moderately strong or strong quakes) and of "quakes jumping between two ends" within the aftershock area. The small and large earthquakes of the sequence and the crustal deformation before and after the mainshock seemed to be confined within a graben with recent intense fault depression. The results of deep geophysical prospecting show that the graben is situated just at the slope of a local mantle upheaval. The investigations and interviews indicated that anomalies of groundwater level, water colour, and the apparent unusual behaviour of cattle, poultry and other animals, had appeared before the Xingtai earthquake. In addition, earth light and sound, and electromagnetic wave anomalies were also reported. All these macroscopic anomalous phenomena were observed one or two days before the mainshock. All in all, the Xingtai earthquake took place in a particular tectonic environment, with a complicated characteristic sequence, and with many macroscopic anomalies observed prior to the mainshock. These typical features might force us to consider how to predict the mainshock of M7.2 which occurred 14 days after the first strong shock of M6.8, how to judge the type of such a seismic sequence, and whether the macroscopic anomalous phenomena which appeared one or two days before the mainshock could be regarded as precursors for the prediction of an imminent earthquake.

After the Xingtai earthquake, the Hejian earthquake occurred in March 1967 in Cangzhou, 160 km northeast of Xingtai. The earthquake was felt clearly in the Beijing and Tianjin regions. After that, many observation stations with various sensors were established in the Beijing–Tianjin region. As a result, some data were obtained before the Bohai earthquake of July 1969. The epicenter of the Bohai earthquake was about 200 km from Hejian in a notheasterly direction. For 3 years and more after the Xingtai earthquake, there were successive occurrences of strong shocks with M over 6 in the northern part of North China, and the epicenters tended to migrate gradually northeastward. By examining all the Bohai earthquake characteristics, it was found that there was a mid–term anomalous process with a period of 6 months to 2 years for various observed anomalies of seismicity, geomagnetism, telluric current and so on before the Bohai mainshock. The small earthquakes in the Beijing–Tianjin region were distributed in a WNW–ESE orientation from Huailai, northwest of Beijing, to the Bohai epicentral area, with minor quakes concentrated in strip–like patterns. This strip–like minor quake zone continued to extend ESE during and after the Bohai earthquake, cutting the NNE strip–like zone which appeared at the same time. One or two days before the Bohai earthquake, workers in the Tianjin Zoo found that almost 20 kinds of animals including panda, swan, yak and others behaved strangely, and moreover they reported these to the appropriate authorities 2 h before the mainshock. According to the investigations, numerous aquatic animals in Bohai waters behaved unusually, such as fish shoals disbanding and fleeing in the epicentral area. The characteristics of various anomalies observed before the Bohai earthquake alerted seismological workers to the continuous northward migration of strong earthquakes and led to the proposal that seismological work should be intensified in the southern part of Northeast China. Meanwhile, the idea of mid–term prediction, supported by the data of the Bohai earthquake, spurred us to consider the whole seismogenic process as divided into several different stages. We began to study the relationships between seismic activities and active faults, which suggested the important role played by the WNW fracture zone in the crustal tectonic movements and seismic activities in North China. In addition, biological researchers paid attention to abnormal animal behaviour stimulated by possible high–frequency oscillations or electrical currents occurring prior to earthquakes.

After the Bohai earthquake, the central seismological working group headed by Professor Li Siguang was set up, responsible for organizing the Seismological Office of the Beijing–Tianjin Region.

In early 1970, the M 7.7 Tonghai earthquake occurred in Yunnan Province. As the Sichuan–Yunnan Region is the seismically most active region in Mainland China, earthquake observations and studies were started in the mid 1960′s and it was predicted that the Tonghai area, south Yunnan region, would be a seismic risk area, though the detailed occurrence time could not be given. At that time, since there were no stations for monitoring precursors within several hundred kilometers of this area, no preseismic data were obtained. However, the post–earthquake survey results demonstrated that there were salient anomalies in water level, water quality, water colour, the behaviour of various animals and earthquake lights precursory of the imminent mainshock. This further deepened our impression that macroscopic anomalous phenomena occurred prior to large earthquakes. The Tonghai earthquake featured notable strike–slip faulting, with bedrock dislocations of tens of kilometers in length exposed at the surface. This implied a direct relation between the earthquake and the active fault.

Fifty days after the Tonghai earthquake, about 750 km to the north, the M6.3 Dayi earthquake occurred in Sichuan Province. On December 3 of the same year, an earthquake of M5.5 took place in Xiji county, Ningxia Hui Autonomous Region, about 650 km to the north of Dayi. Though the magnitude of the earthquake was not so great, more than 100 people were killed due to poor local construction of houses. Within a few years after the Xingtai earthquake, earthquakes happened successively in North China, South China (the M6.4 Yangjiang quake in Guangdong Province on 26 July 1969), Southwest China, and Northwest China, which, in consequence, led to three important events in Chinese seismology: (1) Establishment of the State Seismological Bureau, as the State Council in 1970 decided that a national agency be set up responsible for a unified administration of work on earthquake monitoring, prediction and research.(2) Determination of the policy, principles and methods in seismological work. At the National Seismological Conference held in early 1972, the working policy "under the centralized leadership of the Party, to give priority to precautions against earthquakes, bring together professional workers with the masses, use both indigenous and foreign methods, and for all to join forces to wage battle" was set forth in accordance with a series of instructions put forward by Premier Zhou. Meanwhile, an experimental working procedure for prediction was formally proposed for major earthquakes of M over 7, based on a synthesis of work done over the past several years, and on experience gained from earthquake predictions both in China and abroad. Theoretical studies were thus further promoted on the division of the seismogenic process into stages, as were also laboratory experiments.(3) Establishment of the system for an annual National Consultative Meeting on the Earthquake Situation. Since the mid–period precursory anomalies continued for one or two years before a major earthquake, the state of nationwide seismic activity very often varied within a single year, and seismic activity in different regions seemed to be correlated, it was decided at the Seismological Symposium held in Linfen, Shanxi Province in December 1972, that an Annual National Consultative Meeting on the Earthquake Situation be founded. The purpose of the meeting would be to pass review on the nationwide earthquake situation in the coming one or two years, and by collective discussion define areas for more intensive study and coordinate arrangements for the coming fiscal year on a national scale. Also at the Linfen Symposium, preliminary results of studying the general earthquake situation were reported, relationships between the seismicity in China and the world discussed, and the idea was suggested that earthquake activities in China were now at a "high tide" period.

In February 1973, a severe shock occurred in Luhuo, Sichuan Province. Some unique features of the Luhuo earthquake stimulated us to study new problems. For example, the long axis of the meizoseismal area with a seismic intensity of X degrees of the Luhuo earthquake, about 40 km long, overlapped most of the meizoseismal area of IX degrees of the 1968 M6.8 Zhuwo earthquake, and closely linked up the epicentral area of X degrees of the 1923 M7.5 Xialatuo earthquake. Said facts implied the non–strict " in–situ recurrence" of strong quakes, and broke with the old conception that "no major earthquakes will recurt in the same area within a short period after a strong quake has occurred". Such cases are rare in China, but they merit in–depth research on the problem of seismic energy release and reaccumulation in regions of "in–situ recurrence" of strong shocks. In addition, as the Luhuo earthquake took place in the some 300 km long Xianshuihe fracture zone, a surface fracture 90 km long was produced during the mainshock, and the fracture continued to creep in the direction of coseismic dislocation after the mainshock. How was such a large intracontinental strike–slip earthquake generated? Where does the driving agent for intracontinental earthquakes originate? What are the differences between such earthquakes and intracontinental ones? Furthermore, before the Luhuo earthquake, in Guzan, Kangding, 300 km southwest of the epicenter, abrupt pulse–like variations of water radon content were observed. Is it possible that such variations manifest that the source region and its closely related local tectonics have entered into an unstable state before the impending mainshock?

In May 1974, the M7.1 earthquake occurred in the Yongshan mountain area, Zhaotong District, Yunnan Province. There were a few stations for precursory observation in this area. Since no evident surface breaks were caused by this earthquake, and both the azimuths of small earthquake alignments before and after the mainshock and the predominant orientations of isoseismal line extension were all northwest– and northeastward, it was difficult to judge the azimuth for the main fault plane. This phenomenon provided evidence for proving the type of conjugate source dislocation that was suggested later.

Following the Yongshan earthquake, people were on the alert for distinct anomalies occurring in North China in the vicinity of Bohai, and as a result, the Consultative Meeting on the Earthquake Situation in North China and the Bohai Region was convened in June 1974. At the meeting, two estimations were made on whether strong quakes would recur in North China in the near future. The first suggested that based on the historical records of earthquakes, the return interval of strong quakes in North China, was normally long, with a strong shock of M over 6 every 10 years on the average. Since within less than 10 years since 1966 two earthquakes of M over 7 and one with M over 6 had occurred in North China, it might not be possible for large earthquakes to occur in the coming years. The other concluded that the time distributions of strong earthquakes were not uniform and historically there existed "high tide" periods in which strong shocks happened repeatedly. Though such periods were unusual, they were not unprecedented, and furthermore some outstanding anomalous phenomena, particularly in crustal deformation and seismicity in the north Bohai region, have been observed, so attention must be paid to the possibility of a strong earthquake occurrence. Through discussions at the meeting, it was recommended that monitoring work in the Beijing–Tianjin–Tangshan– Bohai Region should be intensified. As approved by the State Council, monitoring and prediction work was also strengthened in the north Bohai area including Liaoning Province. As was described above, the consideration of possible northward migration of strong shocks in North China helped speed up earthquake observation work gradually in Liaoning Province, following the 1969 Bohai earthquake. It was noted in 1972 that the regional seismicity was increasing gradually. After the regional seismogeological survey was completed, and the mantle uplift in the Yingkou area was discovered in 1973, Jinxian, Yingkou, Liaoyang and others were demarkated as seismic risk regions. Leveling measurements at Jinxian station and seismic activities in south Liaoning Province all showed clear anomalies in 1974. These anomalies were clarified and further ascertained by on–the–spot investigations in September 1974. In October 1974, intensive dissemination of earthquake information was carried out, and thousands of temporary amateur observation points were organized. From mid– November, a variety of macroscopic anomalies were first noticed in the Dandong and Xiuyan areas, and the anomalies became evident by mid–December, so a professional group was immediately dispatched to work in the Dandong area. On 22 December 1974, more than a hundred minor earthquakes took place in the Qinwo reservoir area, Liaoyang, the largest being of M4.8, and with this M4.8 earthquake, the possibility of fairly strong earthquakes in south Liaoning was increasingly realized. However, at the National Consultative Meeting on the Earthquake Situation held in January 1975, the Jinxian–Yingkou–Dandong region was still estimated as a large–scale seismic area with no concrete place pinpointed, and it was expected that the magnitude of the forthcoming earthquake would be about 6. Up to the end of January and in early February 1975, large quantities of macroscopic anomalies concentrated in the Yingkou and Haicheng areas and, in particular, from February 1st to the early morning of February 4, about 500 small shocks occurred in the Haicheng epicentral area, with a maximum magnitude of 4.7. In such an unusual situation, the people's government of Liaoning Province decided that prediction of an imminent earthquake be issued in the Yingkou, Anshan and Jinxian areas in south Liaoning, and that at the same time special planned measures shold be taken against the coming earthquakes in this region. The success of the tentative Haicheng earthquake prediction was objectively due to a series of long–, mid–, short–term and immediate anomalous phenomena before the quake, and subjectively attributable to rigorously executed earthquake monitoring and observations, and the precautions taken against the earthquake.

In view of the abundant precursory phenomena observed before the Haicheng earthquake, numerous important subjects were proposed for research. For instance, a large number of foreshocks were monitored before the Haicheng earthquake. How can we tell foreshocks from a normal swarm? How can we apply them to prediction practice? Why did abrupt anomalies occur intermittently one or two months before the mainshock, and migrate toward the epicenter? Which source mechanism was indicated by the consistency in time–space distributions of various macro– and microscopic sudden anomalies? Were the macroscopic anomalies, such as groundwater level, temperature, earth mist and light, and so on occurring in the epicentral area, perhaps related to intense static electrical fields in the lower atmosphere, and possible severe movements of heat–carrying fluids in the earth before the impending earthquake? Moreover, the area in which the anomalies took place before the shock amounted to 500 km in length, with the anomalous points scattered non–uniformly, and very often clustered at the fault intersections (knots) of the tectonic network. How can we explain such phenomena?

On 29 May 1976, an earthquake of M7.3 occurred, and was closely followed by an earthquake of M7.4 in Longling county, Yunnan Provence. Several foreshocks, the biggest being of M5.2, happened dozens of minutes prior to the shock; hence a prediction of imminency was issued locally. However, the extent of seismic risk in the Longling area had been underestimated, because of the obsence of strong historical earthquakes, and because of the nearby Tengchong area, which was considered as a seismic area with M6 earthquakes. From the geological viewpoint, the Longling earthquake originated from an almost completely granite region, and it was generally thought that earthquakes did not easily occur in such areas. On the other hand, thanks to many stations in the East Yunnan region, many anomalies were detected, while there were few stations in the Longling area, so fewer anomalies were observed there. That was why some researchers emphasized the possibility of earthquake occurrence in the East Yunnan region prior to the Longling earthquake. The generation of the Longling shock within the granite body was really a rare case among the major earthquakes of M over 7 in Mainland China. Were there any peculiar seismogenic conditions for such a quake? Furthermore, as the Longling earthquake sequence showed the features of strong swarms, and each strong one in the sequence almost produced double meizoseismal areas, what were the seismogenic and earthquake–producing tectonic characteristics for such a quake?

The great Tangshan earthquake of M7.8 occurred on 28 July 1976, killing about 240000 people, the greatest number of earthquake casualties in this century. At the National Consultative Meeting on the Earthquake Situation in early 1976, it was estimated that an earthquake with M5 or 6 might occur in the Tangshan–Liaoxi region, and intensification of work in this region was recommended. In the May–June period, the Seismological Bureau of Hebei Province sent the Tangshan Earthquake Working Group to conduct field surveys in the Tangshan area, and several group members were killed during the earthquake. Nevertheless, why was no prediction issued as at Haicheng? the reason was that we lacked sufficient understanding of earthquake regularities generally. In addition, there might be other critical reasons for the prediction failure : (1) Mistaken assessment of the basic situation. After the Haicheng earthquake of 1975, a few workers suggested that, based on seismic and meteorological anomalies, there still existed the background for a possible large earthquake, but a majority of researchers doubted the possibility that another strong earthquake of M over 7 would occur in the coming year within several hundred kilometers of Haicheng. Therefore, although there were anomalies in the Tangshan–Liaoxi region, this region was only regarded as a background area for earthquakes of M5 or 6.2. (2) An earthquake of M6.3 occurred on April 6 in Helingeer, Inner Mongolia Autonomous Region, about 530 km west of Tangshan City, and some of the anomalies in the Beijing–Tianjin region died down after this earthquake; meanwhile, another earthquake of M4.4 took place on April 22 in Dacheng, southwest of Tianjin. Therefore, many people considered that the anomalies were associated with these two earthquakes. (3) No large quantities of abrupt anomalies, and no foreshocks, were reported in the epicentral area and its vicinity, hence the failure to expect that the earthquake could come so soon with such a great magnitude. (4)For issuing an earthquake prediction in the Beijing–Tianjin region, it was unavoidable to take the extremely serious social impact into consideration.

The Tangshan earthquake highlighted several problems: Were the differences between the Tangshan shock and the Haicheng one essential or not? Was there a unified seismogenic model? Did different types of earthquakes exist? Were the Tangshan and Haicheng earthquakes maturing separately or simultaneously, with the inter–event distance of 350 km, the interval of 18 months, and the overlapping of precursory anomaly areas? How could precursory anomalies for successive strong shocks be identified? In addition, as only small surface fractures occurred in Tangshan, differential neotectonic movements were not so salient, and no great earthquakes of M over 7 had ever been recorded historically, the seismic intensity zoning in Tangshan was defined as an area of VI degrees. The M7.8 Tangshan earthquake belied the above–said seismic zoning principles, as did the Haicheng and Longling earthquakes. What principles could we use to determine in advance what would be a seismically strong area of high intensity?

With the initial success in the Haicheng earthquake prediction, people became optimistic as regards earthquake prediction. With the failure to predict the Tangshan earthquake, we had to realize that the diversity and complexity of preseismic anomalous phenomena could by no means be explained by one seismogenic model, and were thus forced to take into account various combined precursory characteristics, explore different types of earthquakes, and pursue seismogenic models based on actual observational data.

The M7.2 earthquake occurred in Songpan–Pingwu, Sichuan Province on 16

August 1976. Similar to the process of predicting the Haicheng earthquake, this earthquake prediction took 4 to 5 years of research work and interpretation. In January 1976, at the National Consultative Meeting on the Earthquake Situation, the seismic risk in the boundary area of Sichuan, Qinghai, and Gansu Provinces was announced, especially the possibility of an earthquake with M6 in the Songpan and Maowen areas. Some abrupt macroscopic anomalies appeared in the Dayi, Maowen, Qionglai and other areas in early March, and the anomalies in the Dayi area became more salient in mid–June, so it was declared that an earthquake of about M6 might occur in the central south segment of Longmen Mountain in the coming one or two months. Up to mid–July, no strong earthquakes occurred at all, only a small earthquake of M3.7, and the macroscopic anomalies also died down. The macroscopic anomalies increased again in mid–July, but without large shocks. A third "high tide" of macroscopic anomalies occurred in mid–August, and the anomalous places moved clearly towards the Beichuan and Jiangyou areas. The Provincial Earthquake Precaution and Prevention Headquarters and the Provincial Seismological Bureau issued an urgent cable to the districts, cities, prefectures and counties concerned, and declared a state of earthquake emergency as from August 12, and ordered to prepare for earthquake rescue and relief. The Songpan earthquake occurred on August 16. Among the various short–term and imminent precursors before the Songpan earthquake, the macroscopic anomalies were the most outstanding, occurring three times within 2 or 3 months. How can we then decide which is the last time of anomaly appearance before the shock? With the "high tide" places of macroscopic anomalies moving constantly, how can we judge the final earthquake–producing region? These are the urgent problems for the prediction of imminent earthquakes.

It is now more than 5 years since the Songpan earthquake, and no large earthquakes of M over 7 have occurred in the eastern part of Mainland China. Does this demonstrate that the episodes in which earthquakes of M over 7 occurred repeatedly from 1966 to 1976 is over? So the problem remains, how to determine the beginning and the termination of a certain seismic episode.

During the recent years, which have been basically quiet without strong earthquakes, we have been able to gain some new knowledge on the normal variations of numerous items of observation. This facilitates very much to recognize of further precursory anomalies from earthquake case studies.

From the above discussion, one can get some idea how the nine major earthquakes have led gradually to a deepening of our understanding of earthquakes, and have shown us how to practice predictions. In order to further improve the quality of earthquake prediction, we would like to recommend the following important subjects, deriving from the nine large earthquakes, for detailed research: (1) The variational morphology and the time–space distribution features of earthquake precursors; (2) The division into separate stages of the process for anomaly variations during a seismogenic period, and the mechanisms of anomaly occurrences; (3) The size of a seismogenic area and the earthquake field; (4) The phasic and episodic separations of an earthqake sequence; and (5) seismogenic physics and earthquake types. In addition, it is desirable to formulate a tentative plan and procedure for earthquake prediction, in accordance with our present knowledge of earthquake processes, and with our actual needs, and to test and refine this plan and procedure in practice. The following chapters of this book will discuss these problems in principle.

CHAPTER 2

Characteristics of Earthquake Sequences

Within 11 years, from 1966 to 1976, there were 15 sequences of large shallow earthquakes with M > 7 in China (some sequences including two earthquakes of M > 7) (see Table 2.1), of which four were in Taiwan Province. The relations between the distribution of these earthquakes and the neotectonics is shown in Fig. 2.1. The Arabic numerals in Fig. 2.1 are the serial numbers of the earthquake sequences in Table 2.1.

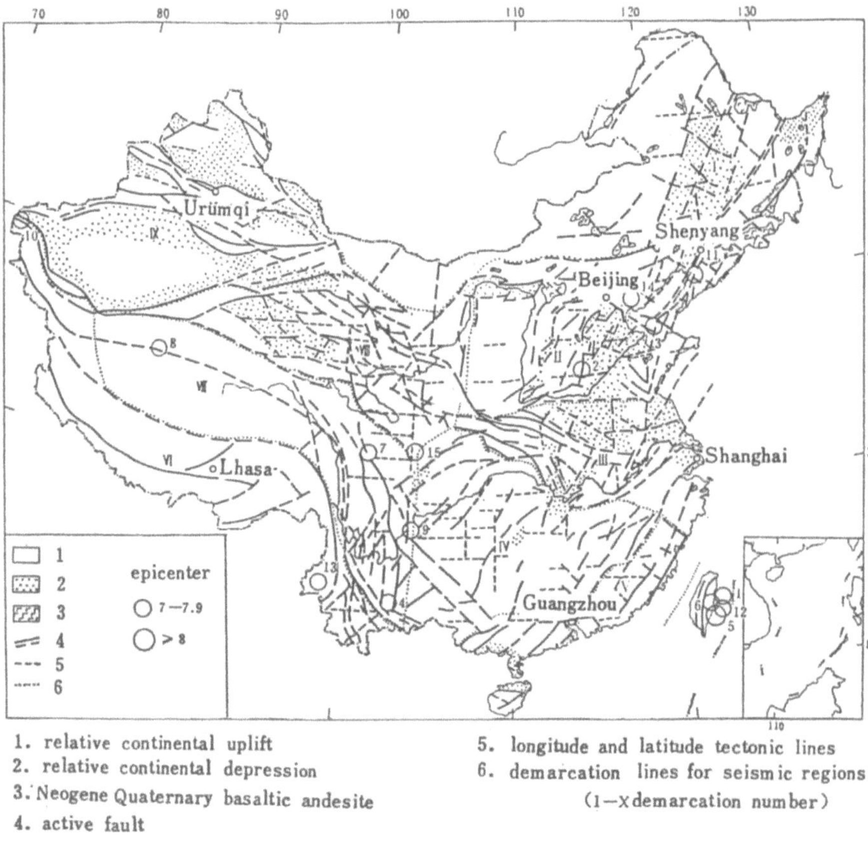

Fig. 2.1. Neotectonics of Mainland China and distribution of recent strong earthquakes

Within the first 77 years of this century, there were altogether 45 sequences of large earthquakes with M > 7 recorded in Mainland China and its coastal areas (except Taiwan Province). These great earthquakes were uneven in their temporal and spatial distribution. The temporal and spatial distribution, frequency and energy release for large shocks ($Ms \geqslant 7.0$) during the period of 1900–1976 are shown in Fig. 2.2. Figure 2.2a shows the time–space distribution of seismic activity, in which the horizontal lines separate time intervals of 11 years. It is shown that: (1) within this century, there were two periods of active seismicity, with 13 earthquakes in the period of 1944–1954, and 11 shocks from 1966 to 1976 (Fig. 2.2b); (2) comparisons of energy released by the earthquakes in different periods show that the energy released in the period of 1966–1976 was fairly low, but that nine of the earthquakes were centered in heavily populated areas. Therefore, the past ten or more years represent a period of remarkably active seismicity in China, which has aroused grave public concern. During this episode, the nine great earthquake sequences occurred mainly in North China and in the Sichuan–Yunnan region of Southwest China.

The serial numbers for these nine earthquakes are 2, 3, 4, 7, 9, 11,13,14 and 15 in Table 2.1. For convenience, they are called the nine great earthquakes in this book.

2.1 Basic Parameters of the Nine Large Earthquakes

2.1.1 Earthquake Parameters and Focal Mechanisms

The occurrence time, epicentral location, depth, magnitude and other parameters of the great earthquakes with $M \geqslant 7$ in the nine earthquake sequences are given in Table 2.1.

The focal mechanism parameters for P waves of the earthquakes with $M \geqslant 7$ are shown in Table 2.2. The nodal plane A in Table 2.2 represents the earthquake fault plane, which was deduced from the data of macroscopic investigations, deformation measurements, geometry of aftershock areas, and wave spectra.

From a comparison of the focal mechanism solutions for the first strong quakes in the nine sequences it can be seen that the ruptures feature the following: (1) Except for the Songpan quake, the nodal planes for most focal mechanism solutions have a fairly large dip (larger than 78°); that is, the fault plane is almost vertical. The fact that the angle between the azimuth of the dislocation axis (X_A) and the strike of the fault plane is less than 5° demonstrates that the dislocation force at the fault plane is nearly horizontal. Therefore, the dislocation type for these earthquakes is principally horizontal dislocation at an approximately vertical fault plane. (2) The four earthquakes in the North China region exhibited either NNE–NE right–lateral strike–slip dislocations (like the Xingtai, Bohai and Tangshan earthquakes), or, on the contrary, a NW–NNW left–lateral strike–slip dislocation (like the Haicheng earthquake). These earthquakes occurred due to dislocations under a NE–SW or nearly EW regional compression. The Tangshan and Haicheng earthquakes also have a tensional normal faulting component. The first four strong earthquakes in Southwest

Table 2.1. Catalogue of strong shallow earthquakes($M_s \geqslant 7.0$) in China from 1966–1976

Serial Number	Date (year,month,day)	Time (hour,minute,second)	Epicentral Location Longitude (E)	Epicentral Location Latitude (N)	Reference Place Name	Magnitude	Focal Depth (km)	Source of Data
1	1966.3.13	00–31–20	$122^{o}.6$	$24^{o}.1$	East Hualian, Taiwan Province	7.8	63	Compiled by SSB, 1977
2	1966.3.22	16–19–46	$115^{o}03'$	$37^{o}32'$	Xingtai, Hebei Province	7.2	9	"
3	1969.7.18	13–24–49	$119^{o}.4$	$38^{o}.2$	The Bohai Sea	7.4	35	"
4	1970.1.5	01–00–37	$102^{o}.7$	$24^{o}.0$	Tonghai, Yunnan Province	7.7	13	"
5 (1)	1972.1.25	10–06–23	$122^{o}.4$	$22^{o}.6$	Central North-East Sea of East Taiwan, Taiwan Province	8		"
(2)	1972.1.25	11–41–24	$122^{o}.1$	$23^{o}.1$	"	7.6		"
6	1972.4.24	17–57–21.8	$121^{o}.6$	$23^{o}.6$	West Dagangkou, Taiwan Province	7.3		"
7	1973.2.6	18–37–08.3	$100^{o}.4$	$31^{o}.5$	Luhuo, Sichuan Province	7.9	17	"
8	1973.7.14	12–51–21	$86^{o}.5$	$35^{o}.3$	North-West Mani, Xizang(Tibet) Autonomous Region	7.3	33	"
9	1974.5.11	03–25–18.3	$103^{o}.9$	$28^{o}.2$	Yongshan, Yunnan Province	7.1	14	"
10	1974.8.11	09–13–59.9	$73^{o}.8$	$39^{o}.4$	South Wuqia, Xinjiang Autonomous Region	7.3		"
11	1975.2.4	19–36–08	$122^{o}48'$	$40^{o}39'$	Haicheng, Liaoning Province	7.3	12	"
12	1975.3.23	15–32–04	$122^{o}36'$	$22^{o}58'$	East Sea of Xingang, Taiwan Province			
13 (1)	1976.5.29	20–23–18	$98^{o}38'$	$24^{o}22'$	Longling, Yunnan Province	7.3	20	Chen Lide et al. 1979
(2)	1976.5.29	22–00–22.5	$98^{o}45'$	$24^{o}33'$	"	7.4	20	"
14 (1)	1976.7.28	03–42–53.8	$118^{o}.2$	$39^{o}.6$	Tangshan, Hebei Province	7.8	16	Compiled by SSB, 1977
(2)	1976.7.28	18–45–31.0	$118^{o}.7$	$39^{o}.9$	"	7.1		"
15 (1)	1976.8.16	22–06–46	$104^{o}06'$	$32^{o}42'$	Songpan, Sichuan Province	7.2	15	Seismological Bureau of Sichuan Province
(2)	1976.8.23	11–30–10	$104^{o}08'$	$32^{o}30'$	"	7.2	22	"

Table 2.2. Focal mechanism solutions

Serial number	Parameters / earthquake	Nodal plane A			Nodal plane B			Axis X_A		Axis Y_B		Axis P		Axis T		Axis N		Nature of nodal plane A	Source of Data
		Strike	Dip	Inclination	Strike	Dip	Inclination	Azimuth	Elevation	Azimuth	Elevation	Azimuth	Elevation	Azimuth	Elevation	Azimuth	Elevation		
2	Xingtai	11°	SE	85°	282°	SW	86°	192°	4°	101°	5°	56°	1°	147°	6°	320°	84°	Right-lateral strike-slip	Research group of earthquake mechanism, SSB, 1973.
3	Bohai	20°	SE	80°	292°	SW	75°	202°	15°	109°	10°	246°	4°	155°	18°	349°	71°	"	"
4	Tonghai	306°	SW	87°	217°	NW	83°	307°	7°	216°	3°	351°	3°	261°	7°	102°	83°	"	Kan Rungju 1977
7	Luhuo	119°	SW	89°	209°	SE	80°	119°	10°	209°	1°	74°	7°	164°	8°	307°	80°	Left-lateral strike-slip	Research group of earthquake mechanism, SSB, 1973.
9	Yongshan	320°	NE	80°	46°	NW	85°	316°	7°	46°	12°	92°	3°	2°	11°	197°	79°	"	Su Sui 1976
11	Haicheng	290°	NE	81°	23°	SE	75°	112°	15°	20°	9°	66°	17.5°	157°	4°	100°	72.5°	"	Gu Haoding 1976
13	Longling ①	101°	NE	78°	14°	SE	72°	104°	18°	11°	12°	148°	3°	56°	21°	274°	68°	Right-lateral strike-slip	Chen Lide 1979
	②	14°	SE	45°	79°	NW	67°	349°	23°	104°	45°	141°	12°	35°	51°	241°	36°	Oblique thrust faulting	"
14	Tangshan ①	41°	SE	85°	309°	NE	70°	39°	20°	130°	6°	86°	18°	346.5°	10°	258°	69°	Right-lateral strike-slip	Qiu Qun 1976
	②	25.6°	SE	61.4°	100.3°	NE	64.3°	190.3°	25.8°	296.6°	28.6°	242°	40.3°	333.5°	1.8°	65.6°	49.7°	Oblique thrust faulting	Zhang Zhili 1980
15	Songpan ①	0°	W	63°	299°	NE	46°	29°	44°	270°	27°	63°	10°	319°	54°	160°	34°	Oblique thrust faulting	Seismological Bureau of Sichuan Province 1979
	②	325°	SW	76°	49°	SE	65°	139°	25°	235°	14°	95°	7°	189°	28°	354°	6°	Right-lateral strike-slip	"

Same serial No. as shown in Table 2.1 Axis X_A: Force axis of dislocation at nodal plane A, Axis Y_B: Force axis of dislocation at nodal plane B, Axis P: Principal compressive stress axis, Axis T: Principal tensional stress axis, Axis N: Intermediate principal compressive stress axis

China (except the Songpan earthquake) feature mainly strike–slip faulting on a WNW fault plane, with a compressional thrust faulting component. The first strong earthquake of the Songpan sequence is dominated by thrust faulting. The said two characteristics are consistent with research results on recent stress fields in Mainland China.

In Table 2.3, other parameters of the first strong shocks of each sequence are shown, including the rupture length, dislocation amplitude, stress drop, earthquake moment, and rupture velocity. These are obtained by inversion from spectral and geodetic data. It can be seen from this table that the rupture length, dislocation and moment seem to be related to the magnitude of the earthquakes; the stress drop varies within dozens of bars; and the rupture velocity varies from 1 km / sec to 6 km / sec. As these parameters are derived by different methods, further comparison and analysis cannot be made.

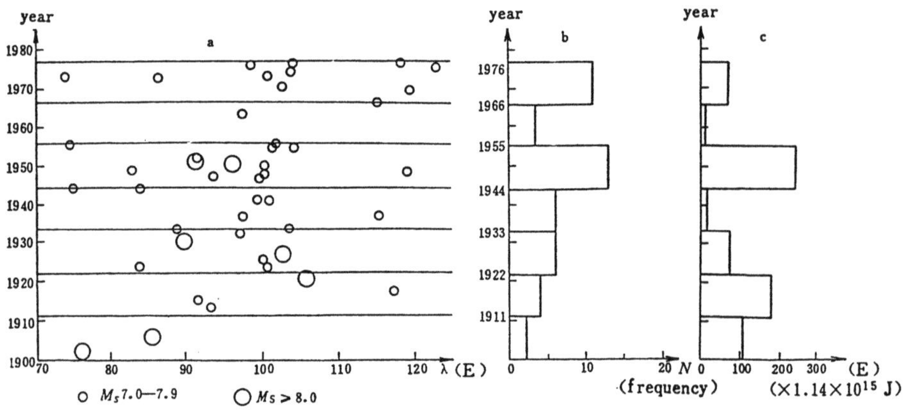

Fig. 2.2. Seismic activity in Mainland China and its offshore regions from 1900 to 1976; *a* time–space distribution of earthquakes Ms ≥7; *b* earthquake frequency; *c* seismic energy release

2.1.2 Isoseismal Contour Maps

The isoseismal contours for the nine earthquakes are shown in Fig. 2.3, in which the solid lines indicate isoseismal contours obtained from field investigations, dotted lines represent isoseismal contours for the second shocks of $M > 7$ in the Songpan and Longling sequences, and the broken lines provide inferred isoseismal contours for reference.

The shapes of the inner isoseismal contours for these earthquakes can be roughly divided into two kinds. The first is a simple ellipse (such as that for the Tonghai, Luhuo, Xingtai, Tangshan and Songpan earthquakes), and is described as "unilateral" in Table 2.3, with the orientation of the major axis close to the long axial direction of the aftershock area. The second is not regular, with two mutually overlapping ellipses (such as the Haicheng, Yongshan and Longling earthquakes), and is depicted as "bilateral" in Table 2.3. Such shapes may imply

Table 2.3. Focal parameters of first shocks with M>7 for large earthquake sequences, and their isoseismal lines

Serial No.	Parameters / Earthquake	Rupture length L km	Rupture width W km	Average dislocation U (M)			Stress drop $\Delta\sigma$ (bar)	Seismic moment Mo (dyne.cm)	Rupture velocity V (km/sec.)	Methods used	Source of Data	Inner isoseismal line	
				Strike	Dip	Total						Shape	Ratio of lenghts for major and minor axes or biaxes
2	Xingtai			1.34	0.50		42	5.9×10^{25}		S wave spectrum	Chen Yuntai 1975	Unilateral	2:1
3	Bohai	34				0.32	9.7	1.1×10^{26}	3.1	Inversed from deformation data	Research group of earthquake mechanism, SSB, 1973.		
4	Tonghai	52	22	2.24	0.14		34	8.5×10^{26}		S wave spectrum	Wang Chunrung 1978	Unilateral	10:1
		54.5(ES) 20.0(NW)	17.5	1.52	0.60		25	7.8×10^{26}	4.3(ES) 2.5(NW)	Inversed from deformation data	Research group of earthquake mechanism, SSB, 1973.		
7	Luhuo	120(NW) 15(ES)		1.71	0.34		24	1.2×10^{27}	5.8(NW) 2.8(ES)	S wave spectrum		Unilateral	8:1
9	Yongshan	66				0.07	74	3.1×10^{26}	1.74(NE)	S wave spectrum	Liu Wanqin 1979	Bilateral	2:1
11	Haicheng	54	24	1.17	0.33		22	5.2×10^{26}	1.3(SE) 1.3(NW)	Body wave spectrum	Lin Banghui 1979	Bilateral	1.5:1
13	Longling	26.2	20			0.46	18.6	0.92×10^{26}	3.4(SE)	"	Yan Fengtong 1978	Bilateral	1.5:1
14	Tangshan	84	34	4.59	0.50		29	4.3×10^{27}		Inversed from deformation data	Chen Yuntai 1979	Unilateral	2:1
		115	24			1.36	12	1.24×10^{27}	2.7	Body wave spectrum	Zhang Zhili 1980		
15	Songpan											Unilateral	2:1

that the occurrence of the earthquakes is controlled by a conjugate set of seismotectonic faults (Ma Zongjin et al. 1980c).

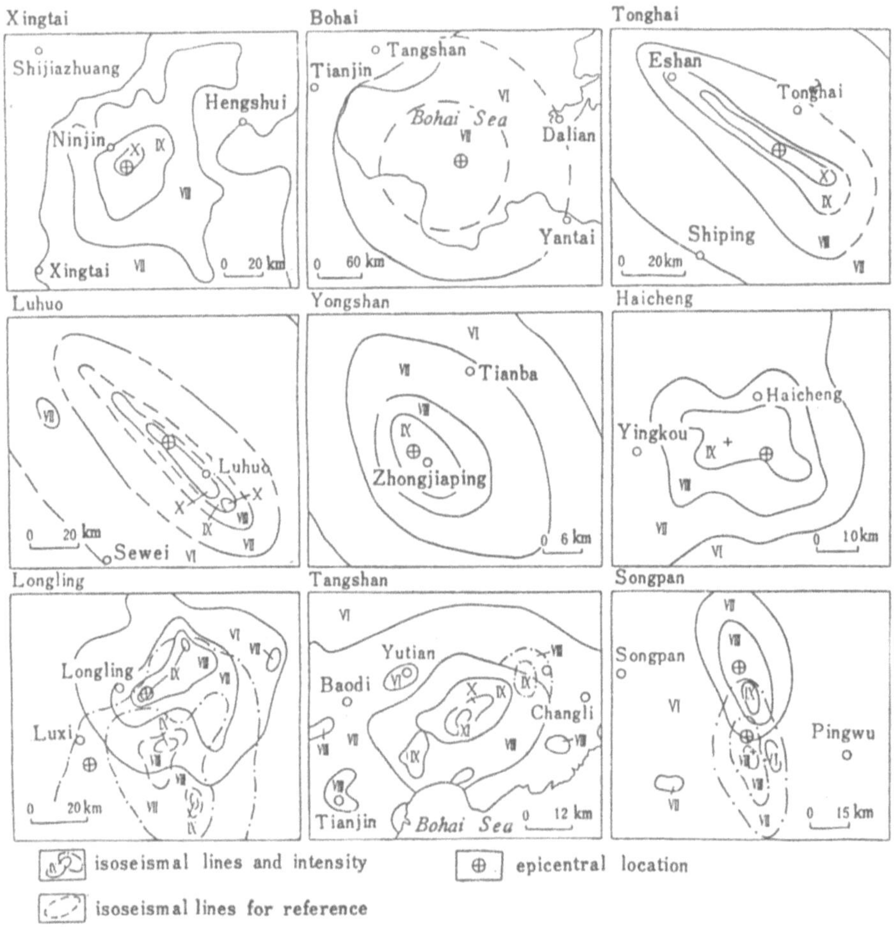

Fig. 2.3. Distributions of isoseismal contours for the nine great earthquakes

In addition, there are some anomalous distributions in the intensity maps of some of the large earthquakes. For the Tangshan earthquake, the Yutian area has a lower seismic intensity, while Tianjin is a high intensity region. The intensity anomalies seem to be associated with local geological structure, seismic wave absorption properties of the medium, and kinematic and dynamic factors due to buildings. Besides, the intensity anomalies are still dependent on detailed field investigations.

2.1.3 Surface Earthquake Cracks

Numerous ground cracks were caused by the Luhuo, Tonghai, Xingtai, Tangshan and Haicheng earthquakes. The crack lengths for the Luhuo and Tonghai quakes are 90 km and 50 km respectively, and the widths for both are 2–3 km. The crack bands consist of a series of oblique and pinnate ground cracks, and the dislocation orientations are in agreement with the focal mechanism solutions. The ground cracks of the Tangshan and Haicheng quakes are distributed widely, and many small fissures are scattered in different azimuths within a range of tens of kilometers, as well as the short crack bands in the meizoseismal area. For example, the major crack band for the Tangshan quake was only 8 km long, the surface dislocation amounting to 1.5 m, and the combined shape of the oblique cracks indicated right–lateral strike–slip faulting for the dislocation. The scale of the major rupture associated with the Haicheng earthquake was less than that for the Tangshan earthquake but the small cracks of the Haicheng quake were distributed widely, not only in the areas of recent unconsolidated sediment coverage, but also scattered in the east Liaoning mountainous area. The distribution of ground cracks associated with the Xingtai earthquake was generally similar to that of the Haicheng earthquake.

The ground cracks associated with the Songpan and Yongshan earthquakes were less developed, and could hardly be identified as tectonic cracks. After the Longling earthquake, the ground fissures were traced, but were not distinct, along the orientation of the original rupture band. The analysis and induction of different ground cracks associated with earthquakes is an important subject for research into seismotectonic classification.

2.2 Features of Aftershock Sequences

After the great earthquakes occurred, the time series, spatial distribution and severity variations of the frequent aftershock activity exhibited a somewhat stable statistical nature, providing rather significant information for studying the stress–strain state and the properties of the medium in the source regions.

Table2.4 gives some parameters characterizing the aftershock activities of the nine great earthquakes, including the length and width of the aftershock area, the predominant distribution of source depth of aftershocks, the activity level (a) and severity ratio (b) of the aftershock sequence, the attenuation coefficient (p) of aftershock sequence frequency, the maximum magnitude (M_1) of aftershock, and the duration (ΔT) of strong aftershock activity.

The following scctions discuss the characteristics of the spatial distributions, the time series, and the severity of aftershock activities for the nine great earthquakes, in an attempt to analyze the similarities and differences in these aspects of their aftershock sequences.

Table 2.4. Some parameters of aftershock activities for the nine great earthquakes

Serial number	Parameters / earthquake	Aftershock area			Magnitude of Maximum Aftershock (M_1)	Magnitude Difference between the Mainshock and the Maximum Aftershock (ΔM)	Aftershock Series			
		Length (km)	Width (km)	Depth (km)			a Value	b Value	P Value	ΔT (Day)
2	Xingtai	90	30	10–25	6.2	1.0	5.56	0.75	0.96	645
3	Bohai	60	25		5.1	2.3	5.34	0.86	0.89	1
4	Tonghai	85	35	4–16	5½	2.2	3.79	0.72	1.15	1
7	Luhuo	110	50		6.3	1.6	3.33	0.48	1.58	3
9	Yongshan	22	10	2–16	5.7	1.4	4.26	0.68	1.27	36
11	Haicheng	70	30	2–12	6.0	1.3	6.00	0.95	0.79	1198
13	Longling	90	45	0–20	6.8	0.6	5.48	0.76	1.45	136
14	Tangshan	145	50	6–18	7.1	0.7	7.04	0.98	0.80	288
15	Songpan	80	25		6.7	0.5	4.57	0.69	1.32	7

2.2.1 Spatial Distribution of Aftershocks

The epicentral distributions of aftershocks are shown in Fig. 2.4 for the nine great earthquake sequences. The starting magnitudes of aftershocks of these great earthquakes are either greater than 3 or 4.

1) Size of aftershock area.

The contours of the epicentral distributions of aftershocks for the nine great earthquakes can in general be simplified as elliptical areas, with the ratio (L / D) of major axis to minor axis being about 2 to 3. The major axes are nearly parallel with the orientations of the earthquake faults.

The epicentral distributions of aftershocks are mainly constrained by the principal rupture directions and earthquake –producing faults in the source regions and are affected by the inter–crossing fractures distributed in the source regions. For example, the aftershock distribution of the Xingtai earthquake was predominantly oriented in N35 ° E, and was extended northwestwardly on different scales, being closely correlated with the characteristic tectonics in North China. Again, during the aftershock process of the Yongshan earthquake, for one week after the mainshock (11 May 1975), the aftershocks were distributed in the direction of N30 ° W, in conformity with the mainshock rupture orientation and the strike of the inner isoseismal line; beginning from June 5 to 17, the distribution of aftershock activity notably turned to the direction of N40 ° E. From Fig. 2.4e it can be seen that overall distribution of the aftershocks is influenced by conjugate fracture zones in the said directions. Therefore, the shape of the Yongshan aftershock zone is irregular.

Professor Utsu, a Japanese seismologist (1961), studied the relation between the length (lkm) of the aftershock area and the magnitude of mainshock (M), obtaining the following statistical expression:

$$\log l = 0.5\,M - 1.8. \qquad (2.1)$$

The statistical equation given by Utsu (straight line) and the data of the nine great earthquakes (open circles) are shown in Fig. 2.5a.

The comparison shows that there seems to be no contradiction between the data of the nine great earthquakes and the expression given by Utsu; that is, the bigger the magnitude of mainshock, the longer the aftershock area.

2) Aftershock area and mainshock position.

Matsuzawa Takeo, a Japanese seismologist (1979), pointed out that for numerous earthquake cases, the mainshock epicenters were very often located at one end of the aftershock zone. This used to be regarded as an important regularity in the theory of great earthquake occurrence. However, the data of the nine great earthquakes show that the mainshock epicenters are mostly close to the centers of the aftershock zones, instead of being near their ends (see Fig. 2.4).

The dominant range of aftershock depth distribution for the six great earthquakes listed in Table 2.3 shows that the aftershock activity is confined to a certain crustal depth range. Meanwhile, it can be seen that the depths for five of

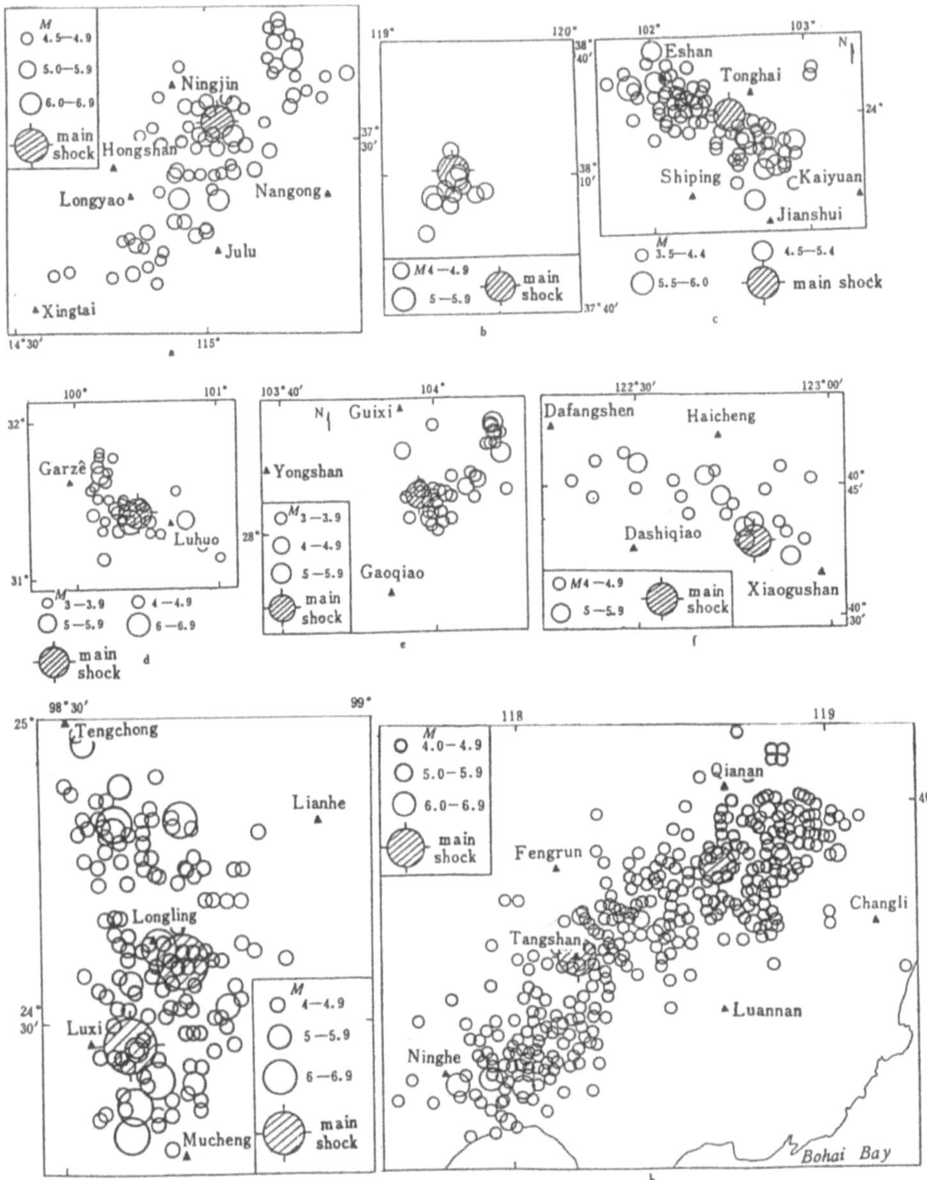

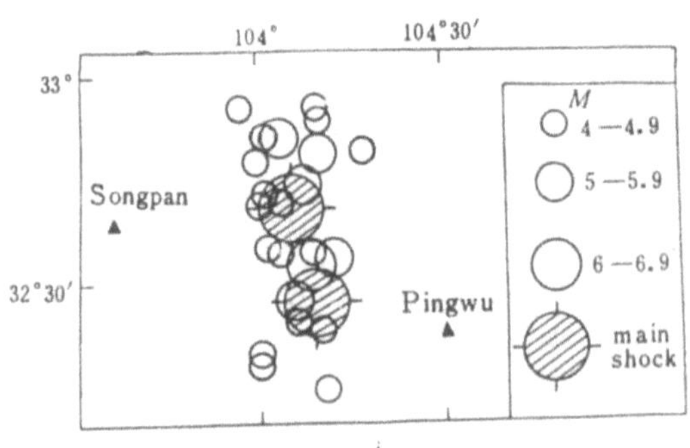

Fig. 2.4. Epicentral distributions of aftershocks for the nine great earthquakes; *a* Xingtai; *b* Bohai; *c* Tonghai; *d* Luhuo; *e* Yongshan; *f* Haicheng; *g* Longling; *h* Tangshan; *i* Songpan

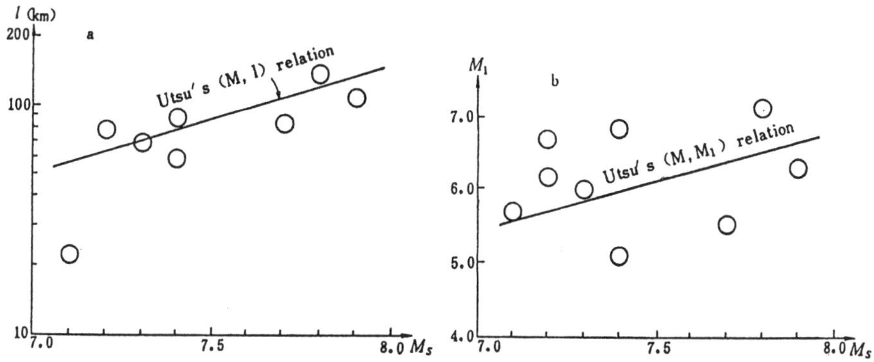

Fig. 2.5. Relations between the mainshocks and the aftershocks for the nine great earthquakes; *a* the length of aftershock area and the mainshock [Eq. (2.1)]; *b* the maximum magnitude of aftershock and the mainshock [Eq. (2.6)]

the mainshocks (Haicheng, Tonghai, Longling, Tangshan and Yongshan) are positioned near the lower boundary of the predominant layer of aftershocks, while the position for the Xingtai earthquake is close to the upper boundary of the aftershock zone. The cross sections of aftershock depth distributions for the Haicheng and Xingtai earthquakes are plotted in Fig. 2.6, with the mainshock positions marked. Therefore, the hypocenters of the mainshocks are located near the boundaries of the elliptical aftershock distributions, but not near their ends.

3) Spatial jumping of aftershock activities.

The regular migrations of strong aftershock activity within the aftershock areas are very distinct in some of the great earthquake sequences. Figure 2.7 shows the profiles of migrations of strong aftershocks ($Ms > 5.7$) for the Xingtai earthquake and of strong aftershocks ($M > 6.0$) for the Longling earthquake, with two characteristics as follows: (a) The epicenters of strong aftershocks leaped to and fro between south and north within the aftershock area, that is the so–called phenomenon "jumping between both ends", and after several rounds like this, the last strong aftershock in the sequence would occur at one end of the aftershock area. This demonstrates that the two extremes of the aftershock area were critical locations of stress redistribution and readjustment after the mainshock. (b) Meanwhile, the focal depths of strong aftershocks seemed to become either deeper (such as the Xingtai earthquake sequence) or shallower (such as the Longling sequence) unilaterally. As the depths were measured with fairly big error, such a tendency needs to be further examined by accurate observations.

2.2.2 Time Series of Aftershocks

Following the great earthquake, the aftershock activity lasts for some time, hence constituting a time series for statistical analysis. Figure 2.8 shows the (M–t) diagrams of $M > 4$ aftershock sequences within a year after the nine great earthquakes.

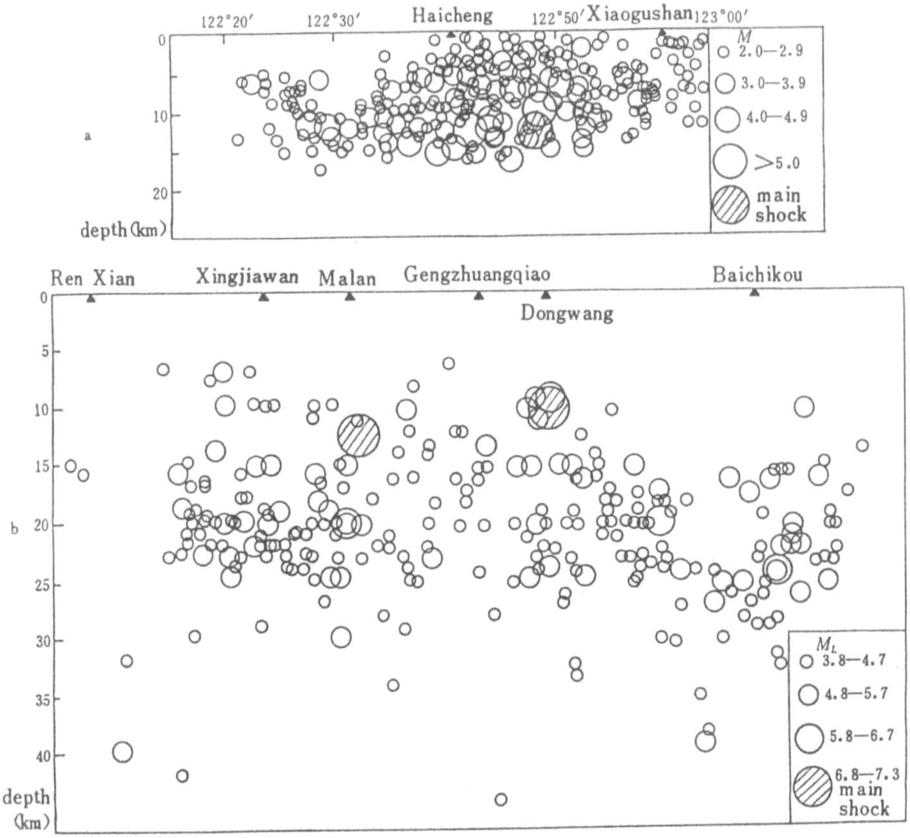

Fig. 2.6. Profiles for focal depth distributions of aftershocks; *a* the Haicheng sequence (profile N69 ° W); *b* the Xingtai sequence (profile N35 ° E)

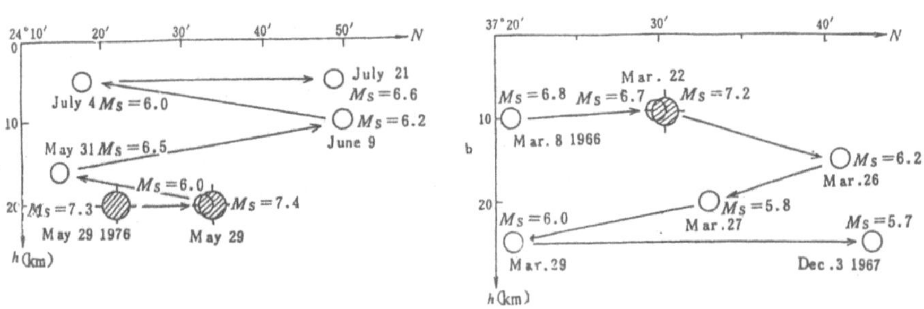

Fig. 2.7. Profiles for spatial migrations of strong aftershocks; *a* the Longling sequence; *b* the Xingtai sequence

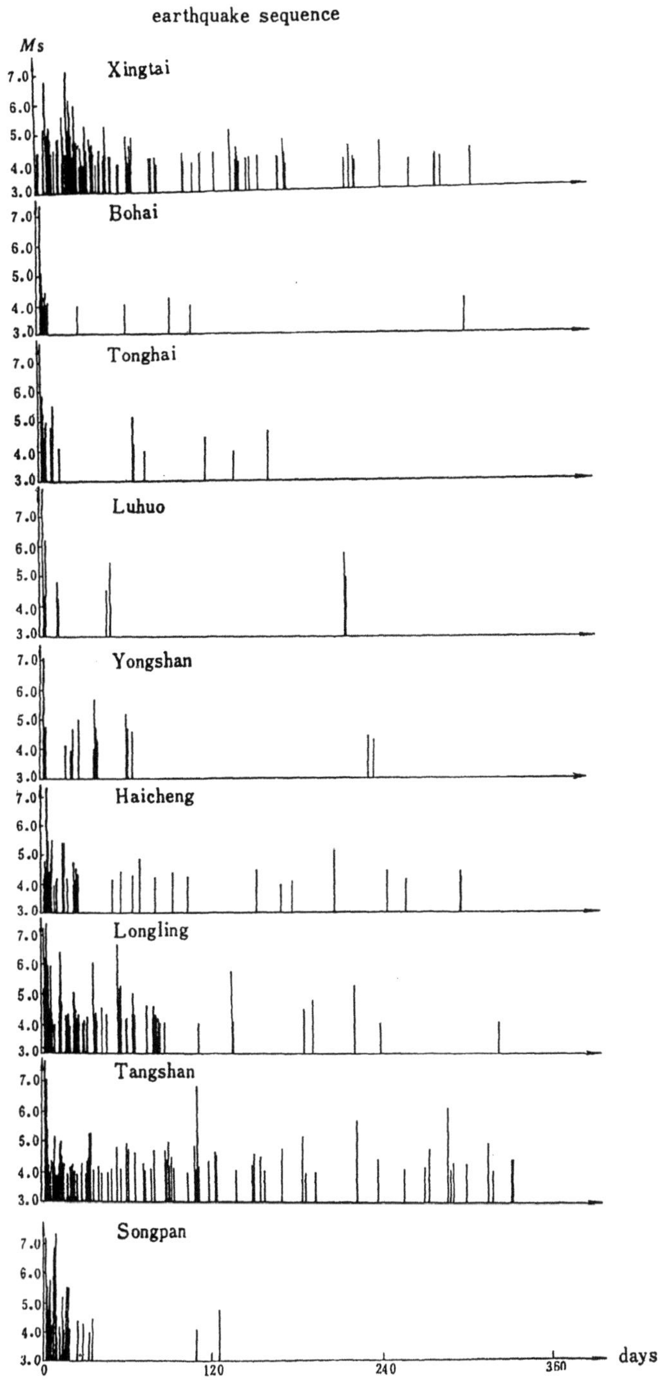

Fig. 2.8. (M–t) digrams of the aftershock sequences for the nine great earthquakes

1) Division of aftershock activities into stages.

Benioff, a U.S. seismologist (1951a), divided the aftershock sequence into two stages, with the first stage related to recovery of compressive rocks, and the second to recovery of shear creeping. Papazachos (1966) confirmed the above–mentioned two stages, while studying aftershock sequences in Greece. He suggested that the first stage would begin right after the mainshock, characterized by energy release of minor intensity aftershocks, and then the major–magnitude aftershocks would be dominant in the stage of shear creeping recovery. On the average, the largest aftershocks in the sequences occurred about 3 days after the mainshock. In addition, Papazachos pointed out that there existed a third stage in some earthquake cases, that is the "late aftershock" period, which depended upon the time length which is needed for stresses to be accumulated continuously till the residual intensity of rocks.

Lu Peiling discussed the stage divisions of the aftershock sequences of six great earthquakes (Haicheng, Bohai, Luhuo, Tonghai, Longling and Tangshan), indicating that the aftershock sequences could be divided into three stages within a year after the mainshock: the earliest stage (within 2 days after the mainshock); the early stage (2–120 days after the mainshock); and the late stage (120 days–1 year after the mainshock). The largest aftershocks in the sequence all occurred in the earliest stage (Fig. 2.8). There were also fairly strong aftershocks during the other two stages, and the magnitude difference between these aftershocks and the largest aftershocks in the sequence might not exceed one. Lu Peiling also suggested that there seemed to be " latest aftershock activity" 1 year after the mainshocks.

2) Quasi–periodicity of strong aftershocks.

Gu jicheng (1979) and Wang Biquan et al. (1979) discussed separately the "waiting time" and quasi–periodicity of medium–strong aftershocks in the sequences. They defined the time interval ($t_i - t_{i-1}$)between two successive strong aftershocks as the waiting time (Δt_i) for the following strong aftershock, that is:

$$\Delta t_i = t_i - t_{i-1}. \tag{2.2}$$

They studied many case histories, including the nine great earthquake sequences, and found that the "waiting time" satisfied the following statistical relation:

$$\log \Delta t = A \log t + B. \tag{2.3}$$

which implied that the waiting time for strong aftershocks increased with the extension of their occurrence time (t); that is the occurrence time for the whole sequence of strong aftershocks featured quasi–periodicity.

3) Duration of strong aftershock activity.

The quasi–periodic regularity of strong aftershocks undoubtedly provides a valuable reference for forecasting the occurrence time of strong aftershocks. As the whole, strong aftershock sequence is completed in a limited time period, it is necessary to further study the cutoff time of strong aftershock activity, or to discuss how long strong aftershock activity lasts.

Fist of all, let us set two restrictions as follows: (1) for the aftershock sequences, the magnitude limit (M′) of strong aftershocks is assumed to be M′ > M (mainshock) − 1.7. This means that if the magnitude difference between any aftershock in the sequance and the mainshock is less than 1.7, then that aftershock would be defined as a strong aftershock in the sequence; (2) during the entire sequence, the time difference between the last strong aftershock and the mainshock is called the duration (ΔT) of strong aftershock activity in the sequence.

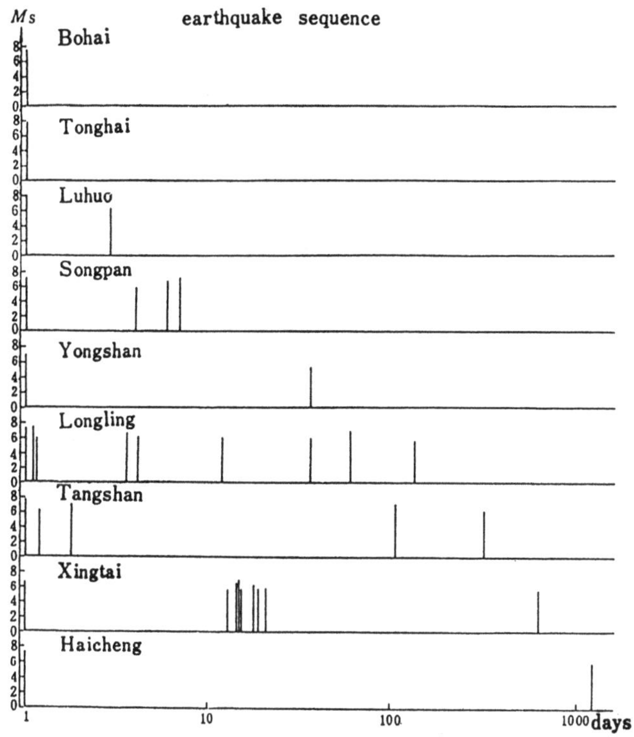

Fig. 2.9. (M−t) diagrams of strong aftershock sequences for the nine great earthquakes

Data on the time series and the duration for strong aftershock activities in the nine great earthquake sequences are shown in Fig. 2.9 and Table 2.4, respectively. All data were taken before December 1979. Analysis of these data suggests the following:

First, the durations (ΔT) of strong aftershock activity for the nine great earthquake sequences differ greatly from each other. For example, the duration of the Haincheng sequence amounted to 1198 days, about 40 months; whereas for the Bohai and Tonghai sequences, no strong aftershocks occurred within the above–mentioned restrictions. For convenient statistics, the duration of both the Bohai and Tonghai sequences is taken as 1 day.

Second, there seems to be no linear relationship between the duration of strong aftershock activity and the size of mainshock magnitude (M mainshock).

Third, the durations for the five great earthquakes in the Yunnan–Sichuan region are comparatively short, less than 5 months, while the ΔT for three of the great earthquakes in North China (except the Bohai sequence) seems longer, exceeding 10 months, with the longest about 40 months.

4) Attenuation of monthly frequency of aftershock activity.

In aftershock studies, the attenuation of aftershock frequency (N(t)) with time is an important statistical regularity which was discovered long ago. In 1894, Ohmori Fusakichi, the famous Japanese seismologist proposed a statistical regularity in which the aftershock frequency attenuated exponentially. In 1961, Utsu suggested a modified equation for hyperbolic attenuation of aftershock frequency as follows:

$$N(t) = \frac{A}{(t+c)^{p}}. \tag{2.4}$$

When $t >> c$, $N(t) = A / t^{P}$ or

$$\log N(t) = \log A - P \log t . \tag{2.5}$$

where N(t) is the number of aftershocks in time period t; A is the number of aftershocks for the first statistics in the first time unit after the mainshock; P is the frequency attenuation coefficient.

We have studied the attenuation process of monthly aftershock frequency for the nine great earthquake sequences. The starting magnitude for aftershocks is about 2, with slight deviations for different sequences owing to the observation status. The time interval for statistical research is longer than the above–mentioned duration of strong aftershock activity (for the purpose of discussing the relation between ΔT and P), and enough data for valid statistics are sure to be gathered in this time interval.

The bi–logarithmic coordinates in Fig.2.10 show the attenuation processes of monthly aftershock frequency for the nine great earthquakes. These figures can be separated into two kinds: (1) the entire process shows a linear attenuation, and the attenuation coefficient (P) is derived from fitting a straight line by the least square method; (2) there are evident inflexion points in the attenuation process; in these cases (indicated by solid circles in the figure), the best–fit line is obtained using data prior to the inflexion points and the attenuation coefficient (P) is derived from this. The attenuation coefficients are listed in Table 2.4.

From Fig. 2.10 and Table 2.4, the following preliminary deduction is made:

a) The P values for the nine great earthquake sequences can be divided into two kinds: for the first, $P<1$, such as the four great earthquake sequences in North China; for the second, $P>1$, like the five great earthquake sequences in Southwest China.

b) For the first kind of sequence (North China, $P<1$) inflexion points exist in the rate of attenuation of aftershocks for three of the great earthquakes (Xingtai, Haicheng and Tangshan sequences). It should be noted that the time of the " inflexion points" is in agreement with the termination time of strong aftershock activities for these sequences. In other words, the attenuation of monthly aftershock frequency speeds up after the strong aftershock sequences are over.

The attenuation coefficients (P) of monthly aftershock frequency and the duration (ΔT) of strong aftershock activity for the nine great earthquake sequences are shown in Fig. 2.11. It can be seen that the P values are larger if the ΔT of the great earthquake sequences are shorter, whereas the P values become smaller if ΔT gets longer, exhibiting an inverse relation between ΔT and P. However, it must be mentioned that the Bohai earthquake is an exception.

2.2.3 Intensity of Aftershocks

1) Magnitude of the largest aftershocks.

Utsu (1961) also studied the relation between the magnitude (M_l) of the largest aftershocks and that ($M_{mainshock}$) of the mainshock, deriving the following statistical formula based on a great number of Japanese earthquakes:

$$M_l = 1.45\, M_{mainshock} - 4.7. \tag{2.6}$$

This illustrates that the magnitude of the largest aftershocks becomes larger with a larger mainshock magnitude. Figure 2.5b gives the graphic solution of Utsu's equation and the corresponding data (open circles) for the nine great earthquakes, and the fit between the two is not very good. This may mirror discrepancies between seismic activity in different regions.

2) a and b values of aftershock series.

In Seismicity of the Earth and Associated Phenomena written by Gutenberg and Richter in 1954, it was noted that, on average, more weaker earthquakes occurred than stronger ones, satisfying the following expression:

$$\log N(M) = a - bM, \tag{2.7}$$

where N(M) is the number of earthquakes with magnitude $>M$. This demonstrates another significant statistical regularity for seismic activity. Such a relation is also embodied in aftershock series. In this formula, the b value describes the proportional relation between strong earthquake frequency and weak earthquake frequency for overall earthquake activity within a given space–time range, and the a value represents a measure of the overall level of seismic activity.

The a and b values for the nine great earthquake sequences are listed in Table 2.4; they are obtained from aftershocks ($M \geqslant 3.5$) occurring within one year after the mainshock. It can be seen from Table 2.4 that the a and b values for the five great earthquake sequences in the Yunnan–Sichuan region of Southwest

earthquake sequences

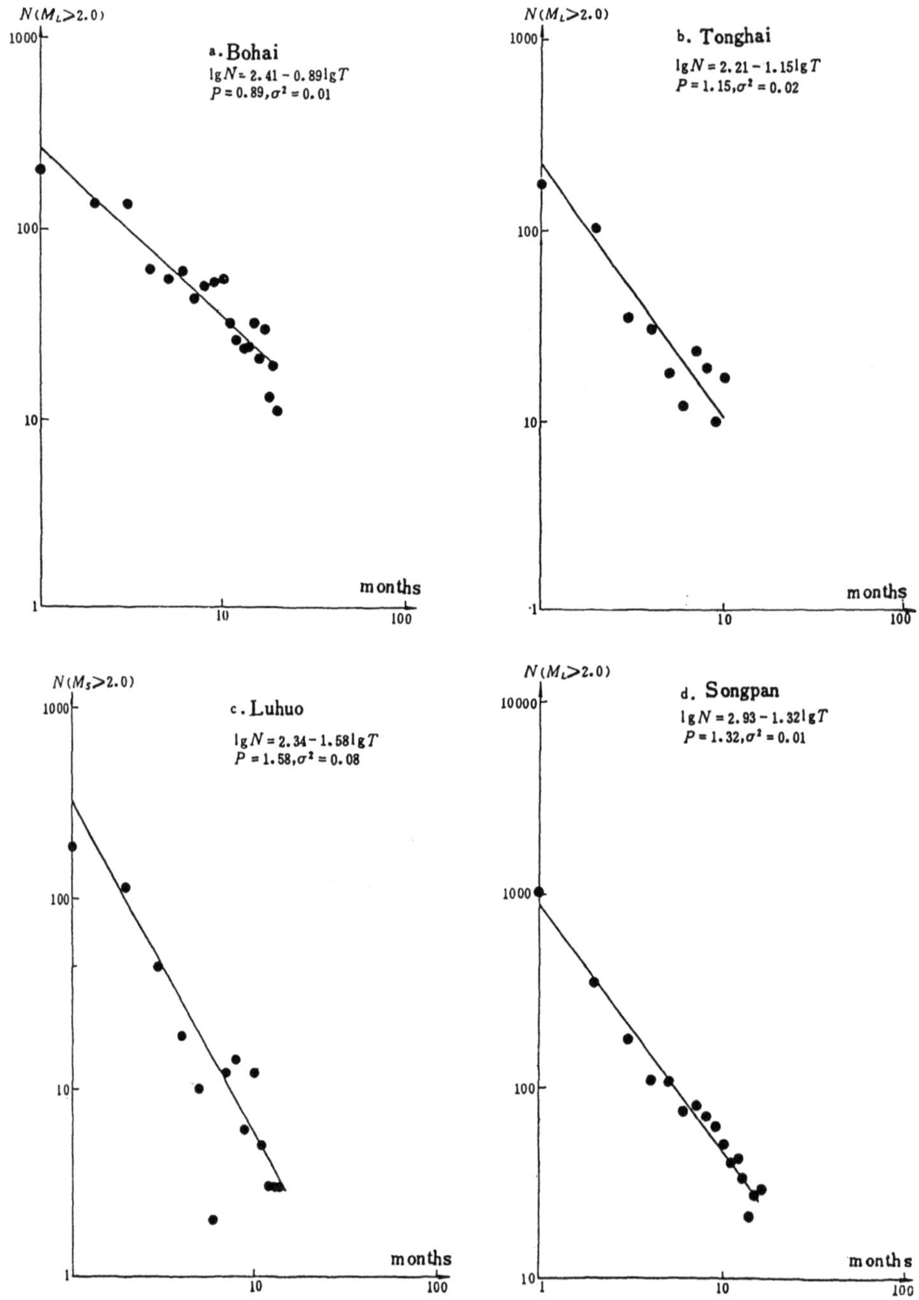

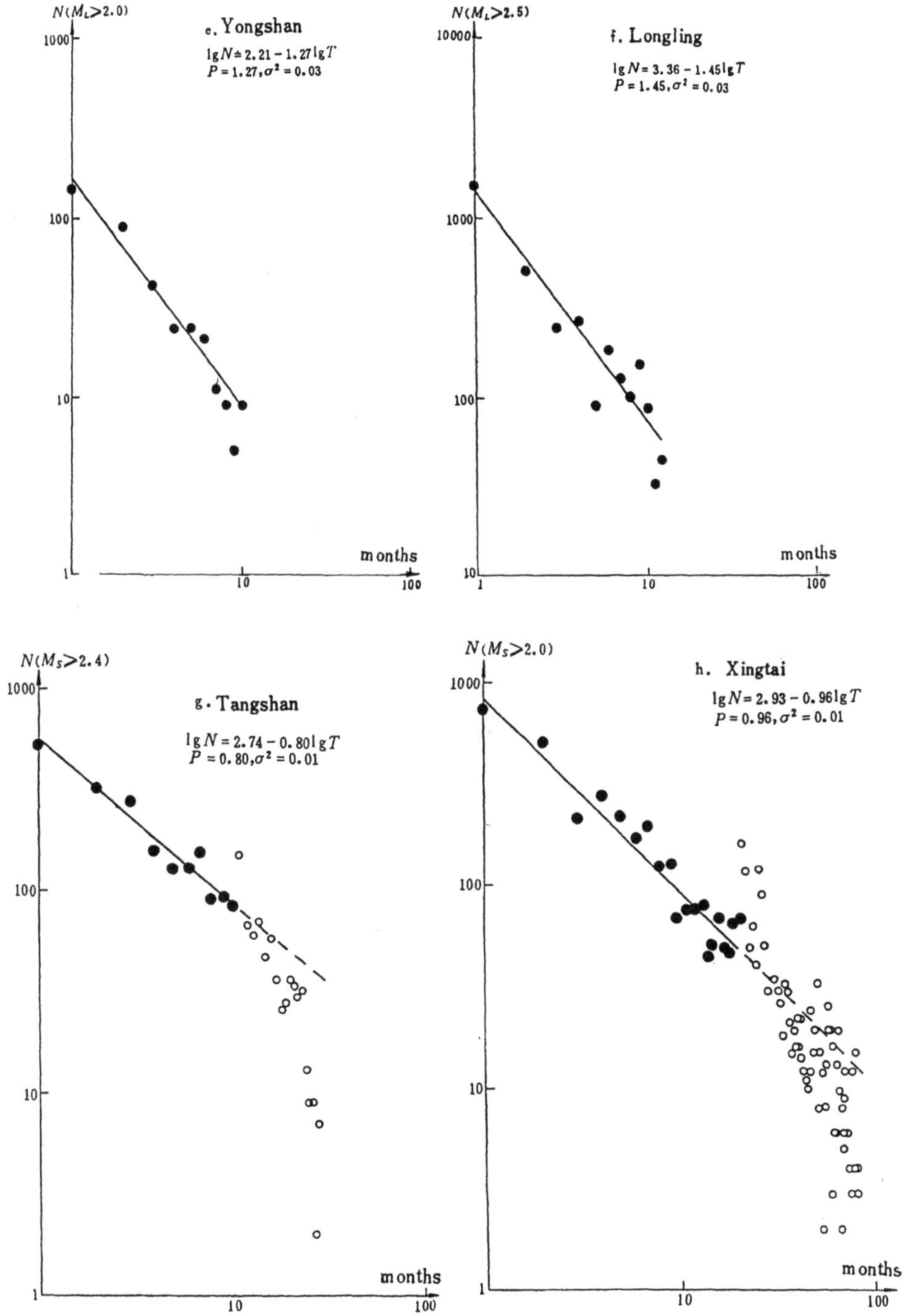

N(ML>2.0)
e. Yongshan
lgN= 2.21 - 1.27lgT
P = 1.27, σ² = 0.03
months
N(ML>2.5)
f. Longling
lgN= 3.36 - 1.45lgT
P = 1.45, σ² = 0.03
months
N(Ms>2.4)
g. Tangshan
lgN = 2.74 - 0.80lgT
P = 0.80, σ² = 0.01
months
N(Ms>2.0)
h. Xingtai
lgN= 2.93 - 0.96lgT
P = 0.96, σ² = 0.01
months

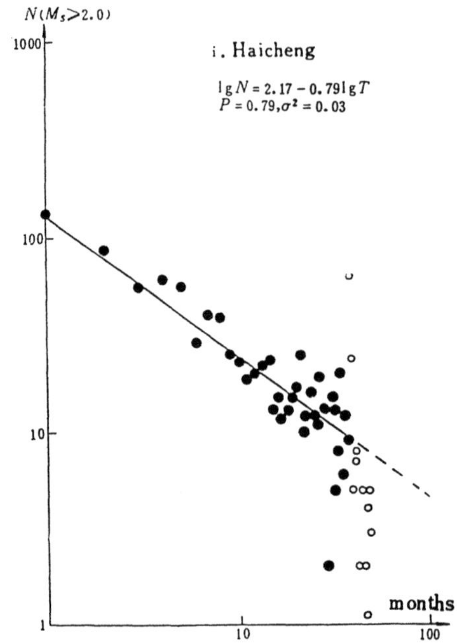

Fig. 2.10. The attenuation of monthly aftershock frequency for the nine great earthquake sequences

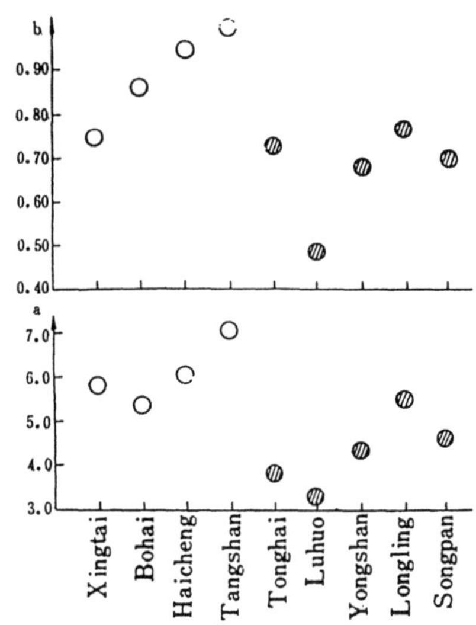

Fig. 2.12. Regional discrepancies of a and b value for the nine great earthquake sequences

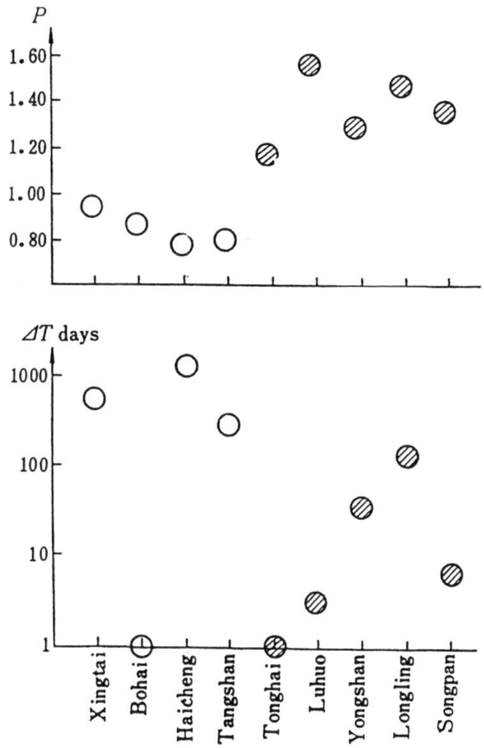

Fig. 2.11. ΔT and P of the nine great earthquake sequences

China are systematically smaller than those in North China. The differences between them are given in Fig. 2.12, which also indicates that variations of aftershock frequency with different strength for the five great earthquake sequences in southwest China are not as large as those in North China.

3) Locations of latest strong aftershocks.

In principle, this section should be included under "Spatial distribution of aftershocks". However, since the concepts of "latest" and "strong aftershock" had not then been introduced we have preferred to put it here.

In the light of the aforesaid stage division of aftershock series, only the Haicheng and Xingtai earthquake sequences included latest strong aftershocks. The locations of these aftershocks have certain characteristics that merit mention.

Three years after the Haicheng mainshock (4 Februry 1975), on 18 May 1978, a latest strong aftershock of M 6 occurred in a "seismic gap" within the aftershock area.

Figure 2.13 shows the evolution of the spatial distribution of aftershocks ($Ms \geqslant 4.0$) for the M 7.3 Haicheng earthquake sequence.

The epicentral and depth distributions of aftershocks with $M > 4.0$, which occurred during the period between the mainshock and the latest strong aftershock of M6 (4 February 1975–17 May 1978) are shown separately in Fig.

2.13a and b. During this period, the spatial distribution (solid circles in the figure) of the early aftershock activity (within 20 days from February 4 after the mainshock to February 24) was not homogeneous, displaying a relatively quiescent "aseismic segment", about 18 km in width (Fig. 2.13a). At both sides of this "aseismic segment", the focal depth of the aftershocks ranged from 7 to 17 km (Fig. 2.13b).

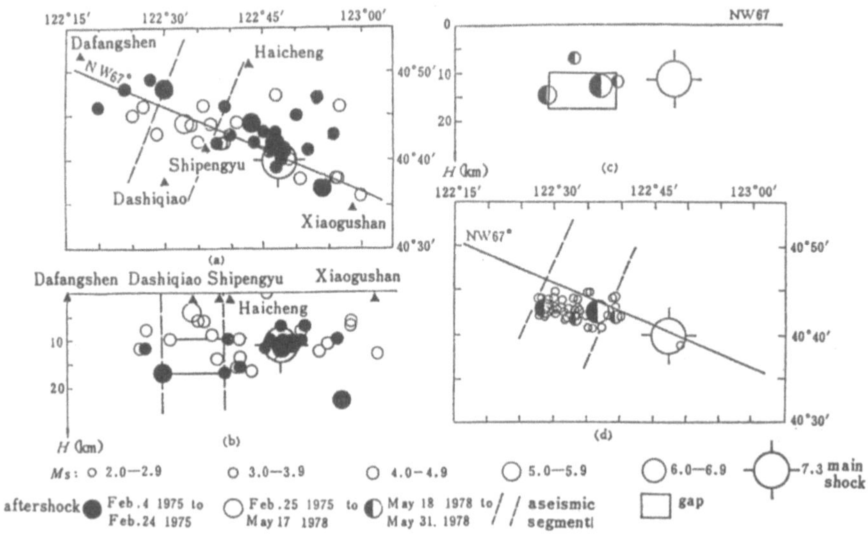

Fig. 2.13. Aftershock "gap" and latest strong aftershock activity of the Haicheng sequence; *a* epicentral distribution of aftershocks(4 February 1975–17 May 1978, Ms> 4.0); *b* depth distribution of aftershocks (4 February 1975 –17 May 1978, Ms>4.0); *c* depth distribution of the strong aftershock of M 6 of 18 May 1978 and its aftershocks (Ms >4.0); *d* epicentral distribution of the earthquakewith M 6 and its aftershocks (18 May –31 May 1978, Ms>2.0)

After the early stage, during the long period from 5 February 1975 to 17 May 1978, six aftershocks with focal depth from 0 to 10 km (Fig. 2.13b) occurred within the "aseismic segment", besides the aftershocks (open circles) that were still active on both sides of this segment. Therefore, up to 17 May 1978, a relatively calm "gap" was formed within the three–dimensional space of aftershock activity, as shown by the framed line in Fig. 2.13b.

The latest strong aftershock (Ms = 6.0) occurred at the east end of the "gap" on 18 May 1978, and its own maximum aftershock (Ms = 5.1) took place at the west end. Also, almost all aftershocks (Ms>2.0) occurred within this "gap" area (Fig. 2.13c and d). Cao Tianqing (1980) used the spectral method to study the nature of the aftershock area for the Haicheng earthquake sequence, and suggested that the aforesaid "gap" might be a "barrier" within the source region.

In addition, the latest strong aftershock (3 December 1967, Ms = 5.7) for the Xingtai earthquake sequence occurred at the north end of the aftershock area, and its focal mechanism apparently differed from that of the mainshock (Fig. 2.14).

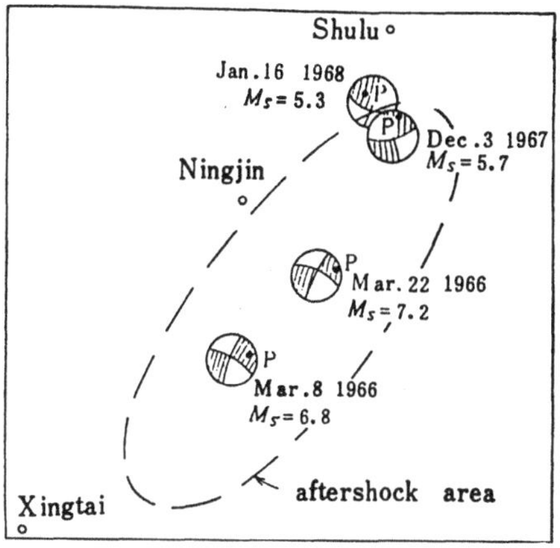

Fig. 2.14. Location of the latest strong aftershock for the Xingtai earthquake sequence —— emergent point for the principal compressive stress axis

4) Classification of earthquake sequences.

Conventionally, shallow earthquake sequences have been divided qualitatively into three types, namely the isolated type (without foreshocks and aftershocks, or with very minor ones); the mainshock type (with large quantities of aftershocks, and some with foreshocks); and the swarm type (without a notable mainshock throughout the entire sequence). The nine great earthquake sequences have all been accompanied by aftershock activities to different extent. It is thought that the Bohai, Tonghai, Luhuo, Yongshan and Haicheng earthquake sequences can be considered as the mainshock type. The Xingtai earthquake sequence may belong to the swarm type, while the Longling, Songpan and Tangshan earthquake sequences, each having two strong earthquakes of $M > 7$, can be termed as the double mainshock or multiple mainshock type.

Since the generation and development processes of great earthquake sequences are closely related to the nature of the regional seismogenic structure, the state of the medium in the source region, and stress–strain variations, we feel that great earthquake sequences should be classified on the basis of some comprehensive general principle, rather than simply from the seismic activity itself. Currently, this problem is far from being solved, and will be dealt with again in Chaps. 3 and 13.

The above-mentioned comparison and analysis of the characteristics of earthquake parameters and sequences for the nine great earthquakes have shown that there are numerous common features, as well as some differences which cannot be neglected, between the two groups of earthquakes in North China and in the Yunnan–Sichuan region of Southwest China. The common features are that the fractures in the source regions are predominantly caused by strike–slip faulting; the mainshocks are mostly located at the bottom of the aftershock area; and the size of the aftershock area is proportional to the magnitude of the mainshock. These features perhaps show that there are similarities in seismogenic conditions and environment for great earthquakes in both regions. The differences are mainly the parameters characterizing the aftershock series. In general, the a, b and ΔT values for the five great earthquake sequences in Southwest China are smaller or shorter than those in North China, whereas the P value is greater than that in North China. These facts may indicate that there exist remarkable discrepancies between the two regions in stress–strain readjustment processes and the rheological properties of the medium during seismic energy release.

CHAPTER 3

Tectonic Environments of the Earthquakes

Shallow–depth tectonic earthquakes are manifestations of local tectonic movements of the crust, and are basically mechanical phenomena. The seismogenic preparation of a strong earthquake seems to be associated with large–scale regional diastrophism, and even with material motions within the mantle, while the concrete conditions for strong earthquake generation depend principally upon the physical properties, structure and state of stress of the crust.

Among the nine great earthquakes, eight of them occurred within the midcrust depth range between 5 and 20 km, with the exception of the Bohai earthquake, which was deeper. Therefore, it is desirable to study the relationship between various layers within the crust and focal–depth distributions. Seismogeological investigations have initially determined that the earth–quake–producing positions of the nine great earthquakes are related to certain tectonic conditions, expressing reactivations of large visible fracture zones, movements of concealed faults or growth of fault zones. In this chapter, beginning with the regional crustal tectonic environments, and particularly emphasizing the controlling roles for earthquake occurrence of crustal layers and active faults, the seismogenic tectonic conditions of the nine great earthquakes and the characteristics of their coseismic dislocations are discussed. We then explore their similarities and regionalized characteristics, so as to further analyze the anomalous phenomena before, during and after the earthquakes.

3.1 Crustal Layering and Tectonic Deformation of North China

3.1.1 Crustal Layering

The Brigade of Geophysical Prospecting, State Seismological Bureau, and Plant 646 of the Ministry of Petroleum cooperatively conducted deep seismic sounding for the "Yuanshi–Jinan" profile starting from Yuanshi County, Hebei Province, through the Xingtai earthquake–stricken area, to the suburbs of Jinan City, Shandong Province. The preliminary analysis was completed in 1971, giving detailed crustal layers and petrographical properties. On this basis, Teng Jiwen et al.(1974, 1975) further analyzed the sounding data, and proposed a layered velocity model for the central part of North China in 1974 and 1975. Making use of the achievements of this research, and taking into account the layering and tectonic phenomena expressed by outcrops of deep crustal layers on the surface, an outline of the crustal layers is shown in Fig. 3.1, together with a succinct description of each layer.

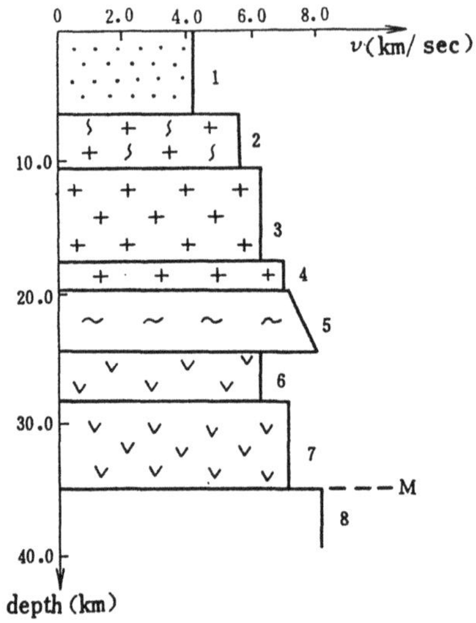

Fig. 3.1. An outline of crustal layers for North China; *1* sedimentary layer; *2* metamorphic crystalline layer; *3* and *4* granite layer; *5* intermediate layer of undetermined lithologic character; *6* and *7* basaltic layer; *8* uppermost layer of upper mantle; *M* Moho surface

The first layer shown in Fig. 3.1 corresponds to unconsolidated Cenozoic sediments and also sediments of the Palaeozoic, which contain many pores and fluid, with very dense stratified planes within them. These latter are prone to small–scale folding and rupturing similar to that exposed at the surface. Such folds and ruptures are very often confined within this layer. The second layer mainly consists of ancient metamorphic rocks, with intrusions of both new and old magmatic rock bodies. This layer features obvious plastic and rheological deformations during metamorphic flexible foldings before the Palaeozoic but, after it became the regional metamorphic rock body, this layer was featured with thick tabular deformation as the Palaeozoic sedimentary base, hence becoming the core of a large–scale bending uplift. The third and fourth layers correspond possibly to granitic ones. Within the regional granitic bedrocks exposed at the surface, there are no thin layered structures, but coarse or finely crystallized tabular textures. And they feature overall thick tabular deformation similar to that of the second layer, particularly with salient brittle broken deformation. The fifth layer is an intermediate one of high seismic velocity, being the transition zone between the granitic and the basaltic layer. The sixth and seventh layers correspond to the basaltic layer, and have no outcrops at the surface for comparison.

As for the differential expression of tectonic deformation in the above–mentioned various layers, more detailed information can be obtained from the "Yuanshi–Jinan" profile (Fig. 3.2).

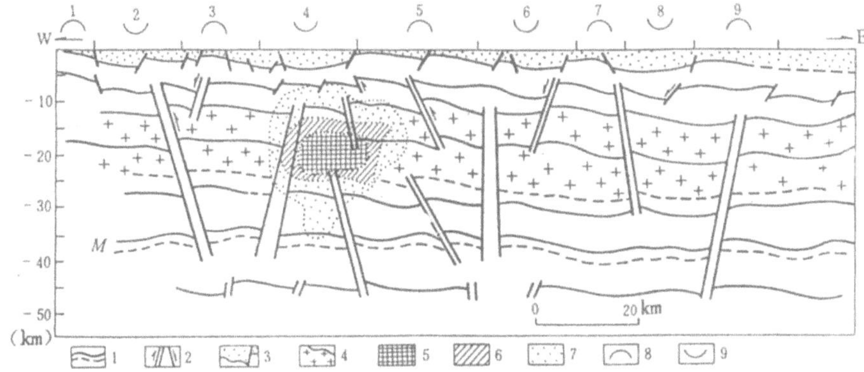

Fig. 3.2. Sketch of the Yuanshi–Jinan crustal seismic sounding profile; *1* velocity layering plane; *2* fracture zone; *3* sedimentary layer; *4* granitic layer; *5* area of dense aftershocks of Xingtai sequence; *6* secondary dense aftershock area; *7* area of few after–shocks; *8* shallow–depth convex structure; *9* shallow–depth concave structure;
(1) Yuanshi convex; (2) Zhaoxian depression; (3) Longyao convex; (4) Shulu depression; (5) Hengxin Convex; (6) Nangong depression; (7) Minghua Town convex; (8) Gucheng depression; (9) Wucheng convex

In general, the various crustal layers exhibit both wide and slow depression and bending, which shows semiwave–like convex and concave structures cut by numerous fractures. There seems to be no major difference in tectonic morphology among the various layers, but there are slight variations in layer thickness, possibly due to the deformation of inhomogeneous bodies and the motion of materials. The maximum variation of the sedimentary layer thickness seems to be mainly associated with uneven uplifting and descending of shallow–depth fault blocks. The thickness of the metamorphic rock layer is also fairly varied, especially near the sides and ends of fault zones. The thickness of the granitic layer is the most constant. The thickness of the basaltic layer is different at the top and flanks of mantle upheaval. From the above, it can be seen that the rigidity of the granitic layer is greater than that of the others, and hence consists of more elastic deformation.

The states of development of fractures are important expressions of the mechanical characteristics of various layers. From the distributions of fractures exhibited by the "Yuanshi–Jinan" profile, the following understandings can be gained: first, we doubt that fractures cutting the sedimentary layer within the upper crust are normal faults, and we do not think they connect with the deeper fractures. Some fracture channels must exist from the deeper crust to the surface, because the in–situ volcanic rocks erupted at many fault–depressed basins in North China. However, most surface fractures are indeed different from deeper fractures in dip, from the overall situation of occurrences of shallow fractures, and the relation between the two is rather perplexing, featuring partly direct connection, and partly disconnection. The fractures confined within the granitic and matamorphic layers of the mid– crust reveal that there were fractures developed within these two layers themselves. These fractures seem to be relatively concentrated where the Xingtai earthquake occurred, at the west flank of the mantle upheaval. Such tectonic environments may serve as the concrete conditions for the seismogenic preparation and production of the Xingtai earthquake. There were few fractures confined within the basaltic layer of the lower crust. If the fracture developments described above are real (and not simply artifacts of the seismic sounding), it can be suggested that there are major differences in the development of fractures within the upper (sedimentary), middle (granitic and metamorphic) and lower (basaltic) parts of the crustal profile, with fractures developed primarily in the mid–crust in which most continental shallow tectonic earthquakes occur. Therefore the conception of a "seismogenic competent" layer or "earthquake competent" layer (or "earthquake–prone" layer) has been put forward.

3.1.2 Earthquake Competent Layer

When analyzing the seismotectonics of North China in 1975, we indicated that the composite body of the regional metamorphic bed and the regional granitic bed might be the principal physical layer as well as the main layer for the preparation and occurrence of earthquakes. We call this the "seismogenic competent" or "earthquake competent" layer. The earthquake competent layer implies that: (1) when the crust deforms as a multiple plane–layered body, the intermediate layer with the largest elasticity modulus controls the deformation of the combined body; (2) the competent layer is the major stratum for accumulating and releasing elastic energy; (3) the occurrence of earthquakes is directly related to the tectonic status of the earthquake competent layer(Ma Zongjin 1984a).

The focal depths of the nine great earthquakes were constrained by the earthquake competent layer. Furthermore, such competent layers may play significant roles in the preparation and occurrence of most shallow earthquakes. Xu Ji studied the focal depths of earthquakes with $M > 6.5$ within this century in Mainland China as well as individual large historical earthquakes (68 earthquakes). He showed that almost all of these earthquakes occurred within the crust, particularly in the middle crust.

Shao Xuezhong et al. (1980) studied converted wave profiles of eight natural earthquakes in the Beijing–Tianjin–Tangshan region, and by comparing the relationship between the focal depth and the crustal layering (Fig. 3.3), they suggested: "almost all strong earthquake sources are located at certain specific positions within the granitic layers of the region". The so–called specific positions are segments of the granitic layers cut by large fractures. The sources of both large and small earthquakes are mostly centered at the boundaries where fractures intersect the upper and lower planes of granitic layers, and this is possibly associated with the local motions of crustal layering planes near fracture zones.

Obviously, our present knowledge on the competent layer is still very limited, and some of our interpretations may be incorrect owing to insufficent data. Nevertheless, in order to better understand seismic source regions, we present the above discussion as a working model, and use it as a basis on which to improve gradually our over–simplified assumptions about source regions.

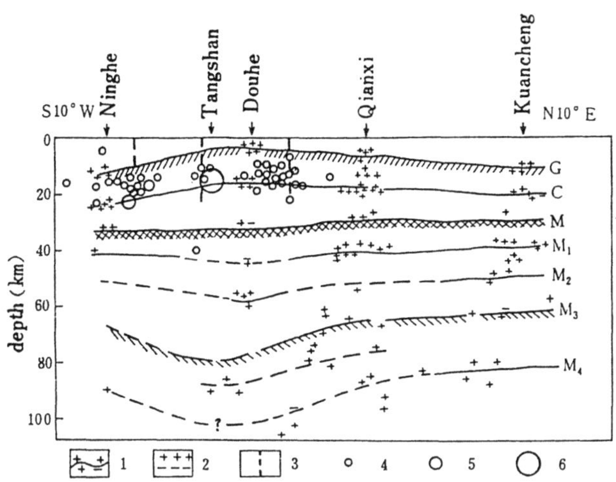

Fig. 3.3. Converted wave profile of crust for Tangshan region; *1* converted planes; *2* converted points, positive and negative; *3* speculated faults; *4* Ms<6.0; *5* Ms=6–6.9; *6* Ms=7–7.9

3.2 Regional Tectonic Features of the North China Seismic Region

The North China seismic region is the most seismically active region in the eastern part of Mainland China.

The limits of this region are shown in Figs. 2.1 and 3.4. This seismic region is within a Palaeozoic platform. Since the mid Mesozoic, the region has experienced the severe Yenshan movement, resulting in tectonic motion and magnetic activity dominated by compressive folding and fractures. In the late period of the Yenshan movement, many large and small fault–depressed basins and fault block uplifts occurred, with large–scale eruptions of basic magma, due to the evident role of gravity equilibrium adjustment. Since the late Neogene, new regional compressive, rhaegmageny and tensile fracturing movements have taken place, accompanied by large–scale rising and descending movements and, by then, the magmatic activity faded and was centered at the intersection parts of major active fault zones.

In order to understand the relationship between seismic activity and geologic structure, and between recent seismicity and the tectonic framework in North China, a map (Fig. 3.4) of geological structures and historical earthquakes in North China, and a map (Fig. 3.5) of epicenters of earthquakes with $M_s \geqslant 3$ from

1966 to 1980 have been compiled. They are based on data from geological structures, airborne magnetic surveys, regional gravity measurements, oil prospecting and historical earthquakes in North China.

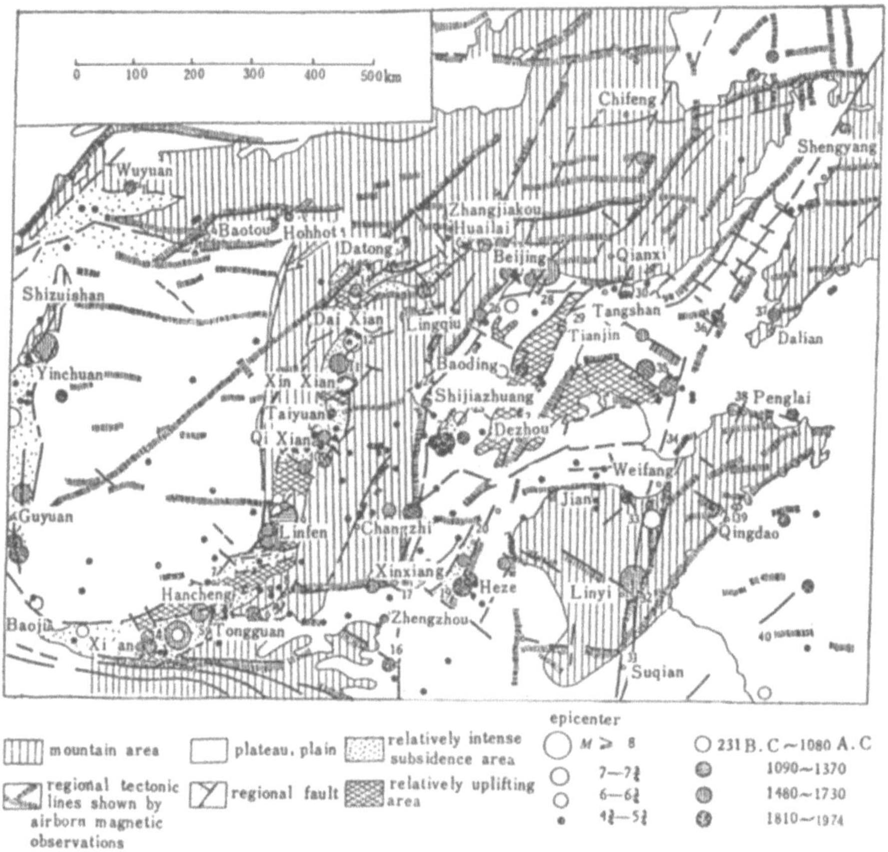

Fig. 3.4. Distribution of geological structures and historical earthquakes in North China

From these maps, we suggest that the distribution of strong earthquakes with M > 6 in North China is basically controlled by three tectonic directions which have also constrained the recent four great earthquakes in North China.

3.2.1 NE–ENE Tectonic Seismic Zone

Figure 3.4 shows some remarkable phenomena, indicating a NE alignment of strong earthquakes, with four clear seismic zones from west to east:

Fig. 3.5. Epicentres of earthquakes with M>3 from 1966 to 1980 in North China

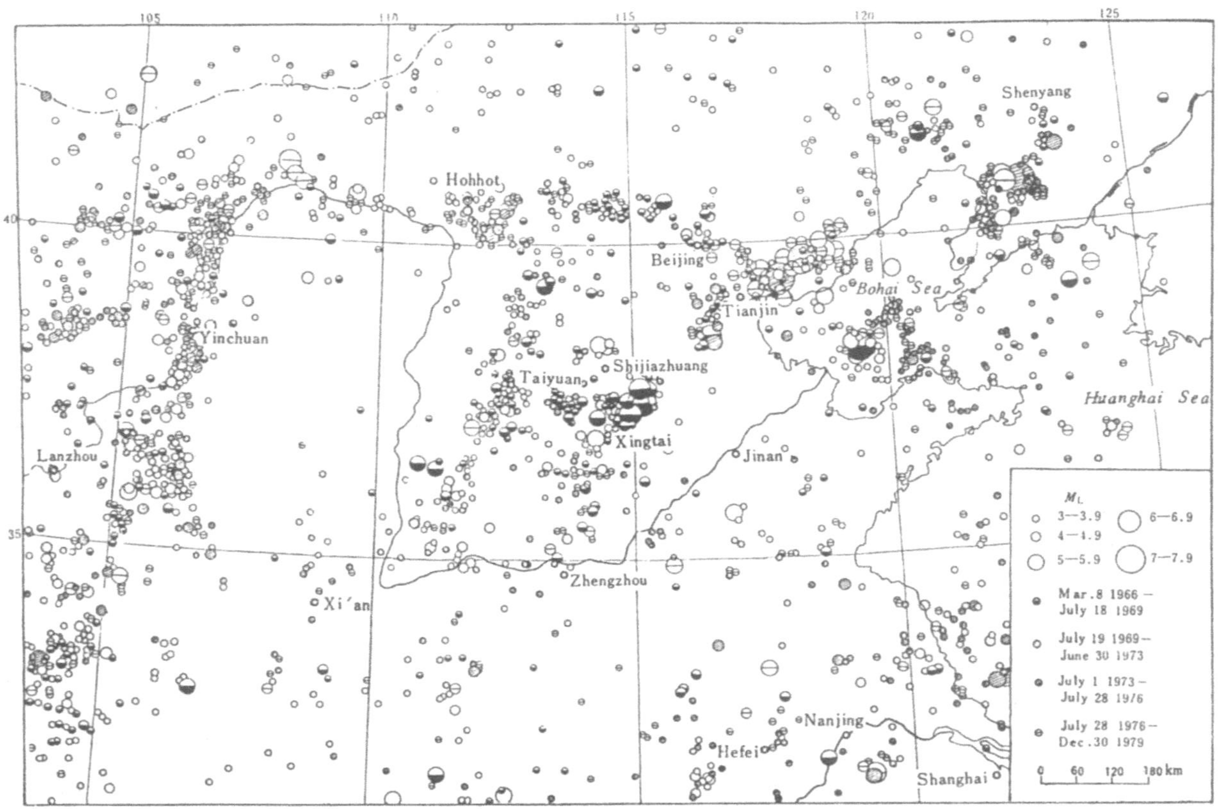

First, the Shanxi seismic zone, which starts from Xi'an in the south, through Taiyuan to Datong and Huailai in the north. The zone is controlled by a series of graben structures near the axis of the great anticline of Shanxi plateau; hence it is also called Fen–Wei graben seismic zone.

Second, the North China plain seismic zone, which is divided into east and west branches, restricted by the east and west marginal tectonic zones of the North China depression. The west branch starts from the Xinxiang–Tangyin graben in the south, via the Xingtai–Shulu graben, to the Sanhe graben in the north; the east branch begins at the Heze graben in the south, through the Changdong fracture zone, to Tangshan in the north.

Third, the Tanlu seismic zone, which begins at Linyi, Shangdong Province in the south, and runs through the Bohai Sea to Haicheng, Liaoning Province in the north. It is generally constrained by the Tanlu fracture zone and its northward extension.

Fourth, the Huanghai (Yellow Sea) seismic zone, which is widespread, possibly due to poor epicentral locations. From the dense distribution of small earthquakes in Fig. 3.5, however, it can be seen that the NE seismic zone can be still identified. Geophysical data on the sea floor are limited, but it appears that the important NE tectonic zone may exist, generally parallel with the seismic zone.

3.2.2 NW–WNW Tectonic Seismic Zone

In the past, little attention has been paid to the existence of regional fracture tectonics and seismic zones in this direction. From analysis of small earthquakes recorded at regional seismic stations after the Xingtai earthquake, it was learned that the NW seismic zone existed, and further seismogeological work and (later) satellite images gradually deepened our knowledge of the tectonics in this direction. The tectonics feature a zone of rather short fault segments which can be connected intermittently, and epicentral distributions along this direction also indicate similar features. Some of the tectonic features in this direction cut through the NE tectonics, while others are truncated by them. The junctions of these two sets of tectonics play a significant role in controlling strong earthquake locations. There are three clear WNW seismotectonic zones in the North China seismic region:

a) Huailai–Beijing–Tianjin–Yantai seismic zone;

b) Daixian–Shijiazhuang–Dezhou–Weifang seismic zone;

c) Qixian–Anyang–Linyi seismic zone.

NW–NWW tectonics also appears in the Xingxiang area, to the north of Zhengzhou City, where it plays a role in controlling the distribution of moderate and small earthquakes. Cenozoic basaltic eruptions (Fig. 2.1) are sometimes seen at the junctions between these tectonic zones and the NE zone, showing that these NW–WNW tectonic zones are also deeply incised crustal fractures. As well as having fault movements of their own, these fractures also dominate the uplift and downdrop of a series of NWW faulted blocks, as described in paragraph 3).

3.2.3 EW Tectonic Seismic Zone

The relationship between the EW tectonics and earthquakes at first seems less obvious than that in the NE zones, but both may in fact play a greater role in controlling earthquake occurrences. From Figs. 3.4 and 3.5, we see that the south and north boundaries of the North China seismic region are in nearly EW direction, and there are distinct EW alignments of strong earthquakes within this region. Three main seismic zones are described as follows:

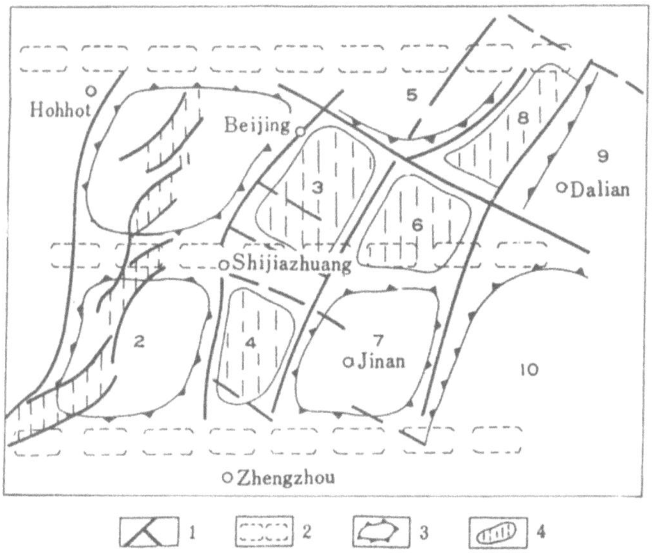

Fig. 3.6. Sketch map of the crustal tectonic framework of North China; *1* Fault block upheaval area of north Shanxi Province: *2* fault block upheaval area of south Shanxi Province; *3* fault block depression area of central Hebei Province; *4* fault block depression area of south Hebei and north Henan Provinces; *5* fault block upheaval area of east Hebei Province; *6* fault block depression area of Bohai; *7* fault block upheaval area of west Shandong Province; *8* faut block depression area of Liaodong Bay; *9* fault block upheaval area of Liaodong; *10* fault block upheaval area of west Shandong Province;

(1) fractures; (2) EW tectonic zone; (3) fault block upheaval area; (4) fault block depression area

1) The south marginal seismic zone of the Yinshan EW tectonic zone runs from Wuyuan in the west, through Huhehot and Beijing to Haicheng in the east. This seismic zone features radically different occurrences of earthquakes to the south and north parts of it (Fig. 3.5). The tectonics of southern marginal Yinshan

is very complicated, with some intense EW fault depression basins, and with some intersections between the large EW regional fractures and the NE and NW structures, which give rise to complex fault depression basins. Most of the strong earthquakes are located at the edges of these active basins.

2) The southern boundary seismic zone of the North China seismic region runs from Xian in the west, through Heze and Linyi and finally to the Yellow Sea. The earthquakes to the south and north of this zone are also radically different (Fig. 3.5). Geophysical explorations have shown that a buried EW tectonic zone may exist along this direction.

3) The EW seismic zone of the North Central China seismic region, runs from Taiyuan in the west, via Shijiazhuang and Dezhou, to Jiaodong Bay. This zone shows a clear EW negative anomaly in the airborne magnetic charts (Zhu Ying 1979), with apparently different anomaly distributions to the south and north. It also seems to mark a change in the distribution of small earthquakes. There is a striking structural contrast between the recent depression in central Hebei Province and in Bohai to the north, and the Shangdong Upheaval to the south in the east part; and also a striking contrast between the Wutaishan uplift in northern Shanxi Province and the Qinshui syncline in southern Shanxi. This demonstrates that this seismic zone has a controlling influence on the deeper crust, and also plays some role in the seismicity.

The three tectonic seismic zones intersect each other, forming a very complex network, and they also delineate the framework for dividing the North China crust into fault blocks, as shown in Fig. 3.6.

This tectonic framework controls the strong seismicity of North China. However, there are differing opinions on the dynamics of the present tectonic movements. Some researchers emphasize the thermodynamics of the mantle uplift, and some stress the general extension of the crust; whereas we have suggested that both compressive and torsion tensile stresses exist simultaneously.

3.3 Hypocentral Tectonics of Four Strong Earthquakes in North China

Using the regional tectonic framework discussed above as a basis, this section will outline the tectonics of the Xingtai, Bohai, Haicheng and Tangshan earthquakes (Fig. 3.7). The following describes these earthquakes in order of their tectonic complexity:

3.3.1 Bohai Earthquake

Geophysical explorations and drilling performed by the Ministry of Petroleum in the Bohai area confirm that the Tanlu fracture zone extends into the Bohai. The fracture at its east side passes through the west of Miaodao and is called the west Miaodao fracture near Haicheng (Fig. 3.7; 1), and the fracture at its west side reaches the central Bohai and is called the central Bohai NE fracture (Fig. 3.7; 2). To the west of the central Bohai NE fracture, the Cenozoic sediments are very thick, of more than 4000 m. The local central Bohai fault block upheaval has been developed in these sediments, and the central Bohai–Yantai

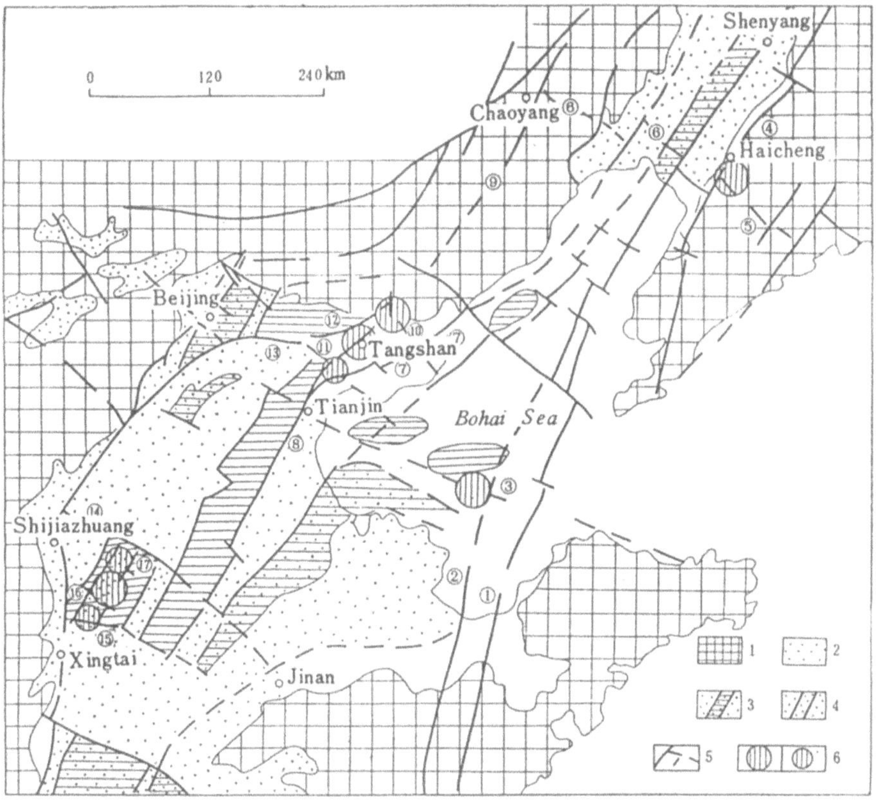

Fig. 3.7. Tectonics of the epicentral region of four strong earthquakes with Ms > 7 in North China from 1966 to 1976; (1) West Miaodao fracture;(2) Central Bohai NE fracture; (3) Central Bohai–Yantai NW fracture; (4) Haicheng–Jinzhou fracture; (5) Xiaogushan–Xiuyan buried fracture; (6) Yinkou– Chaoyang buried fracture; (7) Changli fracture; (8) Cangdong fracture; (9) Chaoyang–Jianchang fracture; (10) Luanxian NW fracture; (11) Ninghe fracture; (12) Qianan NE fracture;(13) Baodi fracture; (14) Shenxian NW fracture; (15) Longyao NW fracture; (16) Shulu NE fracture; (17) Julu–Baichikou fracture; [1] mountain areas of basic rocks; [2] quarternary overburden areas; [3] uplifts (horsts) in overburden areas; [4] depressions (grabens) in overburden areas;[5] large fractures; [6] epicenters (M = 7 or 6)

WNW fracture at the south side of the upheaval is the east segment of the Huailai–Beijing–Tianjin–Yantai tectonic seismic zone mentioned in the previous section. The seismogenic tectonics of the 1969 Bohai earthquake hypocenter might be the contact position between the central Bohai–Yantai fracture and the central Bohai NE fracturre. The occurrence of two nodal planes for the focal mechanism solution and the orientation of small earthquake distributions before and after the mainshock agree with the strike of these two fractures, hence indicating the role played by both fractures. Therefore, it can be considered that the Bohai earthquake resulted from a conjugate seismogenic structure.

3.3.2 Haicheng Earthquake

Before the occurrence of the 1975 Haicheng earthquake, attention was paid to the revived activity of the Haicheng–Jinzhou NE fracture (Fig. 3.7; 4), but the existence of the WNW fracture was not noticed, since the local fracture along the resultant earthquake crack zone could not be identified from regional geological and gravity surveys. However, 1 or 2 months before the mainshock, large quantities of macroscopic anomalies appeared in the Dangdong and Xiuyan areas, and then the anomalies spread WNW towards the epicentral area, so that people began to suspect the existence of a buried NWW fracture. Post–earthquake investigations revealed a set of intermittently connected fractures in the piedmont geomorphology along the ESE Ximuhe course from Haicheng to Xiaogushan. This fracture might spread south–eastwardly to the Xiuyan and Zhuanghe areas, which is the presently supposed Xiaogushan–Xinyan fracture (Fig. 3.7; 5). By comparing leveling data obtained in 1958 and 1970, it can also be seen that consistent bending occurred in the contours of elevation variations of this fracture (Fig. 6.5a), showing the modern activity of this tectonic line. It should be noted that recognition of such buried fractures before an earthquake is very difficult in the absence of sufficient data on recent crustal movements.

The Xiaogushan–Xiuyan fracture and the Haicheng–Jinzhou fracture meet at a " T" junction, but the Haicheng earthquake occurred on the Xiangushan–Xiuyan fracture, about 20 km away from the junction (Deng Qidong et al. 1976). Therefore, the Haicheng earthquake occurred possibly due to the dislocation of the only locked segment of the Xiaogushan–Xiuyan fracture. Small earthquake activities and other anomalous phenomena during the seismogenic process of the Haicheng earthquake, however, have elucidated the continuous activity of the Haicheng–Jinzhou fracture, and a great number of sudden anomalies also took place prior to the mainshock along this fracture, revealing its weakening and creeping. It is suggested that the Haicheng–Jinzhou fracture together with the Xiaogushan–Xiuyan fracture played a common role in the seismogenic process of the Haicheng earthquake. The Xiaogushan–Xiuyan fracture does not cross the Haicheng–Jinzhou fracture, but to the west of the Haicheng–Jinzhou fracture geomorphological and geological data suggest an intermittent fracture zone through the Liaohe plane, that is, the WNW Yingkou–Chaoyang fracture (Fig. 3.7; 6). About 1 year before the Haicheng earthquake, a zone of small seismic activity appeared roughly along this WNW direction. From Fig. 3.7, it can be seen that fractures (6) and (5) belong to the same regional buried fracture zone, which, in general, crosses the Haicheng–Jinzhou fracture, and exactly along which the two major axes of the isoseismal contours of the Haicheng earthquake extended. Therefore the Haicheng earthquake is similar to the Bohai one, resulting from conjugate seismogenic structure and dislocation movement.

3.3.3 Tangshan Earthquake

The source region structure and the regional tectonic background are very complicated for the Tangshan earthquake which is located at the boundary of several regional tectonic zones. The NE Cangzhou upheaval zone extending NE to Shanhaiguan meets the EW tectonic zone at the piedmont of Yanshan spreading eastwardly from Baodi, and forms a fairly wide boundary in the Tangshan area. This forces the NE tectonic line to become ENE. In terms of regional features, the Tangshan area is situated just to the north of the junction between the Zhangjiakou Yantai WNW fracture zone and the west Liaodong Bay fracture (Shanhaiguan–Changli–Cangzhou), that is at the apex of the southern fault block of east Hebei Province (Fig. 3.6; 7).

The seismotectonics of the source region features Tangshan rhombic fault block structures (Guo Shunming 1977), with the Shanhaiguan–Changli fracture in the southeast (Fig. 3.7; 9), the Luanxian fracture in the northeast (Fig. 3.7; 10), the Ninghe fracture in the southwest (Fig. 3.7; 11), and the (speculative) NE Qianan fracture in the northwest. From data on geology, geophysical prospecting, geomorphology, neotectonics and deformation, especially preseismic anomalies in water level, and on the dynamic features of aftershock activities in the Ninghe and Luanxian area, the EW and NE fractures have been proved to exist within the rhombic fault block. They intersect near Tangshan City, possibly providing the seismogenic tectonic conditions for the M7.8 Tangshan earthquake. All in all, the tectonics of the Tangshan source region exhibits complex movements of the fault block with multiple intersecting fractures.

How deep do these fractures reach into the crust? Seismic sounding and converted seismic wave studies have revealed that there are indeed deep fractures cutting the lower crust in the Tangshan area, but there is no evidence for these fractures reaching the surface. It can be seen from the profile (Fig. 3.3) given by the wave conversion method, that the basaltic bed between the C and M planes varies in thickness, with local thickening at the core of the upheaval structure, but without fractures. The thickness of the granitic layer between the G and C planes does not change much; this is the seismic competent layer, in which the speculated fractures are confined. Therefore, it can be imagined that the observed shallow depth fractures might not cut the deeper crust, but their activities, along with the movement of the fault block itself, nevertheless indicate the tectonic deformation features of the subcrustal earthquake competent layer. The upward bending uplift of the competent layer, and the downward bending and local thickening of various upper mantle layers below the M plane, together with the tectonic deformation of the crust, imply that a tectonic zone exists within the lithosphere.

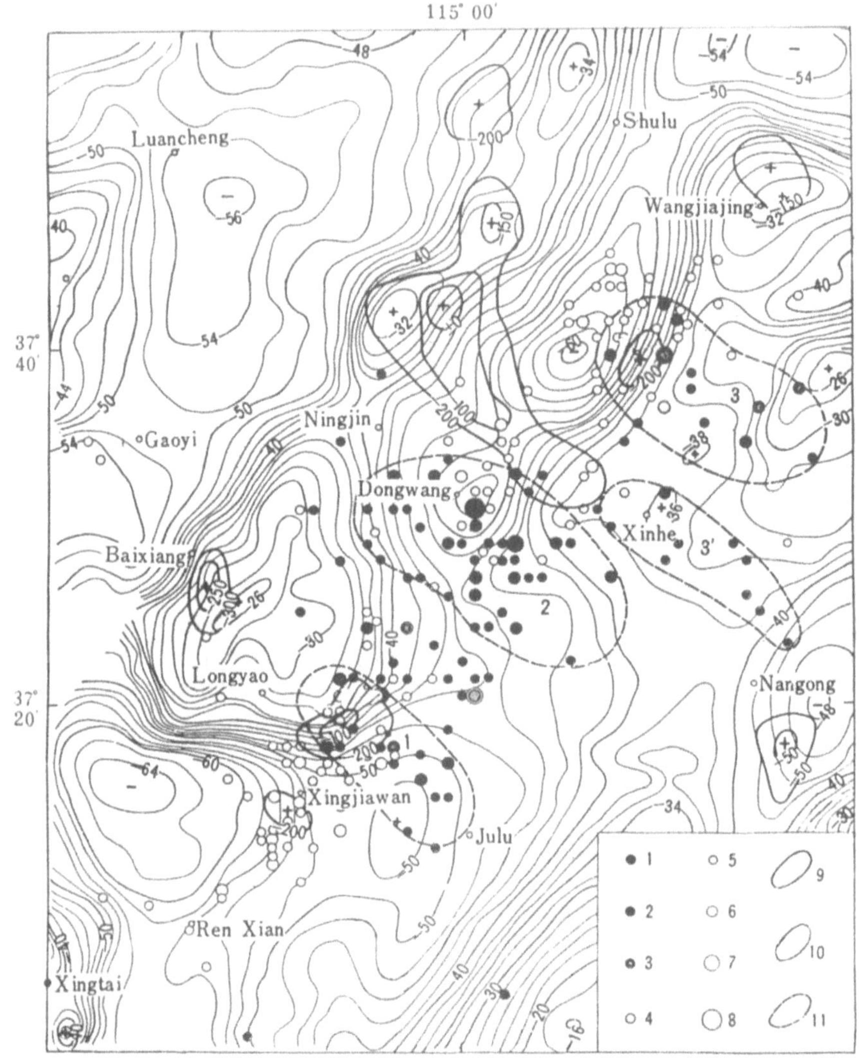

Fig. 3.8. Gravity, geomagnetism and epicenters in the Xingtai seismic area; *1* epicenters (from 6 March to 21 March 1966; *2* epicenters(from 22 March to 26 March 1966); *3* epicenters (from 27 March to 6 April 1966); *4* epicenters(from 7 April 1966 to June 1968); *5* M=4.0–4.9; *6* M=5.0–5.9; *7* M=6.0–6.9; *8* M>7.0; *9* magnetic highs; *10* gravity contours (mgal); *11* aftershock distribution of each strong earthquake in the sequence

This may be the deeper tectonic background of the Tangshan earthquake area, and may also be representative of the North China region.

3.3.4 Xingtai Earthquake

The NE depressed tectonic zone in the North China Plain is not homogeneous or uniform; it has been cut by several WNW fractures, forming interlaced uplifts and depressions. An outstanding fault block uplift zone from Shijiazhuang to Jinan (Fig. 3.6) is dissected by a set of NE fractures which developed as a series of horsts and grabens during the Cenozoic, as given by the "Yuanshi–Jinan" profile in Fig. 3.2. From shallow geophysical prospecting and the thickness distribution of the Cenozoic sediments, it is ascertained that fractures exist at the north and south boundaries of this fault block upheaval, with the Shenxian fracture to the north (Fig. 3.7; 14), and the Longyao–Julu fracture to the south (Fig. 3.7; 15). Four strong earthquakes of the Xingtai earthquake sequence (March 8 earthquake of M6.8; March 22 earthquakes of M 6.7 and 7.2; March 29 earthquake of M6.0) and their aftershocks are basically located within the Shulu graben range with the Baichikou–Julu fracture at the east (Fig. 3.7; 16) and the Shulu fracture at the west (Fig. 3.7; 17). As mentioned previously, the deep crustal structures of the Shulu graben are fairly complicated, as seen from gravity and geomagnetic data (Fig. 3.9). The geomagnetic highs show local intrusions of magnetic bodies, which are distributed along both sides of the graben. The epicenters of earthquakes with $Ms \geqslant 4$ from 6 March 1966 to June 1968 are given in Fig. 3.8. They are generally distributed in a NE orientation, but four of them with $M > 6$ and their aftershocks are aligned NWW, showing the probable existence of a WNW fracture.

The gravity contours and magnetic anomaly areas may also support the above–mentioned extrapolation. Generally speaking, the source region of the Xingtai earthquake seems to be a narrow and small NNE tectonic body, which is dissected by several major WNW fractures. The seismogenic and developmental process of the Xingtai earthquake has revealed the complicated strain and overlapping dislocations of the small fault blocks in this tectonically complex area.

3.4 Regional Tectonic Features of the Sichuan–Yunnan Seismic Region in Southwest China

The Sichuan–Yunnan seismic region is the meizo–seismically most frequent region in Mainland China. Within the first 80 years of this century, there have been 12 sequences of strong earthquakes with $M \geqslant 7$. The epicenters of earthquakes of $M \geqslant 4$ since 1970 and those of earthquakes with $M \geqslant 7$ (including historical earthquakes) before 1970 are given in Fig. 3.9. Four seismically active zones in NS, EW, NE and NW orientations are shown, as are several quiet areas of various shapes and sizes. The distribution patterns of these earthquakes are rather similar to the post–Cenozoic tectonic framework and there are clear geological expressions of almost every active zone (Li Ping 1979).

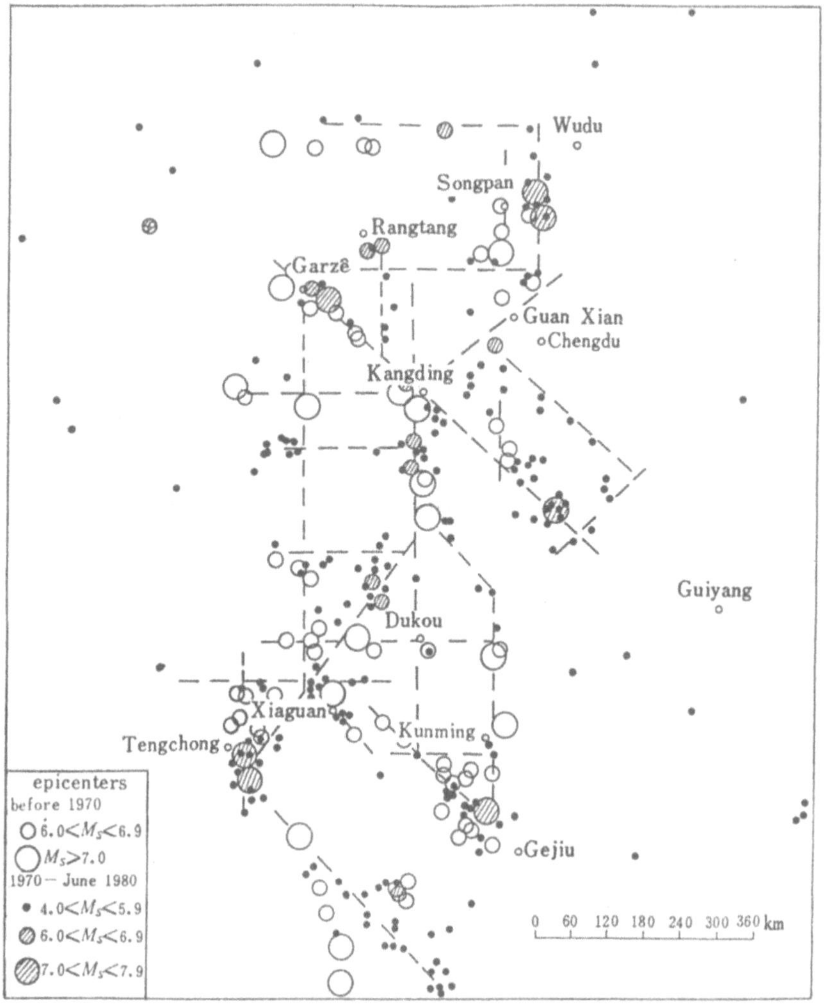

Fig.3.9. Epicentral map of earthquakes in the Sichuan–Yunnan region (earthquakes of M> 4.0 since1970; earthquakes of M> 7.0 before 1970)

The Sichuan–Yunnan seismic region has a very complex geological history. In short, a line starting from Guanxian, through Kangding, Xichang, and Xiaguan, to Gejiu, divides the Palaeozoic and Mesozoic geosynclinal area to the west from the Palaeozoic platform area to the east, but the platform area close to the line has undergone intense tectonism. This seismic region is basically formed from several fault blocks with different crustal thicknesses that result from the Indo–Chinese and Yanshan movements. During tectonic movements since

Cenozoic, the fault blocks have mainly embodied internal deformation, including compression, dislocation and extension; and horizontal movements have governed the differential rise and fall of the relief, and the development of some hundred tectonic basins of large and small scale and of different tectonic natures.

Yan Xianfu compiled an outline map of crustal thickness in Southwest China mainly based on regional Bouguer gravity observations. It shows striking differences in crustal thickness for various fault blocks, and in addition, crustal thickness differs greatly on either side of several major tectonic lines in NS, EW and oblique orientations.

By comparative analysis of the above data, we suggested that the present tectonism of the region is characterized by uniform body strain of the block–combined complex and by relative motions of interblock fractures. This led us to compile an outline map of the crustal structures in the Sichuan–Yunnan region (Fig. 3.10). The main structures of the region are as follows:

3.4.1 Central Sichuan Fault Block (A)

This block looks like a rhombic biconcave lens with a thin crust at its center and a thick crust around its peripheries. The central part is the concave area of Sichuan Basin, which lies above an area of mantle upheaval, with crustal thickness of 39 km. The crustal thickness is about 45 km for the peripheries and normally the block has a thickness of about 42 km. The thickness varies, mostly at the NW and SW boundaries of the block. The seismicity is very weak in the interior of the block, with a few earthquakes of magnitude up to 4 located along two fractures in the NE direction.

3.4.2 Central and East Yunnan Fault Block (B)

This is a compressed rhombic slab, short in the EW direction and long in the NS. The crustal thickness is generally about 47–48 km, becoming thinner toward the NE and SE boundaries, and thickening suddenly toward the northwest boundary. The well–known Red River fault is just located at the southwest boundary, with slight variations in thickness on either side of the fault, and only with local areas of different thickness. Within the center of this fault block, there is a NS zone of crustal thickening, which is the geologically famous Kangding–Yunnan geoaxis, called by us the Kangding–Kumming longitudinal (NS) tectonic zone (λ_E). Within this zone, the block has been divided naturally into two triangular slabs, the Central Yunnan (B_W) and East Yunnan (B_E) slabs. The longitudinal zone is very ancient, the east and west slabs having experienced different geological processes since at least the Proterozoic era. The massive Permian basaltic eruptions cover only the east slab of the block (B_E), while large–area land facies accumulations of the Jurassic and Cretaceous occur almost entirely to the west (B_W), similar to those on the central Sichuan fault block. Differential movements of the east and west blocks have continued to this day.

3.4.3 Northwest Yunnan–West Sichuan Fault Block

This is an almost triangular fault block, whose east "apex" is truncated by the longitudinal tectonic zone (λ_E) to have a trapezoid with a short NS side. The crustal thickness is generally 64 km, with its greatest change at the Kangding–Xichang side, where there is a steep "cliff", with 10 km difference on either side. The crustal thickness of the southeast segment near Xichang has also varied considerably. At the west boundary of the fault block, there is the Batang zone of thinning in a NS direction, with a variation amounting to 2 km. The northeast boundary of the fault block is the Xianshuihe fracture zone.

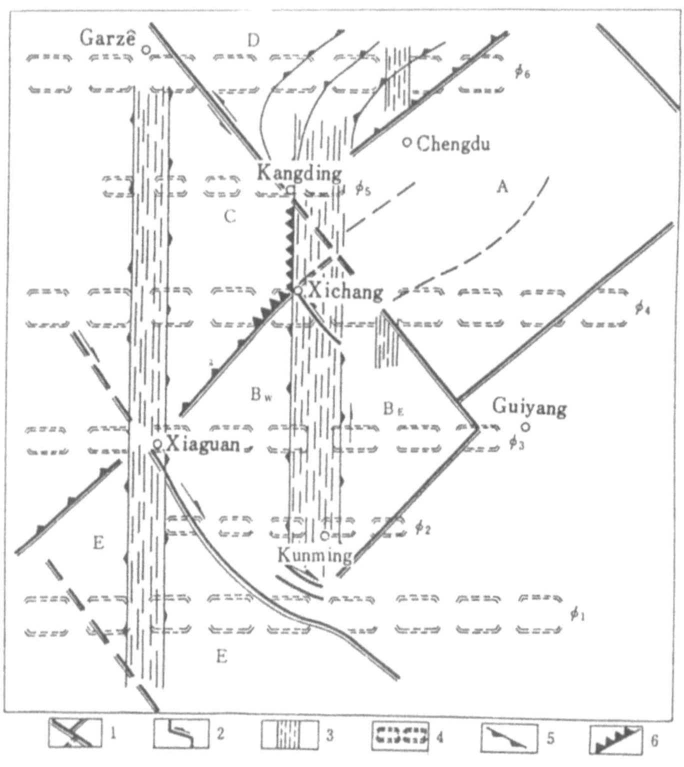

Fig. 3.10. Outline map of crustal structure in the Sichuan–Yunnan region; *A* Central Sichuan fault block; *B* Central and East Yunnan fault block; *C* Northwest Yunnan–West Sichuan fault block; *D* North–west Sichuan fault block; *E* Southwest and South Yunnan fault block; λ_E Kangding–Kunming longitudinal tectonic zone; λ_W Ganzi–Xiaguan longitudinal tectonic zone; φ latitudinal tectonic zone; *1* block boundary; *2* dislocation orientations; *3* longitudinal tectonic zone; *4* latitudinal tectonic zone; *5* compression boundary; *6* intense compression boundary

3.4.4 Northwest Sichuan Fault Block (D)

The Longmenshan fracture zone is the southeast boundary of this fault block, which has a continuous 20 km variation in crustal thickness. This zone of thickness gradient joins the Northwest Yunnan–West Sichuan fault block transversely, thus causing significant deviations in crustal thickness on both sides of the eastern segment of the Xianshuihe fracture zone, and so showing the difference between the southeast and northwest segments of this zone. The crustal thickness of the block usually ranges from 62 to 64 km. Its north boundary might reach the north edge of the Kunlun Mountains, so it is not a rhombic fault block.

3.4.5 Southwest and South Yunnan Fault Block (E)

There is a northwest–oriented zone of varying crustal thickness from Jinghong to Luxi in this block, and possibly there is a fracture zone dissecting the crust. The crust between this zone and the Red River is an elongated plate–like body of 45–47 km thickness, which tends to thin southwestwardly. The Xiaguan–Luxi line seems to be the northwest boundary of the block, and the geology suggests that a northeast–directed deep fracture may exist along a line from Baoshan to Luxi to Ruili. However, the southeast boundary of the block cannot be recognized clearly. A NE fracture may exist between Linchang and Yuanmou, and the minor earthquakes from Gejiu to Simao are sometimes aligned NE. However, these have not enogh to show a definitive southeast boundary for the block. As the block is greatly affected by the longitudinal and latitudinal tectonic zones, the crustal thickness changes very much, both longitudinally and latitudinally.

3.4.6 Kangding–Kunming Longitudinal Tectonic Zone (λ_E)

This complicated NS tectonic zone lies approximately within 102 ° –103 ° E running from Maerkang, to the north of Kangding, via Xichang, to Kunming and Tonghai. It has undergone long and multiple tectonic movements. A longitudinal crust–thickening zone has been produced, with a thickness of 55 km in the north and 53 km in the south and there are gradient zones of sudden crustal thinning at both sides, the Xiaojiang fracture zone to the east and the Anninghe fracture zone to the west. From the surface geology, the zone has been considered to be confined between Kangding and Kunming in several segments. However, if one takes into consideration the overall crustal thickness variations of the zone, it extends as one entity into the interior of the Northwest Sichuan fault block from south to north and does not terminate at Xichang, nor at Kangding. It possibly spreads further north beyond Maerkang, and even through the Qinling Mountain range. The thickness of the tectonic zone itself also shows clear fluctuations, which are aligned roughly NE in enechelon fashion. The zone has undergone many periods of folding, fracturing, metamorphism and magmatic activity, and evidence of multiple tectonism (such as compression, extension, offset, uplift and descent) remains in the valley areas in which the fractures of both the east and west sides of the zone lie (Teng Dezen 1978).

3.4.7 Ganzi–Xiaguan Latitudinal Tectonic Zone (λ_W)

This is a most remarkable tectonic zone or tectonic boundary which occurs at the east Shanjiang and Transverse Range. It is a most clear cutzone, which runs southwardly into the Southwest Yunnan fault block, and northwardly into the Northwest Sichuan fault block. The Ganzi–Litang segment of this tectonic zone is a zone of crustal thinning, while the Litang–Xiaguan segment overlies a wide slope of varying thickness, and the Xiaguan–Gengma segment is a zone of crustal thickening in a nearly NS orientation. This tectonic zone has also underwent a very complicated history, particularly during multiple magmatic activities in Mesozoic and Cenozoic, when the dynamic metamorphism was even more outstanding.

Taking a larger–scale view, the above two longitudinal tectonic zones ($\lambda_E$and λ_W) appear to be the edges of a much wider longitudinal tectonic zone, which runs through China and Mongolia (Ma Zongjin 1984b). The crust between these east and west sides shows axial concavo–bending deformation in a NS orientation. This is due to east–west compression, as shown clearly in Fig. 3.10. In addition, to the east of the λ_E zone, there may exist another narrow longitudinal tectonic zone (Fig. 3.10) which extends intermittently from Songpan, Northwest Sichuan to Mabian, Zhaotong and Yongshan at the boundary between Yunnan and Sichuan, which is associated with the recent strong Songpan and Yongshan earthquakes.

3.4.8 Latitudinal Tectonic Zone (φ)

From analysis of geological structures, crustal thickness variations, epicentral distributions and the uplift zone of the modern dividing crest between Yunnan and Sichuan which was given by Li Xianggen (1977), we have come to understand that there are a series of latitudinal tectonic zones which have been developed to different extent from south to north in the Yunnan–Sichuan region. The following six main zones can be noted: Gengma–Gejiu (φ_1); Baoshan–Kunming (φ_2); Heqin–Dongchuan (φ_3); Zhongdian–Xichang (φ_4); Litang –Kangding (φ_5); and Ganzi–Maowen (φ_6). These latitudinal tectonic zones have seldom expressed themselves directly in continuous extended fractures, but principally as zones of narrow concavo–bending deformation or of crustal thickness variation. Influenced by these tectonic zones, other structures in different directions have been forced to bend and distort. The zones also control the locations of strong earthquakes, which are aligned in an EW direction, and migrate in an orderly manner. Some of the latitudinal tectonic zones are distinct only within wide and extensive longitudinal tectonic zones, but the majority extend far on either side. These cannot be described in further detail because of the limited space at disposal in this section.

To sum up,the fault blocks and block boundary fracture zones intersect the latitudinal and longitudinal tectonic zones to form a rather complex and regular tectonic framework. And the strong earthquakes have occurred at the tectonic intersections and block boundaries. That is the basic contour of regional geological structures in this region.

3.5 Hypocentral Tectonics of Five Large Earthquakes in the Sichuan–Yunnan Region

Descriptions of the hypocentral tectonics follow in order of their complexity:

3.5.1 Luhuo Earthquake

This earthquake produced co–seismic cracks of typically strike–slip form, and investigations of seismic intensity and aftershocks further demonstrated that the earthquake occurred within the Xianshuihe fracture zone (Fig. 3.11). The nearly vertical dip of the fracture, and the linear distributions of basic and ultrabasic rocks along it indicate that this is a deep crust–cutting fracture zone. The relics of new and old tectonic movements at both sides of the zone, and offsets of rock bodies and streams demonstrate that this is a left–lateral strike–slip fault that has developed over a long period of time. This fracture zone is not a single fault, but a zone of several parallel fractures, with a width of 5 km or so. Different direction fractures intersect the Luhuo, Daofu and Qianning segments,

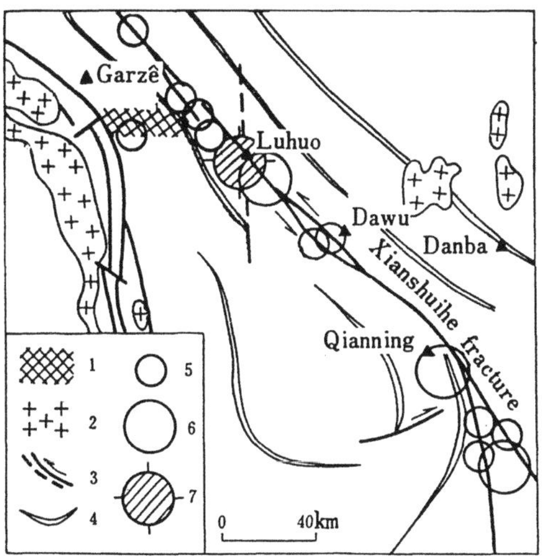

Fig. 3.11. Sketch of hypocentral tectonics for the 1973 Luhuo earthquake; *1* recent intense uplifting zone; *2* granite; *3* fault; *4* folding axis; *5* historical epicenters(M = 6.0–6.9); *6* M = 7.0–7.9; *7* Luhuo epicenter

forming complex tectonic points along the strike–slip fault zone, where small–sized tectonic basins have developed. These complex tectonic points have controlled the locations of several strong earthquakes, and these are the tectonic conditions for high–intensity earthquakes to repeat there. Small–scale NS and NE fractures cross the zone near Luhuo, and a latitudinal tectonic zone crosses where the north apex of the Northwest Yunnan–West Sichuan fault block is also located. This may be the tectonic basis for the frequent strong seismicity in this region.

3.5.2 Tonghai Earthquake

The Tonghai earthquake occurred at a unique tectonic position at the south apex of the East and Central Yunnan fault block (B). This lies within both the Kangding–Kunming longitudinal tectonic zone and the Gengma–Gejiu latitudinal tectonic zone, and is therefore prone to stress accumulation and release due to tectonic movements from various origins. Hence this region has one of the highest frequencies of earthquakes of M over 6 in Mainland China (Fig. 3.12). There have been six quakes with $M > 6$ within this century.

The Tonghai earthquake occurred in the Qujiang fracture zone, producing a co–seismic crack zone of 50 km in length, with a maximum right–lateral horizontal dislocation of 2.3 m, about ten times the vertical dislocation. The Qujiang fracture zone measures 120 km in length, and its eastern and western extremes do not extend beyond the corresponding eastern and western boundaries of the Kangding–Kunming longitudinal tectonic zone. This illustrates that the formation and movement of this tectonic zone is related to the dynamic trend of the Kangding–Kunming tectonic zone. Furthermore, the Qujiang fracture zone is bent southwardly, consistent with the bending of the Red River fault zone, as both are right–lateral displacements. This also indicates that both zones are affected by the Gengma–Gejiu latitudinal tectonic zone. In addition, the elevation of the Tonghai zone is about 2000 m on average, but drops suddenly to 800 m southwardly towards the Red River region. Zhu Chengnan et al. (1978) analyzed the height variations of Neogene sediments and terraces and suggested that altitude changes ended during the Quarternary, whereas the Qujiang fracture zone began its right–lateral slip in the early Pleistocene (so far the total displacement is of 5 km), causing elevation variations at the same time. From the above analysis, it is seen that althougth the Tonghai earthquake occurred in a strike–slip fault zone, the tectonic background of its source region seems much more complicated than that of the Luhuo source region. In short, the Tonghai earthquake was due to the movement of a fairly short strike–slip fault zone whose two ends were restricted, and it was accompanied by regional activities like compression, uplift and depression. Therefore, its aftershocks “jumped between both ends” (Zhang Shurong 1980), that is the aftershock activity alternated between Quxi and Eshan, the two ends of the co–seismic zone. This was not the case for the Luhuo earthquake.

3.5.3 Yongshan Earthquake

No obvious ground breaks were caused by the Yongshan earthquake, possibly because of its small hypocentral rupture. According to the spectral analysis of Liu Wanqin (see Chap. 2), the focal rupture was oriented NE, and the total amount of dislocation was only 0.07 m. Before the Yongshan earthquake, small earthquakes were aligned NW and pointed clearly to the epicentral area. However, the mainshock rupture and the 10–km–long aftershock distribution were mainly oriented NE. The isoseismal contours were basically elliptical with the major axes in the NE and NW fracture zones, showing a conjugate double dislocation mode. The epicentral seismotectonics is related to the tectonic environment of the surroundings. From Fig. 3.13 we see that this region is a boundary between numerous regional tectonic lines, across which the NW and NE fracture zones, as well as the longitudinal and latitudinal tectonic zones all pass. This region is also a seismically active one: the M6.5 Daguan earthquake occurred in 1917, and a M > 6 swarm took place near Mabian during 1935–1936.

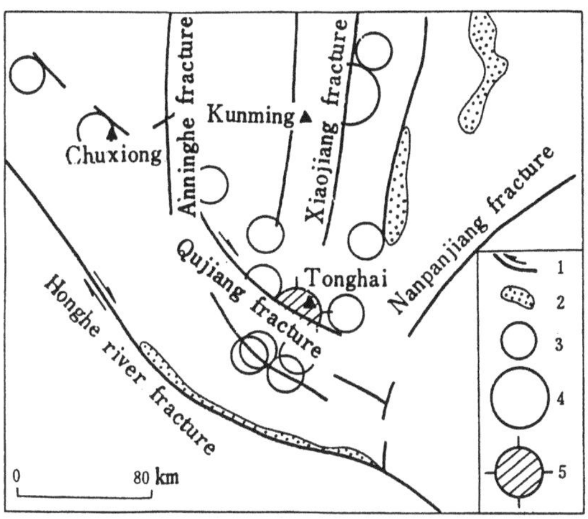

Fig. 3.12. Outline map of seismotectonics of the Tonghai earthquake; *1* fracture; *2* recent sedimentary basin; *3* historical epicenters (M = 6.0–6.9); *4* M = 7.0–8.9; *5* Tonghai epicenter

3.5.4 Songpan Earthquake

The most important neotectonic movement in this region is the presently

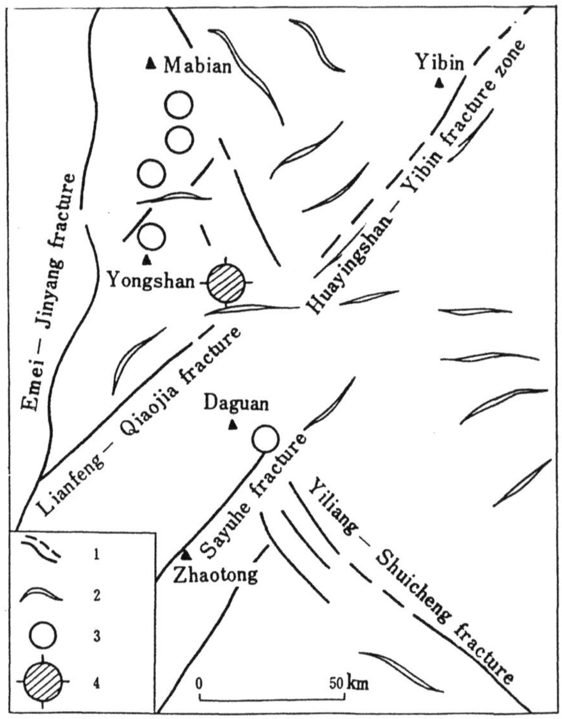

Fig. 3.13. Outline map of seismotectonics of the Yongshan earthquake; *1* fracture; *2* folding axis; *3* historical epicenters(M = 6.0–6.9); *4* Yongshan epicenter

active Xuebaoding upheaval zone (Fig. 3.14). This runs NS at an altitude of 5588 m, and divides the crest between the Mingjiang and Fujiang Rivers and the Minjiang fracture to the west, and the Tiger Tooth fracture to the east. There are many signs of neotectonic movement along both fractures (Li Ping 1979). The Songpan earthquake occurred on the Tiger Tooth fracture. Neither the Tiger Tooth nor the Mingjiang have a visible surface fracture. However, from the overall morphology of surface structures, the Xuebaoding upheaval zone seems to be a subcrustal tectonic zone that took shape as early as the Indo–China movement in the Mesozoic. The focal mechanisms of the two M7.2 earthquakes in the Songpan sequence are of the overthrusting type, with nearly NS strike, but of different dip. This suggests that during the formation of this NS upheaval zone due to EW compression, vertically conjugate faults were produced at the boundary fracture.

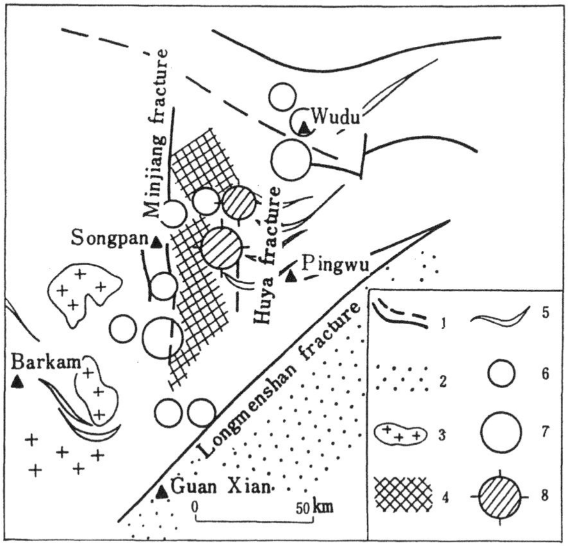

Fig.3.14. Outline map of seismotectonics of the Songpan earthquake; *1* fractue;*2* recent sedimentary basin; *3* granitic rocks; *4* recent intense upheaval zone; *5* folding axis;*6* historical epicenters (M6.0–6.9); *7* M = 7.0–7.9; *8* Songpan epicenter

3.5.5 Longling Earthquake

This earthquake occurred within a Mesozoic granitic body confined by three large fracture zones (Fig. 3.15), the Nujiang River zone, the Wanding zone and the Longling–Ruili zone. Within the triangular area, there are some NNW and ENE fractures and rupture zones, but no large regional faults. Therefore, it might be concluded that this sizeable rock body is a quasi–homogeneous entity. The rupture zones within this body, by analogy with laboratory samples under compression that produce multiple crossing microcracks before macro–fracturing, might generate local stress concentrations before a major quake. Thus the tectonic framework of the rupture zones might determine that each strong earthquake would have an isoseismal area and an aftershock distribution oriented both ENE and NNW (Chen Lide et al. 1979, Liu Zhengrong 1978). This is the case for the four earthquakes of the sequence with $M \gg 6$. The quasi–homogeneous rock body containing ancient rupture zones, acted on by regional dynamic forces, produced interlocking conjugate fractures along the original conjugate fault planes. This is the basic feature of tectonic movement for the Longling earthquake.

To summarize, there are two principal common tectonic conditions for the nine great earthquakes: first, the depth of the competent layers in the crust and their tectonic nature decide the focal depth of the earthquake. The competent layers are basically granitic and partly metamorphic beds in the mid crust, featuring thick platy deformations; secondly, all nine great earthquakes are due to reactiviation of existing fractures or rupture zones. Places where a fracture or

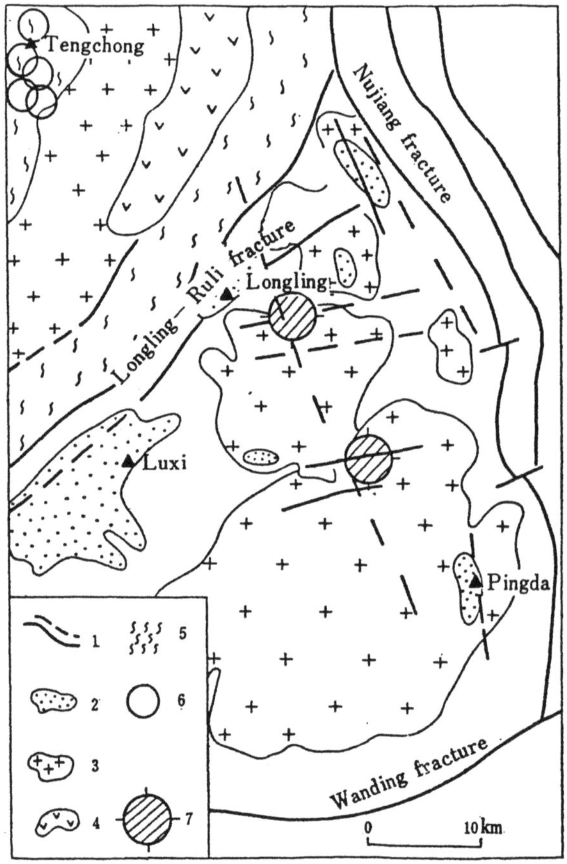

Fig. 3.15. Outline map of seismotectonics of the Longling earthquake; *1* fracture; *2* recent sedimentary basin; *3* granitic rocks; *4* basaltic rocks; *5* metamorphic rocks; *6* historical; epicenters (M = 6.0–6.9); *7* Longling; epicenter

rupture zone cuts a competent layer are the principal tectonic positions prone to local stress concentration, when they have not been only reactivated most recently, but also possess certain locking conditions, hence playing a direct role in earthquake preparation and generation. Large fractures cutting the competent layers are sometimes directly associated with active surface faults, and sometimes indirectly. Therefore, useful speculations on the tectonic movements of competent layers can sometimes be made from the expressions of surface tectonic movements.

Each of the four great earthquakes in North China and the five in the Sichuan–Yunnan region has common regional properties. Relatively speaking, the continental crust of North China is thin, and prone to bending and deforming, which makes it easy for numerous small fault blocks to be developed, and for active boundaries between blocks to vary greatly. Therefore there are various patterns for seismicity in this region, and the subcrustal thermodynamic forces that generate the local mantle uplift speed up the above behaviour. The crust of the Yunnan–Sichuan seismic region is thick, and the independent movements of fault blocks are fairly remarkable and stable, so that internal deformation of the fault blocks themselves determines the relative interblock motions.

The respective characteristics of the nine great earthquakes are, to a large extent, defined by the specific tectonic conditions at the hypocenters. In principle, the hypocentral seismotectonics can be categorized into three types on the basis of the number of fracture zones which control the source region and the manner in which they intersect: single fracturing heave type (strike–slip); double fracturing cross–dislocation type (conjugate); and multiple fracturing overlapping–dislocation type (Ma Zongjin 1980b). There is further discussion on the seismotectonic classification of source regions in Chap. 11.

CHAPTER 4

Seismicity Patterns

Since information on seismic activity contains a rather high signal–to–noise ratio, one of the most basic aspects in earthquake prediction research is to discuss the temporal, spatial and magnitude variations (or so–called seismicity patterns) of moderate and small earthquakes before a major earthquake.

At present, there are various methods for research into seismicity patterns based mainly on the time, space and magnitude parameters given in earthquake catalogues. In this chapter, some features of medium and small earthquake patterns within a range of several hundred kilometers and several years before the nine great earthquakes are discussed. At the end of the chapter, we also discuss the use in earthquake prediction of kinematic and dynamic parameters of seismic waves, such as anomalous seismic velocity and spectral variations.

It is well known that the features of the seismicity patterns before some great earthquakes are very different from others because of the different time–space dimension and magnitude range. It is thus desirable to analyze data within similar time–space–magnitude dimensions while studying the patterns before the nine great earthquakes. Only in this way can the anomalous phenomena prior to the nine great earthquakes be compared, the similarities among them be found, and even more important, the relationships between the different characteristics in different dimensions be understood.

Modern seismic observations in China can be traced to the end of the last century (a seismic station set up in Taibei in 1897), and 27 seismic stations had been established in Taiwan Province, Northeast China and the coastal region of the East Sea by 1940. Beginning about 1956, the Chinese government planned a standard seismic station network and regional networks in Mainland China. Particularly after the 1966 Xingtai earthquake, the construction of seismic stations and earthquake monitoring work have been rapidly developed. The development of seismic station networks in Mainland China is outlined in Fig. 4.1. The open and solid triangles denote 16 standard stations and about 50 regional seismic stations, respectively, that were operational by the end of 1967. At the standard stations, Kirnos–type seismographs ($T_w \leqslant 10s$; dynamic range $V=(1-2)\times 10^3$) are installed. Some stations also operate 62–type ($T_w < ls$, $V=n\times 10^5$),Harlin–type ($T < 1s$; $V=n\times 10^4$), or 513–type ($T_w = 1-10s$, $V=1-100$) seismographs. At most of the regional stations, Vegik–type microseismographs are installed, with a frequency response ranging from $V=10^4$—10^5. The Kirnos–type seismographs are also used in the regional seismic network in Southwest China. The open circles in the figure represent the regional stations (about 74) brought into operation from 1968 to 1971, and the solid circles represent the regional stations (about 110) which were put into operation from 1972 to 1980. Temporary stations are not included in the above–mentioned preliminary statistics. In all there are about 350 seismic stations marked in the figure.

From Fig. 4.1 it can be noted that the majority of seismic stations so far established are east of longitude 100 ° E. With this distribution, the regional moderate and small earthquakes before the nine great earthquakes have been recorded to some extent, and these data have been compiled by various provinces and regions in their earthquake catalogues.

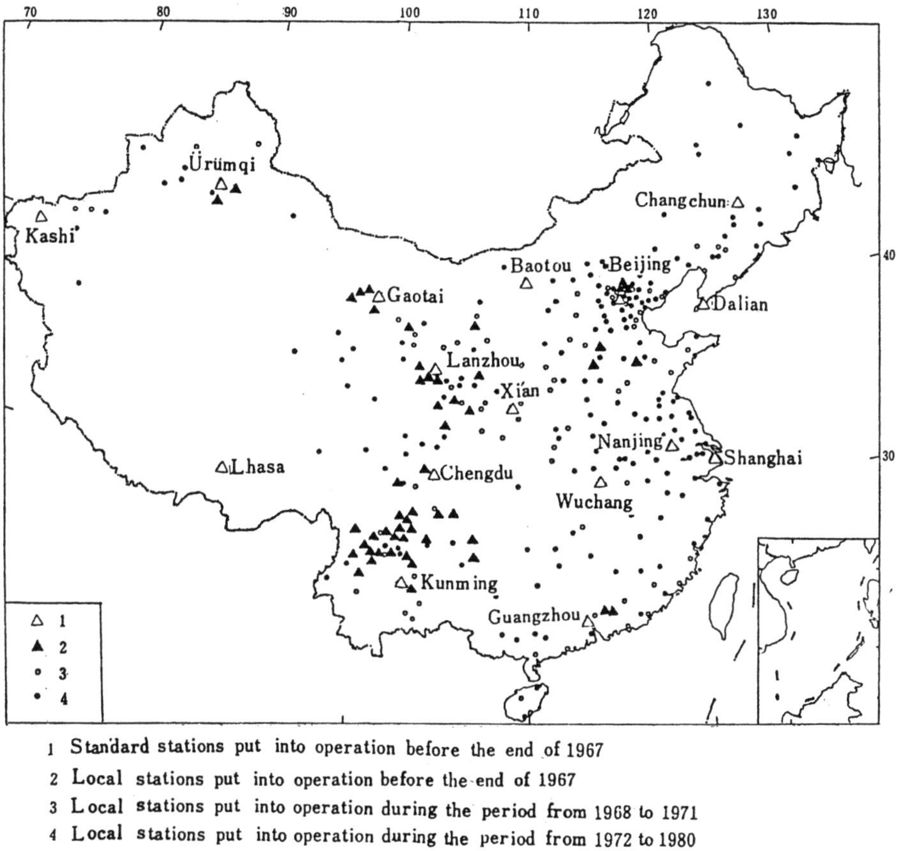

Fig. 4.1. Sketch map of development of seismic stations in Mainland China

4.1 Regional Seismicity Rate and Energy Release

4.1.1 Rate and Energy

First, let us discuss the variation in rate of occurrence and energy release of moderate and small earthquakes within several hundred kilometers of the forthcoming mainshock epicenter, for several years beforehand. According to the Gutenberg–Richter relation:

$$\log E = 11.8 + 1.5\,M. \tag{4.1}$$

the seismic wave energy E (ergs) can be derived. Among the nine great earthquakes, three (Xingtai, Bohai and Tonghai) occurred before 1970, and the data on them are not complete due to limited detectability. Therefore, it is inappropriate to compare these with the others. For the remaining six great earthquakes (Luhuo, Yongshan, Haicheng, Longling, Tangshan, and Songpan), we discuss variations in the 5 years prior to the mainshock. For $Ms \geq 7.5$ earthquakes, the area used is $5° \times 5°$; for $Ms < 7.5$, it is $3° \times 3°$. The earthquake–producing faults and their vicinities are all involved in such areas under research. The areas and minimum magnitudes used are chosen as follows: for the Luhuo earthquake 98°–103° E, 29°–34° N and $M_L = 2.5$; for Haicheng 121°–124° E, 39°–42° N and $Ms = 1.0$; for Longling 97°–100° E, 23°–26° N and $M_L = 2.5$; for Tangshan 116°–121° E, 37°–42° N and $M_L = 2.5$; and for Songpan 103°–106° E, 31°–34° N and $M_L = 2.5$.

Note that in our statistics we do not consider aftershocks of the moderate quakes occurring before the mainshock.

The resulting yearly accumulated rate and energy release statistics are shown in Fig. 4.2, from which the following three points can be illustrated:

Firstly, in general, several years before these great earthquakes, the regional seismicity rate increases, reaching a peak value 1 or 2 years before the mainshock, then remaining high or slightly decreasing. Before the Tangshan earthquake, the rate varied, but it increased every year for 3 years before the mainshock.

Secondly, the pattern of energy release rate can be divided into two types, the first being a variable decrease such as the Yongshan, Longling, and Tangshan earthquakes, with a low energy release in the year before the mainshock; and the second, as shown before the Luhuo and Haicheng earthquakes, being a drastically high level of energy release in the year before the mainshock, in complete contrast to the stable release rate of several years previously.

Thirdly, the quake frequency and the energy release rate do not vary synchronously.

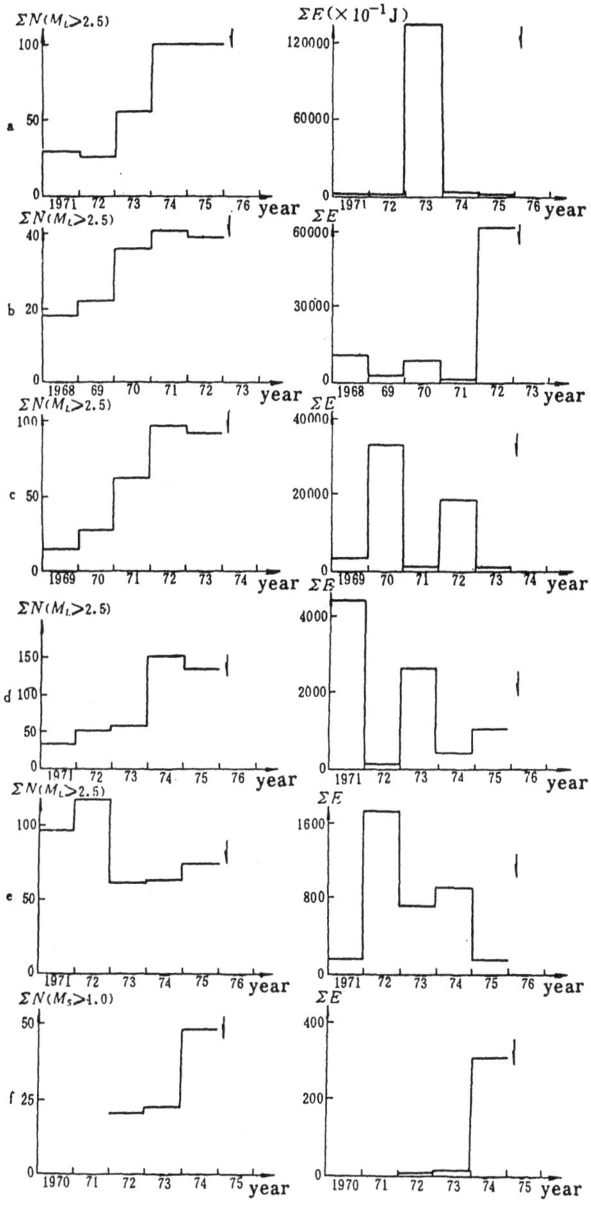

Fig.4.2. Regional seismicity rate and energy release before six great earthquakes: *a* Songpan; *b* Luhuo; *c* Yongshan; *d* Longling; *e* Tangshan; *f* Haicheng (arrows show the occurrence times)

4.1.2 b Value Variations

We now discuss variations in b value; its implications are described in subsection 3 of the first chapter.

Li Quanlin et al. (1978) and the Seismological Bureau of Sichuan Province (1979) calculated the b value variations respectively for the Haicheng, Longling, Tangshan and Songpan earthquakes, and the moderate and small earthquakes within their epicentral areas and vicinities (Fig. 4.3). The regional area for b value computation is as follows: for Haicheng 122° 00′–124° 00′E, 40° 20′–41° 30′N; for Longling 98° 10′–99° 10′E, 24° 00′–25° 00′N; and only the epicentral areas and their close surrounding regions are selected for the Tangshan and Songpan earthquakes.

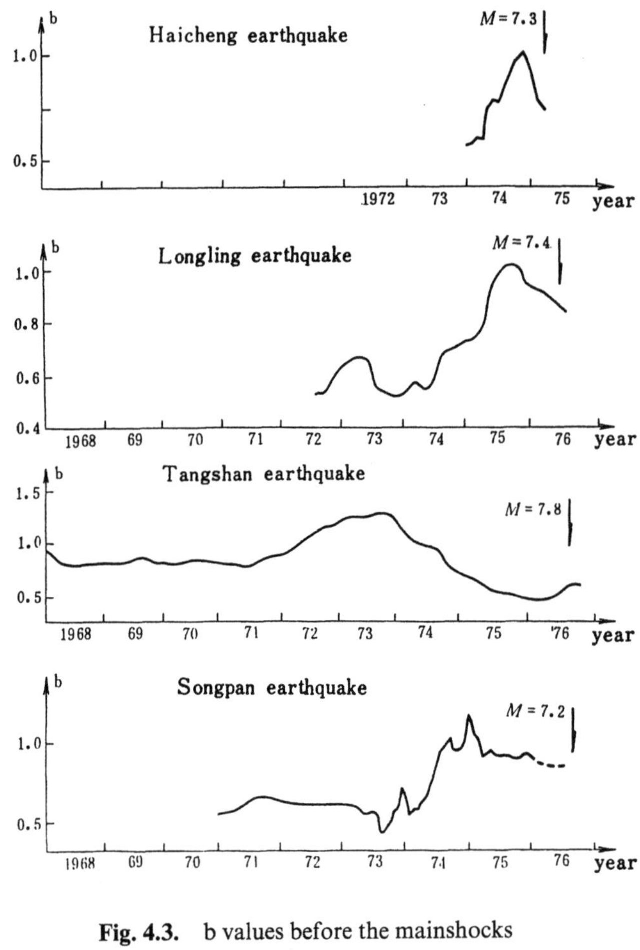

Fig. 4.3. b values before the mainshocks

The b value variations can roughly be separated into two stages:

1) the rising stage, indicating that the number of relatively small earthquakes becomes proportionally larger;

2) the falling stage, showing that the frequency of relatively large earthquakes increases gradually before the impending mainshock.

From fracture mechanics we know that the microruptures in specimens speed up prior to macrofracturing, and small ruptures join each other to become larger ones. The seismicity rate and b value variations discussed above are understandable on this basis.

We studied b value variations in the Beijing–Tianjin–Tangshan–Zhangjiakou region before the Tangshan earthquake. This region is a part of the Yanshan seismic zone, which might be divided into three areas, the Zhangjiakou area (a); the Beijing area (b); and the Tianjin–Tangshan area (c) (Fig.4.4). b values are computed for these areas from 1966 till the Tangshan earthquake (calculations made on data accumulated every 2 years, with magnitudes ranging from 2.0 to 5.0). By comparing the curves for these three areas, it can be seen that the b value in the Tianjin–Tangshan area decreases annually, while the b values in the Zhangjiakou and Beijing areas in the west are comparatively stable, hence mirroring the distinct discrepancies in the crustal strain–stress status in these areas.

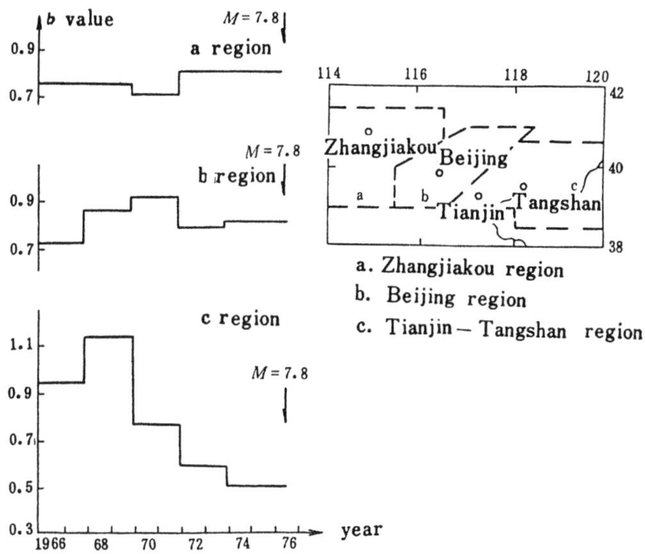

Fig. 4.4. b values in three areas of the Yanshan seismic zone before the Tangshan earthquake

In addition, Huang Deyu et al. researched the spatial distribution of b value in much larger regions before some earthquakes, and suggested that the low b value areas were associated with the locations of forthcoming mainshocks. They showed time–space scanning diagrams of b values before six great earthquakes (Haicheng, Tangshan, Luhuo, Yongshan, Longling, and Songpan), which were plotted by the scanning method of equal earthquake number (B = 50, with scanning window S = 4° × 4°. For instance, the low b value (b = 0.60) area was centered in the Tangshan vicinity, and the Tangshan mainshock occurred at the northeast part of the low b value area.

In this section, different approaches are used to calculate b value; as a result, three kinds of b value results are obtained, but with similar conclusions, e.g., the b values are unusually low within the epicentral area and its surroundings.

4.2 Spatial Distribution of Regional Seismicity

The time–dependence of the spatial distribution of seismicity will now be discussed.

Table 4.1. Band parameters of seismic activities before seven great earthquakes

Parameters / Earthquakes	Strike	Duration of formation (months)	Total length (km)	Minimum magnitude (M_L)
Bohai	NW	14	500	
	NNE	14	400	3.0
Luhuo	NW	10	550	4.0
	NE	15	480	
Yongshan	NW	13	350	3.0
	NE	12	300	
Haicheng	NW	13	1000	3.0
	NE	7	350	
Longling	NW	8	550	4.0
	NE	9	400	
Tangshan	WNW	38	500	3.0
	NE	38	550	
Songpan	NW	31	480	4.0
	NE	18	420	

4.2.1 Band Distribution of Seismic Activity

Within a certain period before the large earthquakes, the regional moderate and small earthquakes were often distributed in narrow bands, forming a very distinct seismicity pattern. Chen Zhangli et al. (1984) studied the formation and characteristics of these bands before seven great earthquakes. The band parameters are given in Table 4.1 and the band–shaped distribution of seismic activity before the Luhuo earthquake is shown in Fig. 4.5.

From Table 4.1 and Fig. 4.5 the following points can be obtained:

1) Before these great earthquakes, the regional seismicity is concentrated in nearly conjugate bands, and the mainshock is located at their intersection.

2) The bands are usually several hundred kilometers long.

3) The bands are generally formed 1–3 years before the mainshock.

Comparing this with Fig. 4.1, it can be seen that most of the bands appear within a year either side of the peak seismicity rate.

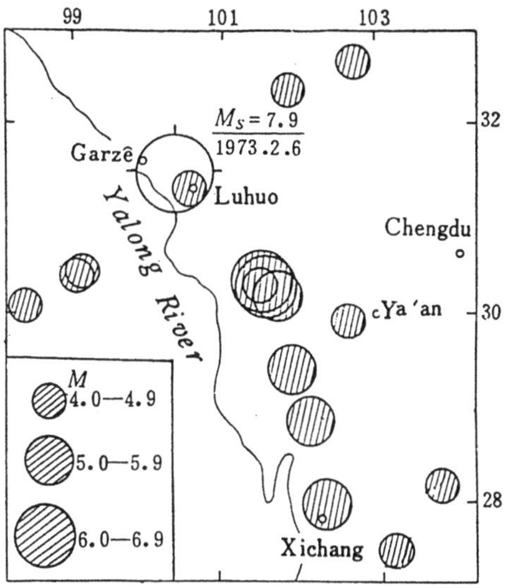

Fig.4.5. The band–shaped distribution of regional seismic activity before the Luhuo earthquake

4.2.2 Regional Seismicity Migration

Over a much longer time scale than that for band formation, the regional earthquake patterns feature dynamic migration. For convenience of discussion, we first divide the region of minor seismicity into three areas (Fig. 4.6).

Area A: the hypocenter of the forthcoming mainshock, which we represent by its aftershock area. The direction and length of the major aftershock area axis are denoted by the shaded circle in Fig. 4.6.

Area B: the near hypocentral area, whose extent is roughly several times that of area A, is delineated by the dotted circle in Fig. 4.6.

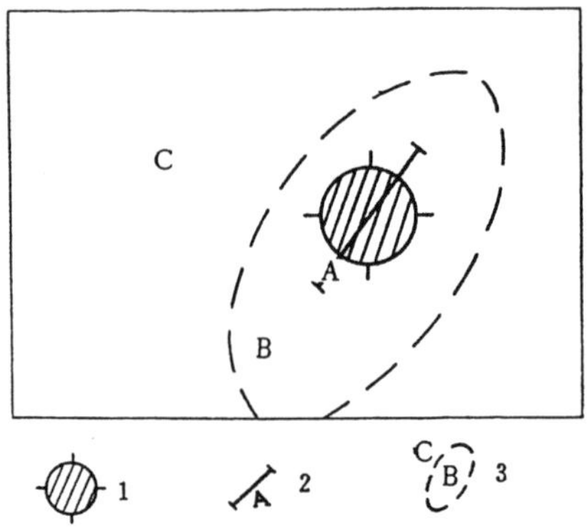

Fig. 4.6. Outline map of division of regional seismicity before a great earthquake; *1* mainshock epicenter; *2* direction and length of major axis of hypocentral area *(A)*; *3* range of focal area *(B)* and range of far focal area *(C)*

Area C: far hypocentral area.

Spatial and magnitude distributions of regional moderate and small earthquakes before seven earthquakes are shown in Fig. 4.7, with seismic activity at different periods marked by different symbols (open and shaded circles). Migration of seismic activity from the far hypocentral area to the near hypocentral or epicentral area was observed before all of the nine great earthquakes except the Songpan quake.

The following characteristics of the seismicity migrations can be inferred:

1) Their duration ranged from 3 to 10 years, which can be separated into two phases:

The first phase, during which moderate and small earthquakes were mostly located at far hypocentral areas (area C), while the seismicity in the nearer areas (B and A) was relatively quiet.

The second phase, during which the earthquakes occurred mainly in areas A and B, whereas the seismicity in area C became quiescent. The mainshocks occurred right after this phase.

The existence of two such phases reveals either that the active and quiescent areas of regional moderate and small earthquakes exchanged spatially, or that these earthquakes concentrated, clustered and migrated towards the mainshock epicenter.

2) During the second phase, the moderate and small earthquakes would continue till months or days before the mainshock (see M–t diagram in Fig. 4.7). The activity at the end of these series either moved into the mainshock epicentral area (A), or approached near hypocentral areas (B), while area A remained calm.

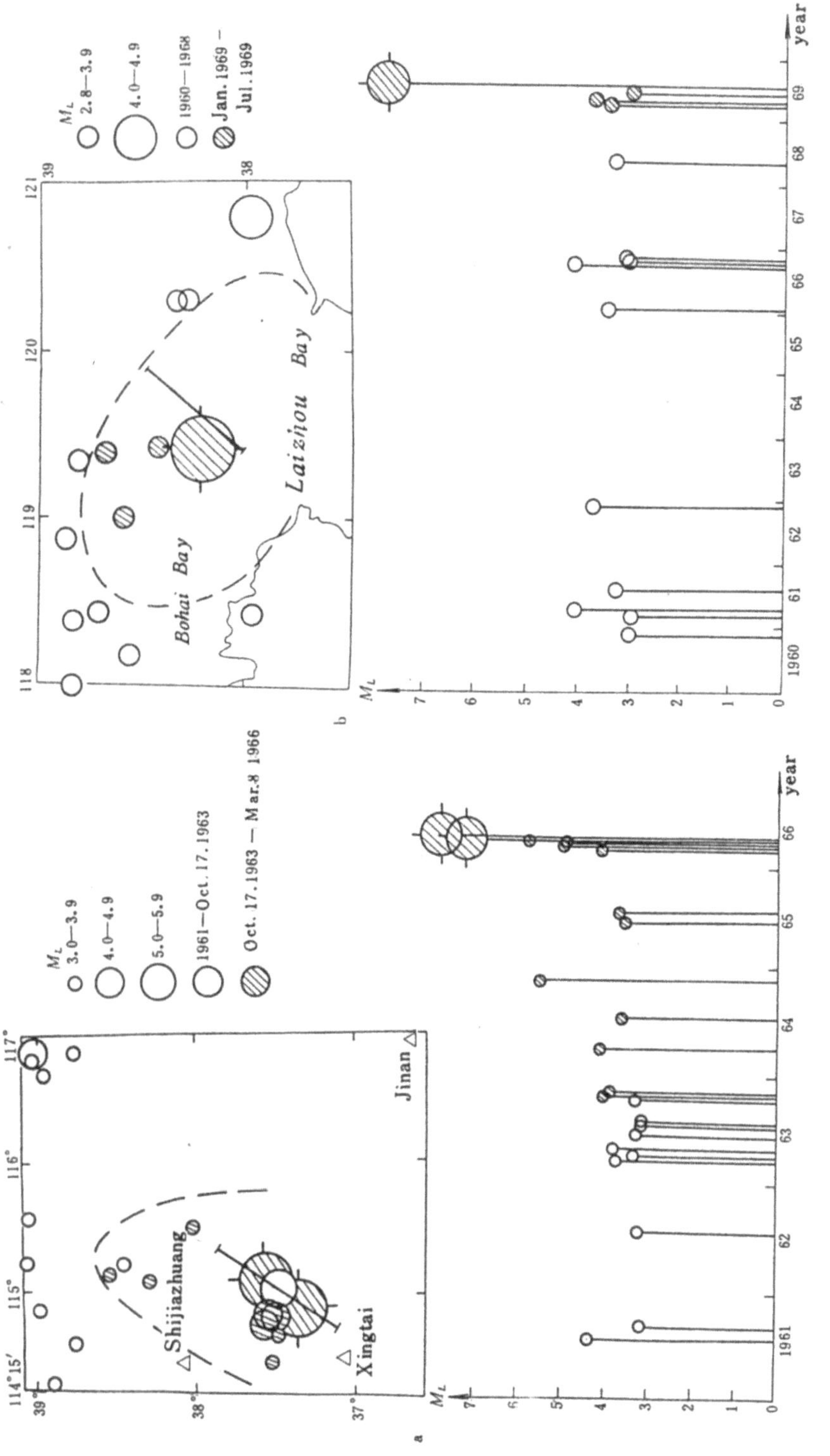
a
114°15′
115°
116°
117°
39°
38°
37°
Shijiazhuang
Xingtai
Jinan
M_L
3.0—3.9
4.0—4.9
5.0—5.9
1961—Oct. 17. 1963
Oct. 17. 1963 — Mar. 8 1966
year
1961
62
63
64
65
66
b
118
119
120
121
39
38
Bohai Bay
Laizhou Bay
2.8—3.9
4.0—4.9
1960—1968
Jan. 1969 — Jul. 1969
1960
61
62
63
64
65
66
67
68
69
0
1
2
3
4
5
6
7

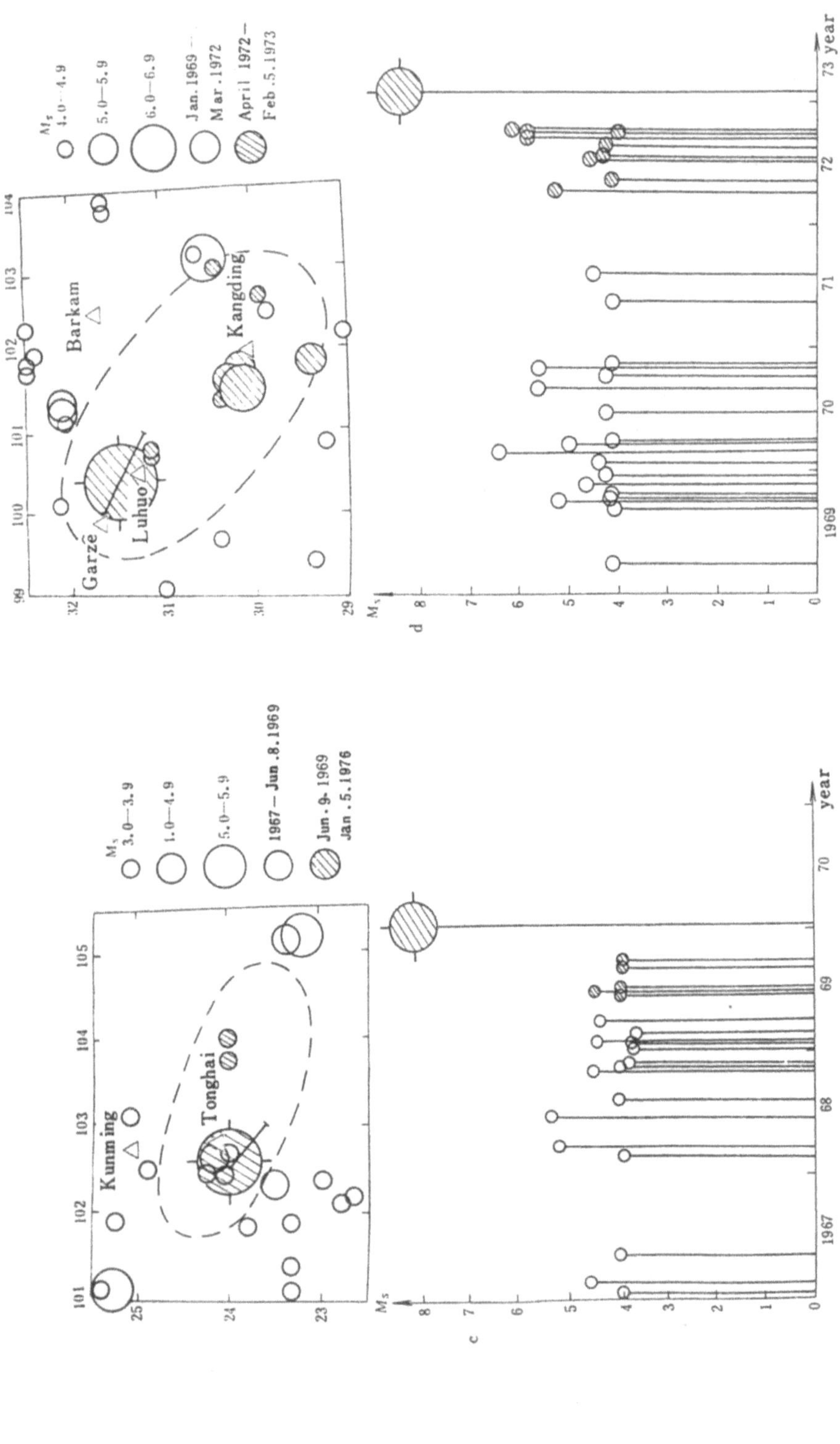
Kunming
Tonghai
M_S
3.0—3.9
5.0—5.9
1967—Jun.8.1969
Jun.9.1969
Jan.5.1976
c
year
1967
68
69
70
Garzê
Luhuo
Barkam
Kangding
4.0—4.9
5.0—5.9
6.0—6.9
Jan.1969—
Mar.1972
April 1972—
Feb.5.1973
d
1969
70
71
72
73 year

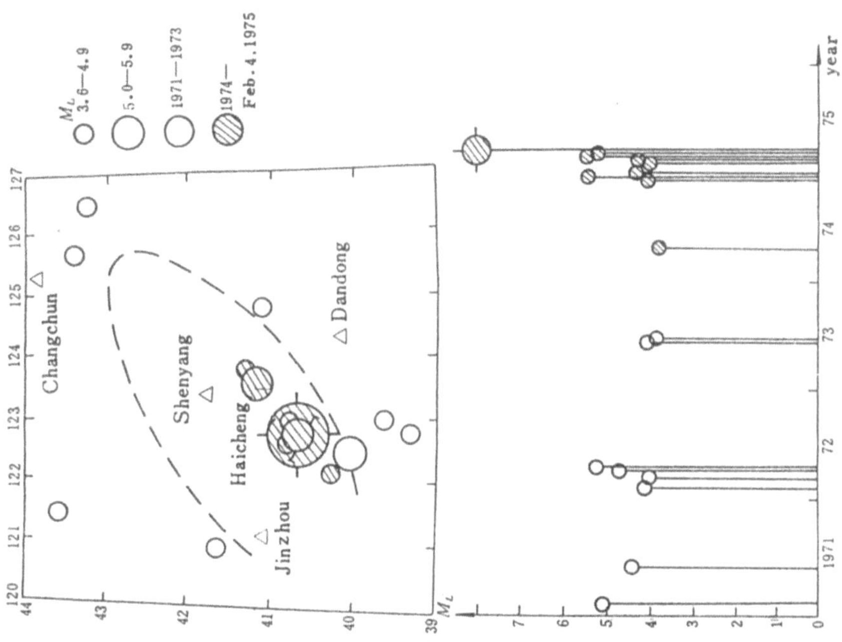

Changchun
Shenyang
Haicheng
Jinzhou
Dandong
M_L
3.6—4.9
5.0—5.9
1971—1973
1974—
Feb.4.1975
year

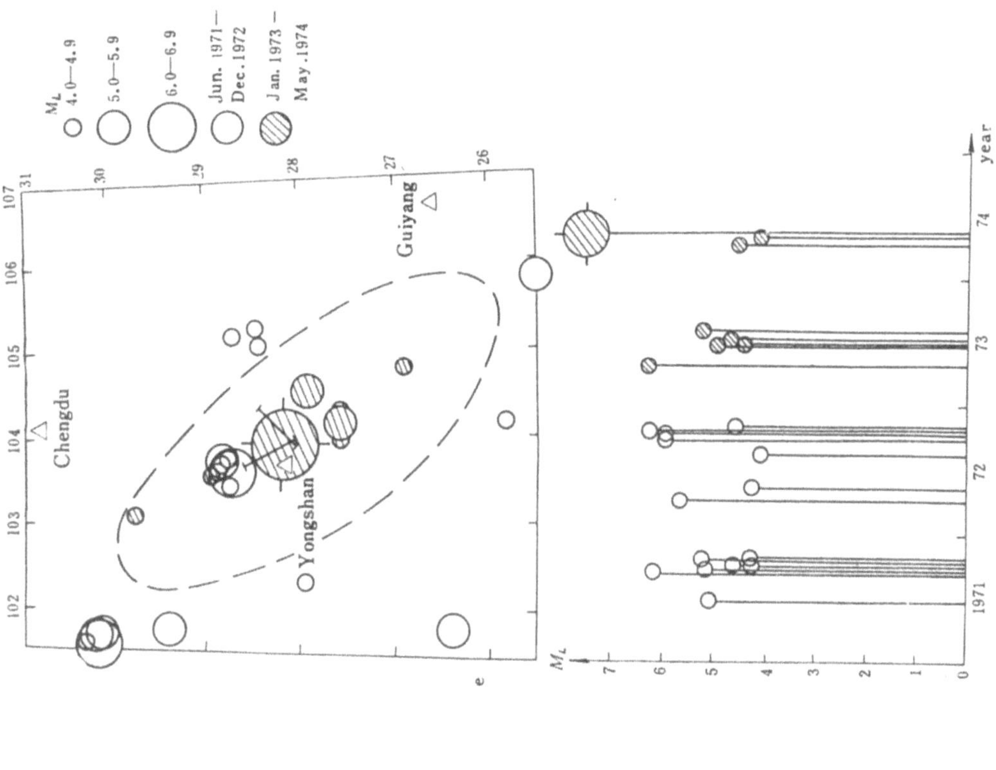

Chengdu
Yongshan
Guiyang
M_L
4.0—4.9
5.0—5.9
6.0—6.9
Jun. 1971—
Dec.1972
Jan. 1973—
May.1974
year

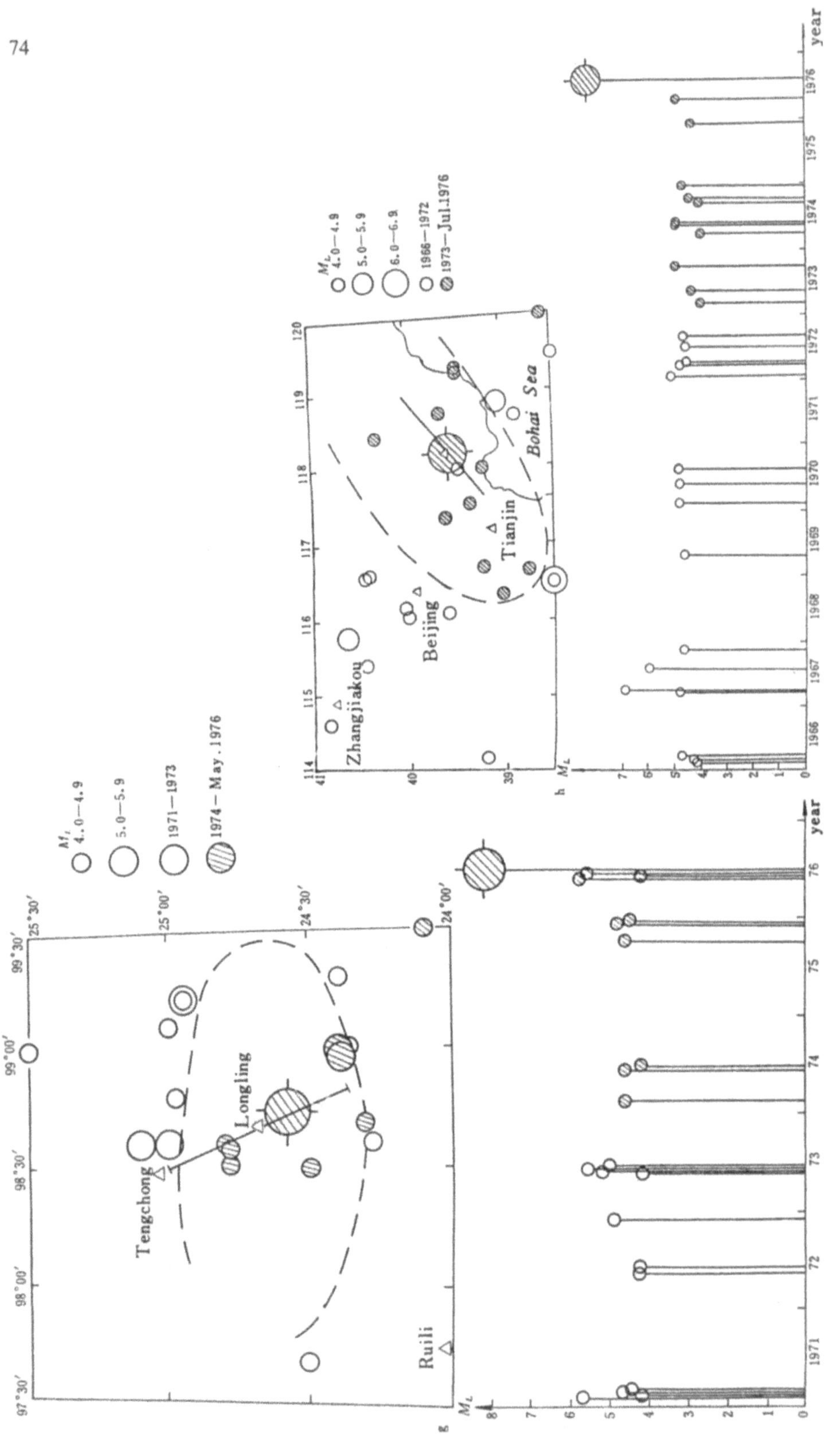
g
99°30′
99°00′
98°30′
98°00′
97°30′
25°30′
25°00′
24°30′
24°00′
Tengchong
Longling
Ruili
M_L
4.0—4.9
5.0—5.9
1971—1973
1974—May.1976
1971
72
73
74
75
76
year
h
114
115
116
117
118
119
120
41
40
39
Zhangjiakou
Beijing
Tianjin
Bohai Sea
4.0—4.9
5.0—5.9
6.0—6.9
1966—1972
1973—Jul.1976
1966
1967
1968
1969
1970
1971
1972
1973
1974
1975
1976
year

Fig. 4.7. Diagram of regional moderate and small seismicity migration before the great earthquakes. *a* Xingtai earthquake; *b* Bohai earth– quake; *c* Tonghai earthquake; *d* Luhuo earthquake; *e* Yongshan earthquake; *f* Haicheng earthquake; *g* Longling earthquake; *h* Tangshan earthquake

For example, two days before the Xingtai, Haicheng and Longling earthquakes, earthquakes were registered in the mainshock epicentral areas, and these earthquakes are sometimes direct foreshocks (treated further in Sect. 4.3). However, the sequences of moderate and small earthquakes ended several months before the mainshocks for the Tangshan, Luhuo, Bohai, Tonghai and Yongshan earthquakes. When the sequences stopped, the seismicity was generally located in the near hypocentral area (B), and not the mainshock epicentral area. This suggests that these cases are quiescent before the impending mainshock.

3) The overall distribution of earthquakes during the second phase is mostly consistent with the strikes of regional active tectonic zones. For the Xingtai, Haicheng and Tangshan earthquakes, the overall distributions of epicenters during the second phase are directed NNE, in agreement with the typical strike of active zones in North China; while the Yongshan, Luhuo and Longling earthquakes are distributed along NW or WNW orientations, asssociated with the features of boundary fracture zones in the crustal fault blocks of West China.

Migration of regional seismicity before great earthquakes has been studied by numerous other researchers.

4.2.3 Seismic Gaps

Some time ago, quiescent regional seismicity patterns (gaps) were discovered by the seismological community. In 1965, Fedotov indicated such phenomena. No. 3 Research Section of the Institute of Geophysics, Academy of Sciences of China studied the regularity of strong earthquakes in North China in 1970, and suggested two types of seismic gap, the first being "blank areas" among bands of strong earthquakes, where mainshocks might occur, and the second being "seismic gaps" of low seismicity formed in the epicentral area of a forthcoming great earthquake and outside of which the level of moderate and small earthquakes was high for several years before the mainshock. In 1979, Mogi also divided " seismic gaps" into two similar kinds, the first being the seismicity pattern evolved over a very large spatial and temporal scale, associated with the range of a seismic zone and the repeat time of seismic activity. In this section, we shall only discuss the second kind of seismic gap. The initial research on this kind of gap was aimed specifically at long term earthquake prediction.

The gaps studied by No.3 Research Section were mostly of 50 years or longer duration. For some of the nine great earthquakes in this book, seismic gaps of several decades duration are also found (Wu Kaitong 1976, Wei Guangxing et al.1978).

Long and short period earthquake prediction requires the identification of an abnormal seismicity gap within years before the mainshock. If larger magnitude earthquakes are used to define a "gap" in a given area within a short period, the difficulty is that the number of earthquakes is not enough. If smaller magnitude earthquakes are used, it is also very difficult to identify the gap, since the probability of small earthquake occurrences (both in time and space) is fairly high, and the size of the gap small.

The seismic gap pattern for the Tangshan earthquake is shown in Fig. 4.8. The epicentral distributions of earthquakes with $M_L > 2.0$ in the Tangshan area from 1970 to 1973 are given in Fig. 4.8a, and basically the distributions are uniform. The distributions of small earthquakes before the mainshock from 1974 to July 1976 are shown in Fig. 4.8b. It is apparent that the seismicity decays drastically, and a gap is formed in the forthcoming mainshock epicentral area, as delineated by the dotted line in Fig. 4.8. Such a gap formed in the medium term is much smaller in size than that generated in the long term. A similar gap appeared before the Songpan earthquake.

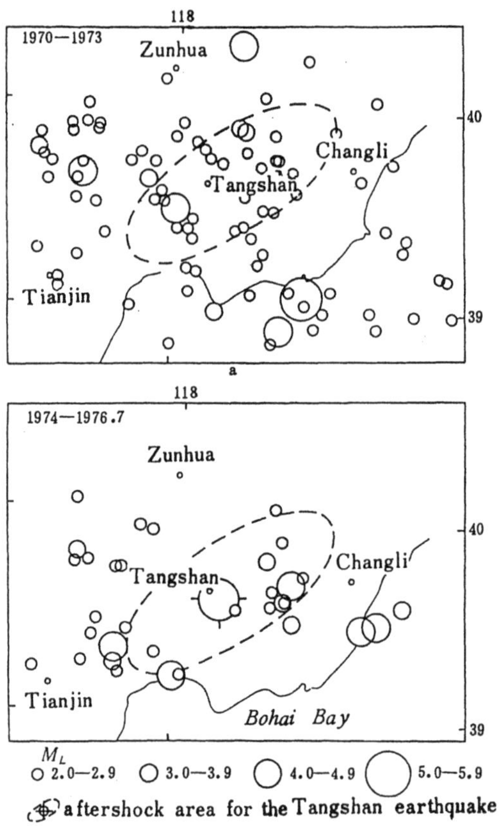

b

Fig. 4.8. The seismic gap (medium–term) before the Tangshan earthquake

Now we shall discuss the evolution of the seismic gap for several years before the Haicheng earthquake (see Fig. 4.9).

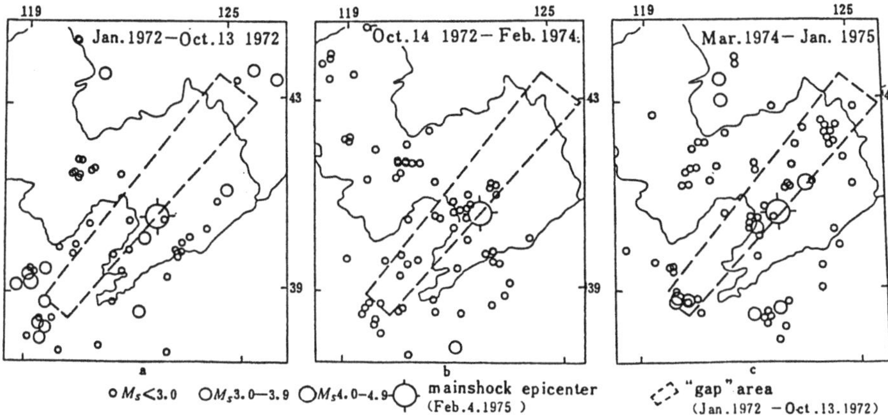

Fig. 4.9. Evolution of seismicity gaps in Liaoning Province before the Haicheng earthquake

The distributions of earthquakes from January 1972 to 13 October 1972 are shown in Fig. 4.9a, indicating a NE seismic gap in conformity with the northward extension of the Tanlu fracture zone. Then, from 14 October 1972 to Feburary 1974, seismic activity moved into the south part of the gap (Fig. 4.9b), and from March 1972 to January 1975, it changed to the north part of the gap (Fig. 4.9c). Finally the Haicheng earthquake occurred on 4 February 1975. These seismicity patterns illustrate that there might be a certain period before the mainshock in which earthquakes are reactivated in the gap area.

Up to now we have described three kinds of spatial distributions of seismicity (bands, migration and gaps) under different time–space–magnitude scales. These patterns have their own characteristics, and are correlated with each other under certain conditions. For example, the spatial range and magnitude are similar for both bands and migration, except that the migration processes take longer. Further, during the second phase of the migration process, the seismicity in areas A and B is actually distributed in bands. Also, medium and short period seismic gaps are very often the quiescent segments within bands.

4.3 Foreshocks and Swarms

4.3.1 Foreshock Activity

Particular importance has been attached to research on the space–time distributions, magnitude variations and spectral characteristics of foreshock activity, as they may contain information on the approaching great earthquake.

At present, different researchers have very distinctive principles for selecting foreshock activities. For instance, Utsu used foreshocks that occurred within 2 weeks of the mainshock, within a range of 0.5° × 0.5°, and M over 4.5 (for earthquakes of M > 7.0, a range of 1° × 1° is taken). Evison (1977), however, discussed the foreshock series during a rather long period before the mainshock, and he suggested that the sudden increase of earthquakes near the hypocentral region of the forthcoming mainshock marked the beginning of the precursory earthquake sequence (precursory swarm); then, the precursory gap appeared, and the seismic activity remained quiescent until the occurrence of the mainshock. The precursory earthquake sequences for the large New Zealand earthquakes studied by Evison lasted about 10 years. Therefore, the foreshocks defined by Utsu can be termed direct foreshocks, while the precursory swarm studies by Evison can be called early foreshocks.

In this section, we are mainly concerned with direct foreshocks. As noted above, direct foreshocks are only recorded for the Xingtai, Haicheng and Longling earthquakes.

Before the Ningjin, Xingtai, M7.2 earthquake of 22 March 1966 occurred, the M6.8 Longyao earthquake took place on March 8. This was the first strong earthquake of the great Xingtai sequence. A group of nine direct foreshocks with M over 4, and maximum M5.2 (on March 6), occurred in the hypocentral region seven days before this earthquake.

About 5 years before the M7.3 Haicheng earthquake on 4 Feburary 1975, the Shipengyu seismic station, Yingkou City, close to its epicentral region was set up, providing an excellent opportunity for recording the foreshock activity. Only nine small earthquakes of maximum magnitude 1.3 were recorded in the mainshock epicentral area from August 1970 to 31 January 1975. Within 4 days of the Haicheng mainshock, from 1 February 1975, about 500 medium and minor earthquakes occurred in a region 20 km from Shipengyu Station, hence contituting a complete swarm series of direct foreshocks, with a maximum of M4.7. All earthquakes with $M_L > 2.0$ were confined within a range of 5 km, indicating a considerable concentration of foreshocks.

Within 15 min before the Longling earthquake of M7.3 on 29 May 1976, 16 direct foreshocks with a maximum magnitude of 5.2 were recorded at the Tengchong Seismic Station (about 40 km from the epicenter), among which six were felt locally.

Among these three cases, the abundant Haicheng foreshock series provided the best conditions for studying further the characteristics of such series, and the results obtained will be of importance for distinguishing whether a group of swarms might be foreshock activity or a set of independent earthquakes.

The existing research results have shown the following features for the Haicheng foreshock series:

1) Low b value.

Wu Kaitong et al. (1976) pointed out that the b value of the Haicheng foreshock series was 0.56, much lower than that (0.86) of the Haicheng aftershock series. Also, the b value of the Longling foreshock activity was rather low (0.41).

2) Consistent dislocation manner.

The dislocation modes of the Haicheng foreshock series seemed to be relatively stable, as indicated below:

a. There was a well–defined consistency for the P wave onset patterns, and the average focal mechanism solutions of the foreshock series were very close to that of the mainshock (Wu Kaitong et al.1976).

b. The maximum amplitude ratio between direct P waves and S waves (vertical orientation) remained constant, while the amplitude ratio for normal swarms is generally rather random (Jin Yan et al. 1976).

c. There was good conformity for polarization angles of S waves (Gu Haoding et al. 1980).

3) Concentration–quiescence process.

There appeared to be a process of concentration–quiescence–mainshock occurrence for the seismicity rate and energy release of the Haicheng foreshock series. The calm stage appeared 12 h before the mainshock (Wu Kaitong et al. 1976).

4) Earthquake deficiency.

Xu Shaoxie et al. (1981) studied the "earthquake deficiency" phenomena during the late Haicheng foreshock series. He suggested that no earthquakes of M over 0.9 took place from 1300 to 1400 during the interval from 1300 to 1936 on February 4 when the mainshock occurred, illustrating that there was a lack of medium earthquakes during this period. This represents extremely unbalanced behaviour of the b value. The paper compared this earthquake with other normal swarm series, and showed that the magnitude–frequency distributions for the late part of those series was fairly even. Whether the above–mentioned features of the Haicheng foreshock series can be regarded as precursors requires further study on other foreshock series.

4.3.2 Regional Precursory Swarm Activity

Some earthquake histories indicate that precursory swarm activity happens over a range of several hundred kilometers, rather than close to the mainshock epicenter.

Song Liangyu et al. noticed a relationship between regional small swarms and large earthquakes in Yunnan province. They defined the small swarms as: (1) a group of earthquakes with consecutive average frequency $\Sigma n_i / t > 2$ or with the first frequency>4, within a 30 km range, which is called a swarm (t, denoting consecutive days of the swarm, n_i, the ith day frequency); (2) the difference between the biggest and the second large shock $M_{max} - M_1 < 1.0$ with the starting magnitude of 1.5.

From January 1966 to April 1979, about 90 small swarms (aftershock clusters of large earthquakes are not included in the statistics) were recorded within the NE tectonic zone in the West Yunnan region. Figure 4.10 shows the annual occurrence rate of these small swarms indicating that the rate varied greatly, being four times as high in 1974, prior to the May 1976 Longling earthquake, as it has been in the 1960's.

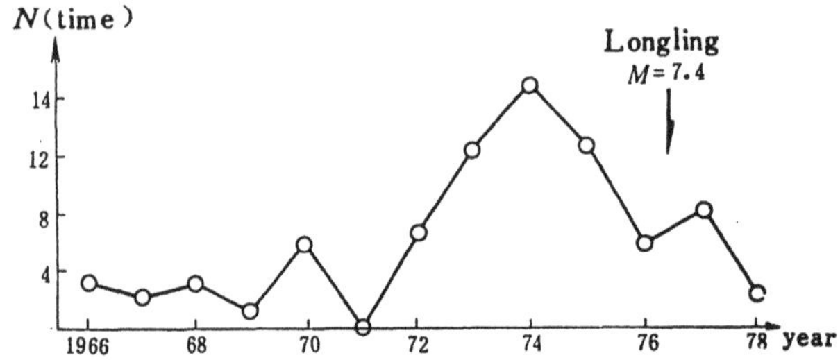

Fig. 4.10. Frequency of small swarms in West Yunnan before the Longling earthquake

Before the Haicheng earthquake, the medium and minor earthquake activity in Liaoning Province and its neighbouring areas was rather low in 1972 and 1973, but became drastic in 1974 (see above), and minor swarms occurred constantly. Altogether seven regional swarms occurred from March 1974 to the time of the Haicheng earthquake. The swarm activities seemed to be associated with accelerated crustal movement of the Liaodong Peninsula, and the last swarm occurred in the hypocentral region, being the direct foreshock series.

4.4 Seismic Wave Anomalies and Spectra

4.4.1 Anomalies of Seismic Wave Velocity

Though during the 1920's some seismologists suggested probing variations of seismic wave velocity within a forthcoming source region, the idea could not be realized because of limited observation accuracy and station distribution. Only since the 1950's have field observations been practical. During the 1960's Soviet scientists observed variations of the ratio of primary wave to secondary wave velocity (V_P/V_S) before some moderate earthquakes in Garm, mid–Asia. (Kondratenko 1962). This aroused great interest in many countries. In the 1970's, a number of velocity and velocity ratio changes within the hypocentral region and its surroundings were reported successively in China, the United States and Japan. These results included studies of variations of V_P/V_S by using arrival times of direct primary and secondary waves (P and S) from local and near earthquakes. The ratio is derived from the Wadati equation. Let the arrival time difference of direct primary and secondary waves be $t_{(S-P)}$, the primary wave arrival time t_p, and the origin time t_o, then the Wadati equation is as follows:

$$t_{(s-p)i} = \alpha(t_p - t_o). \quad (i = 1, 2, \cdots, n) \tag{4.2}$$

where i is station subscript, n is station number, and α is the linear slope of the Wadati diagram. The mean velocity ratio V_P/V_S is related to a by:

$$V_p / V_s = \alpha + 1. \tag{4.3}$$

Therefore, arrival times of direct P and S waves of an earthquake recorded by a set of stations is plotted as a Wadati diagram versus the coordinate ($t_{(s-p)}$), and the slope is measured to obtain the average velocity ratio.

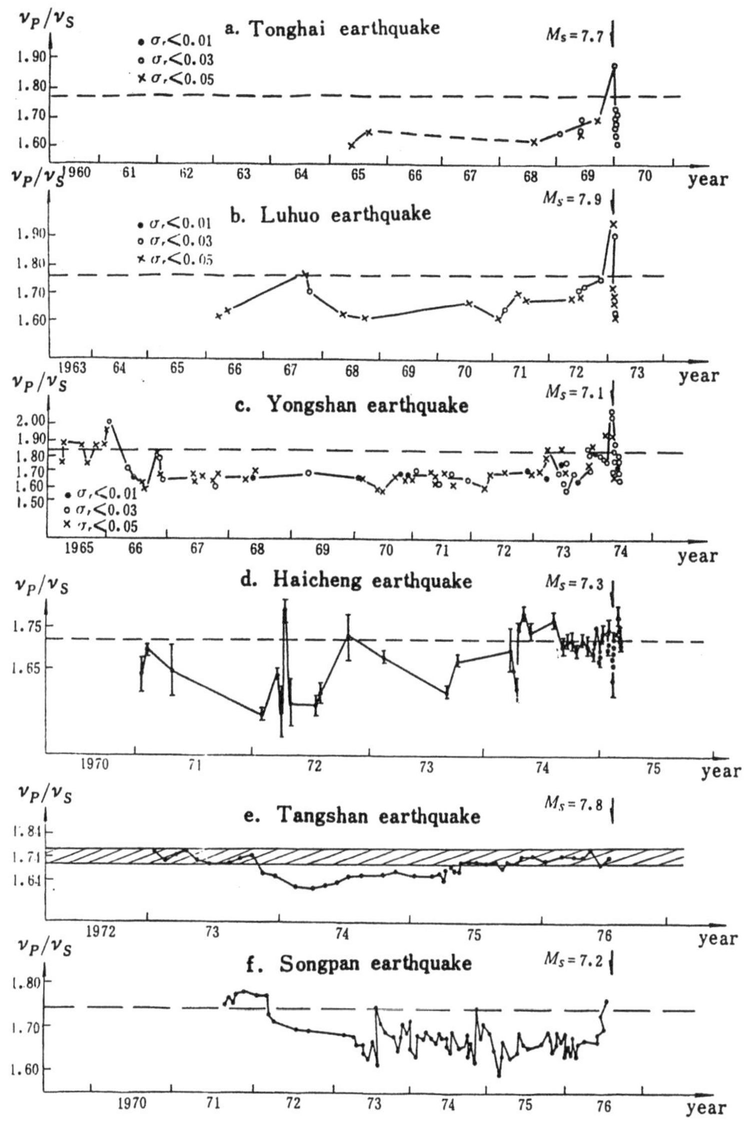

Fig.4.11. Curves of regional average variations of wave velocity ratio before six great earthquakes

The phenomenon of velocity variation within the source region before a large earthquake has been an important aspect in earthquake prediction research. Chinese seismological workers have also done much research on this subject in recent years. Curves of wave velocity ratio variations before six great earthquakes of M over 7 are given in Fig. 4.11, in which σ_r means the data error, and the region used for velocity studies usually includes the epicentral area of the great earthquake and its vicinity.

The curves in Fig. 4.11 suggest:

a) P and S velocity ratio variations before great earthquakes generally display a "bay shape" anomaly. The value of the ratio is suddenly decreased below the normal background variation, and then recovers to the normal level.

b) Before the Tonghai, Luhuo, Yongshan and Songpan earthquakes in Southwest China, the mainshocks followed immediately after the "bay shape" anomaly ended; while before the Haicheng and Tangshan earthquakes in North China, the mainshocks occurred after a period of time (about 1 year) during which the curves kept at a constant level after the "bay shape" finished.

The above two characteristics are frequently noted in other countries.

Many Chinese and other researchers pointed out that the duration of the negative anomaly seems to be proportional to the magnitude of the mainshock. According to this relation, the duration of velocity anomalies for the M7.8 Tangshan earthquake would be expected to last for 10 years or more, but actually the curve in Fig. 4.11e shows that the anomaly duration was only about 3 years. So the above relationship is suspect.

Accurate studies of velocity variations will undoubtedly be very important in understanding the processes in the source region. In order to achieve this end, appropriate distribution of stations and highly accurate observations are required. It is also necessary for frequent small earthquakes to occur near the epicentral area of forthcoming mainshocks, though this is not possible in some areas. Using P wave residuals of distant earthquakes is an appropriate method for compensating this defect (Wyss et al. 1973). Duan Xingbei et al. (1976) adopted this method to study P wave velocity variations within the source region before the Haicheng earthquake, and noted that P wave velocity within this region decreased by 10%–20% within 3 years before the mainshock.

4.4.2 Duration of Seismic Wave Vibration

During the seismogenic processes of great earthquakes, the stress–strain state and physical properties of the medium within the forthcoming source region may change significantly. This may lead to changes in the absorption of passing seismic waves. Work done at the mid–Asia earthquake prediction test site has shown that the durations of seismic wave vibrations from minor earthquakes vary in length before and after some medium–strong earthquakes.

Before the Tangshan earthquake, at Beianhe station in the Beijing region, the DD–1 type seismographs recorded some small earthquakes ($1.0 < M_L < 5.0$) occurring in the hypocentral area and its surroundings, and the duration ratio of their vibrations τ_H / τ_V had obvious anomalies. The Beianhe station is about 190 km from the Tangshan epicenter. τ_H and τ_V denote respectively the durations of horizontal and vertical components of S waves, and the cut–off time taken is when the amplitude has decayed to twice the background noise level. Statistical analyses of Huang Deyu et al. prove that the ratio of τ_H to τ_V is independent of the epicentral distance and the magnitude of small earthquakes.

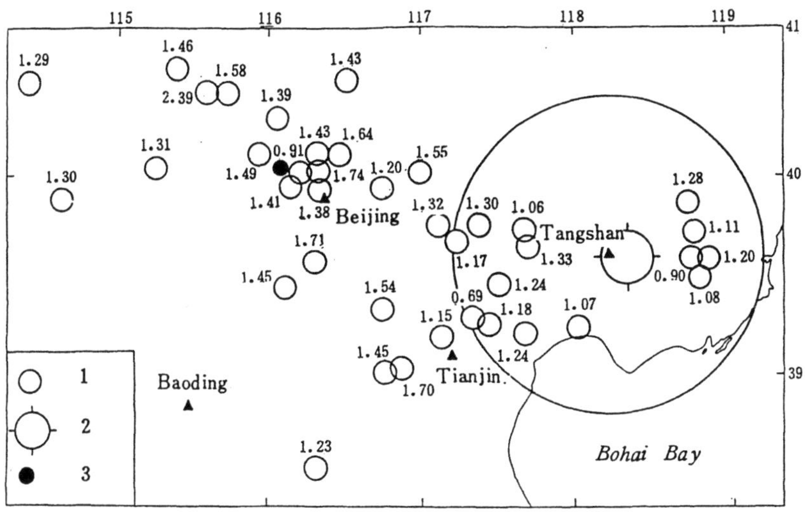

1 epicenters of small earthquakes and τ_H/τ_V values

2 mainshock epicenter 3 Beianhe station

Fig.4.12. Distribution of τ_H / τ_V for small earthquakes (September 1975–July 1976) recorded at Beianhe station before the Tangshan earthquake

Figure 4.12 gives the spatial distribution of τ_H / τ_V recorded before the Tangshan earthquake. Let the radius of the Tangshan epicentral area be 85 km (as indicated by the large circle in the figure), in which 14 minor earthquakes occurred, with an average duration ratio $\tau_H / \tau_V = 1.3$ (rms error, $\sigma_n = 0.17$); outside this area 25 minor earthquakes occurred with $\tau_H / \tau_V = 1.46$ ($\sigma_n = 0.27$). This shows that there were evident differences between the durations of S wave vibrations for minor earthquakes inside and outside the epicentral area before the Tangshan mainshock, with the lower ratio in the epicentral area.

4.4.3 Spectra of Small Earthquakes

The variations of time–space–magnitude patterns and wave velocities of regional moderate and small earthquakes before large earthquakes are undoubtedly important aspects in earthquake prediction research. In addition, attention has been paid to variations in the spectra of small earthquakes.

Zhen Zhizhen et al. (1975) introduced the idea of average dislocation (F) for studying risks of large earthquakes. The expression is:

$$F = M - \log S. \qquad (4.4)$$

where M is magnitude and S is source area (in sq. km) determined from spectral analysis. They suggested that variations in F mainly reflect variations in stress drop.

Figure 4.13 shows the variations in F before and after the Haicheng, Yongshan and Tangshan earthquakes (Zheng Zhizhen et al. 1977, 1978) . For the Tangshan earthquake, small earthquakes with Ms>3.7 are used in Eq. (4.4); for the Yongshan earthquake, small earthquakes with Ms>4.3 are taken; and for the Haicheng earthquake, small earthquakes of $M_L = 2.5$–3.9 are used.

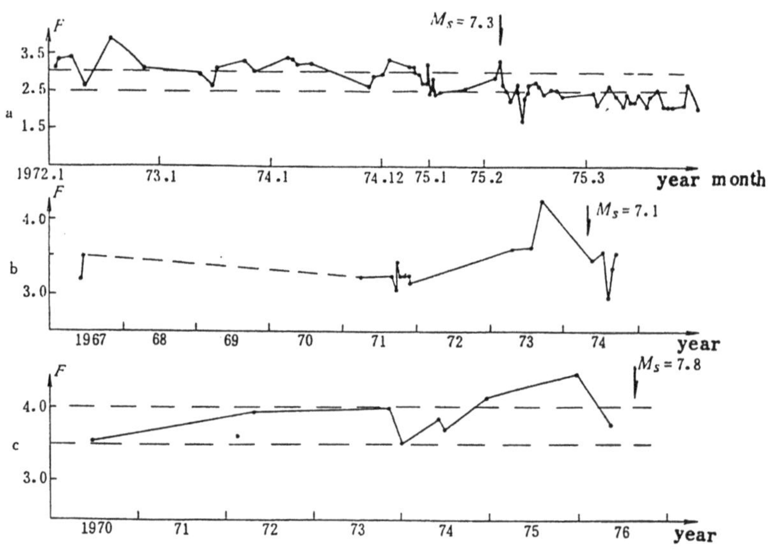

Fig.4.13. Variations of average dislocation *(F)* before great earth–quakes; *a* Haicheng earthquake; *b* Yongshan earthquake; *c* Tangshan earthquake

From these cases it can be seen that a high F value of regional earthquakes may be regarded as an anomalous phenomenon.

Spectral analysis of minor earthquakes occurring before large ones, and the study of the attenuation factor, Q, of the medium are also important areas of earthquake precursor research.

Preliminary analysis of regional seismicity patterns and other anomalous variations before the nine great earthquakes have been made in this chapter, and it is suggested that some of the anomalies can be generalized as follows:

1) The frequency of regional moderate and small earthquakes increases year-

ly several years before the great earthquakes.

2) The regional moderate and small earthquakes are focussed in areas around the future earthquake–producing faults, forming earthquake bands.

3) Migration of regional seismic activity towards the future mainshock epicenters can be observed.

4) There are anomalous changes in the frequency–magnitude relation of earthquakes in the forthcoming mainshock epicentral areas and their vicinities (b value decrease and earthquake deficiency).

5) The seismic wave velocity ratio very often decreases before the great earthquakes.

CHAPTER 5

Anomalous Variations of the Geophysical Field

In this chapter, certain anomalous variations of the geophysical field observed before great earthquakes are summarized. These are mainly geoelectric, gravity and geomagnetic variations in both the deep and the shallow crust, within the epicentral area and its vicinity. Since the observations have only been conducted over a short period, we are still uncertain as to the normal background variations of these anomalies. Therefore, further studies are needed on the relation between the observed anomalies and earthquake occurrences.

5.1 Apparent Resistivity Changes in the Shallow Crust

In 1967, at the Lanzhou Geophysical Institute of the Chinese Academy of Sciences (now the Lanzhou Seismological Institute of the State Seismological Bureau), the Geoelectric Research Group began measurements of apparent resistivity of the shallow crust in the aftershock area of the Hejian and Xingtai earthquakes. They reported that the apparent resistivity changed abnormally by several percent for days to tens of days before certain aftershocks of M4–5 and greater (Qian Fuye et al. 1980). Since then, geoelectric observation networks have gradually been established in some of the principal seismic zones in China, so that long–term variations of apparent resistivity in the shallow crust can be measured continually. At these stations, symmetrical quadrupole devices are used (as in electrical prospecting), with the distance between the poles generally being 1–2 km, giving a detection depth of several hundred metres. Of the nine great earthquakes discussed in this book, several years of observations were carried out before five of them, within a 200 km range of their epicenters. These were the Luhuo, Yongshan, Longling, Tangshan and Songpan quakes, of which the Tangshan data are the richest, being obtained from 14 stations within 200 km of the epicenter. These data (except for the Xushui station) are shown in Fig. 5.1. There are also one or two stations that observed resistivity variations before the Luhuo, Yongshan, Longling and Songpan earthquakes. These variations are discussed below.

5.1.1 The Basic Nature of Apparent Resistivity Changes

Most of the data for the four earthquakes (apart from Tangshan) are shown in Figs. 5.2 and 5.3. The relative variation, $\Delta\rho_s / \rho_s$ of the apparent resistivity is shown, where ρ_s is the mean value of apparent resistivity averaged over many years of observations. For stations with large annual variations, only long–term changes in ρ_s are used in looking for pre–earthquake variations (Qian Jiadong et al. 1985). From these curves, we note the following features.

a) Before these five great earthquakes, the apparent resistivity at stations within a range of 200 km tended to fall for one to several years before the mainshock. The variation ranged from several percent to more than 10%, and the mainshock often occurred close to the time of lowest resistivity.

b) After the mainshock, the resistivity at most stations shows a gradual recovery, but does not completely return to its previous level.

c) At some stations (such as Changli before the Tangshan earthquake, Wudu before Songpan, and Ganzi before Luhuo) the apparent resistivity anomalies could be roughly classified into three phases: (1) the "beginning" stage, during which the apparent resistivity began to drop clearly; (2) the "lasting" stage during which the apparent resistivity kept dropping stably; (3) the "accelerating" stage, during which the apparent resistivity dropped drastically, indicating the imminence of the earthquake. However, such a complete process has only been observed at a few stations so far. Such behaviour possibly contains information on the seismogenic process; in particular, the acceleration stage might be of great importance as a short–term precursor.

5.1.2 Short–term or Imminent Apparent Resistivity Anomalies

The accelerating stage discussed above very often appeared 2 to 3 months before the mainshock. The rate of change at this stage is several times that during the previous stages, so such accelerated variations are thought to mark the change of intermediate–term anomalies into short–term ones. Continuous apparent resistivity data have been obtained at about 20 geoelectric stations within a range of 200 km around the epicenters of the five earthquakes. Of these 20 stations, only the three mentioned above showed the short–term accelerated phase. These stations are close to the epicenters ($\Delta < 100$ km), showing that the short–term accelerated changes might happen principally near the epicenter. Of course, there may be other factors, such as crustal medium conditions and distances to active faults, which determine the advent of short–term anomalies.

Within 1, 2, or even more than 10 days of the mainshock, abrupt, large amplitude variations have sometimes been observed, with the relative variation amplitude reaching a maximum of about 20%. Figure 5.4 shows abrupt imminent anomalies observed before the Tangshan earthquake; these show sudden large amplitude changes, as compared with the several percent intermediate–term anomalies of previous years. In addition, the range in which imminent anomalies are recorded may be greater than that of medium–term anomalies. For example, before the Tangshan earthquake, the medium–term anomalies only occurred within 200 km of the epicenter, while the imminent sudden anomalies were recorded by the Hezi station 550 km from the epicenter. However, the imminent anomalies are distributed unevenly. For the Tangshan earthquake, the Changli station ($\Delta = 75$ km) and the Baodi station ($\Delta =$ 70 km) are both close to the epicenter, yet the former station recorded the imminent anomalies, while the latter did not. This behaviour may be related to the tectonic location, the medium conditions or underground water level changes at the stations.

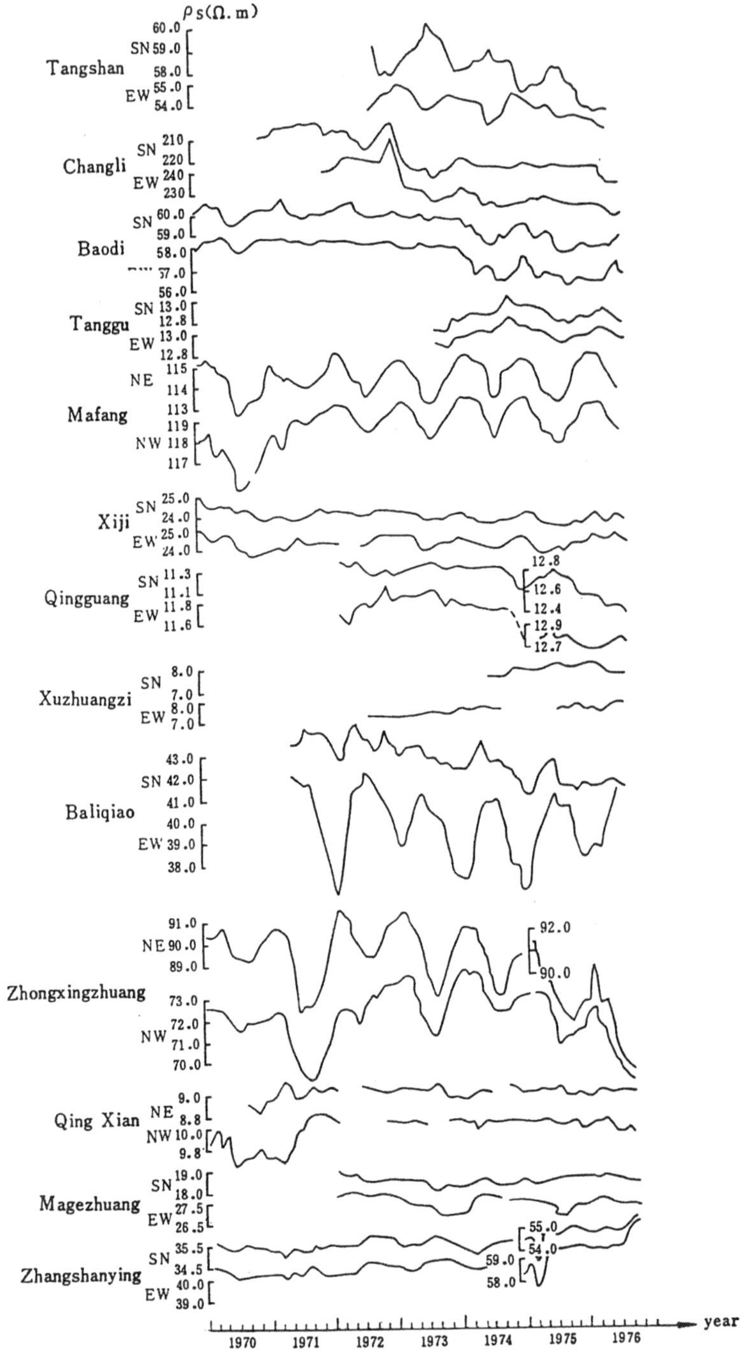

Fig. 5.1. Average monthly variations of apparent resistivity at several geoelectric stations in the Beijing–Tianjin– Tangshan region before the Tangshan earthquake

Fig.5.2. Variations of apparent resistivity before the Songpan earthquake

5.1.3 Relation Between Apparent Resistivity Changes and Earthquake Occurrences

1) Apparent resistivity anomalies may result from stress–hardening effects within the source region and its vicinity during the seismogenic process.

Many experiments on the physical nature of resistivity changes in rocks before large earthquakes have been made, both in China and abroad. Brace et al. (1968) and Brace (1975) found that for saturated or partially saturated granite samples, the resistivity would tend to drop when the loading reached a high stress stage (exceeding one half to two–thirds of the rock rupture strength). The Lanzhou Seismological Institute performed loading and unloading tests in soils, and they observed that apparent resistivity tended to drop during loading, and to recover during unloading. By comparing great earthquake observations with experimental results, it may be qualitatively concluded that the apparent resistivity variations are a result of gradual pre–earthquake stress increases, and rapid post–earthquake decreases.

2) Range and amplitude of anomalies and epicenter.

Only before the Tangshan earthquake were a large number of stations operating, so we will only discuss this case in reference to the spatial distribution of anomalies. More than ten geoelectric stations in the Beijing–Tianjin–Tangshan region recorded the fall in apparent resistivity before the Tangshan mainshock. Contours of anomaly amplitude ($\Delta\rho_s / \rho_s$) are plotted in Fig. 5.5. These show that the anomalous area is about 300 km long from east to west, 100 km wide from north to south, with the epicenter located at the larger–amplitude anomalous area within the anomalous region. As the epicentral distance increases, the anomalous amplitude decreases at different rates in different directions. The contours show two predominant orientations, close to the strikes of the two tectonic zones in the NE and NW directions.

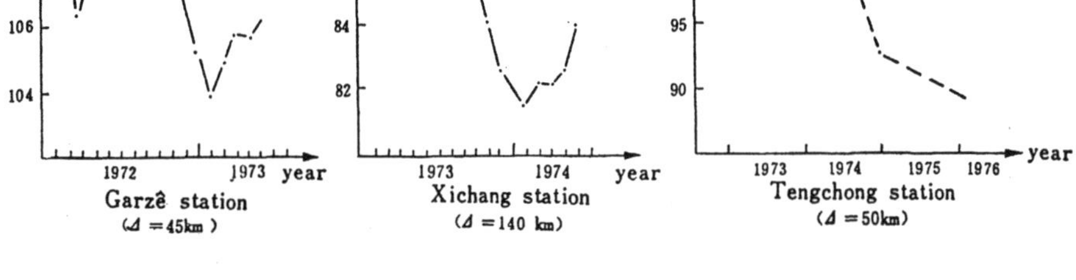

Fig. 5.3. Apparent resistivity variations before the Luhuo, Yongshan and Longling earthquakes

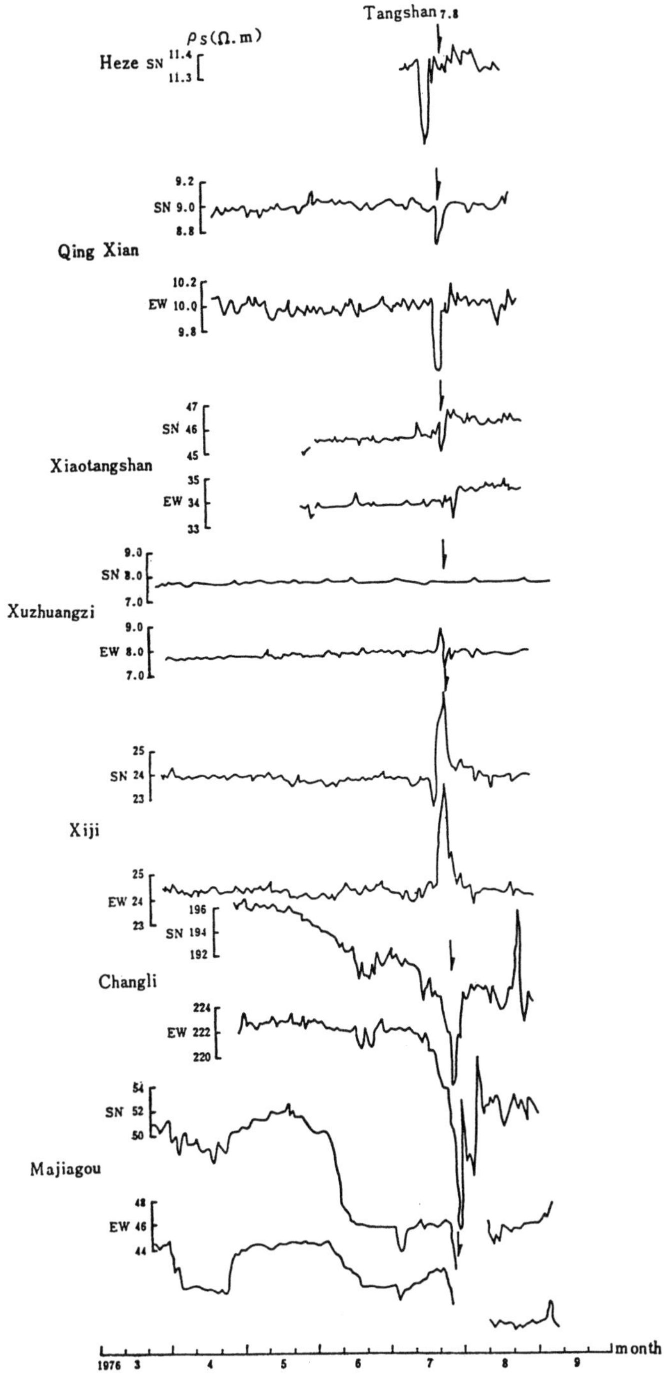

Fig. 5.4. Abrupt imminent anomalies of apparent resistivity before the Tangshan earthquake

In addition, the Analysis Group of the Beijing Seismological Brigade discovered that the time of occurrence of the anomalies depended on epicentral distance, with earlier occurences at stations closer to the epicenter, and later occurrences at stations further away. In summary, the amplitude of resistivity anomalies seems fairly large, and the latter occur earlier within the epicentral area (Beijing Seismological Brigade 1979).

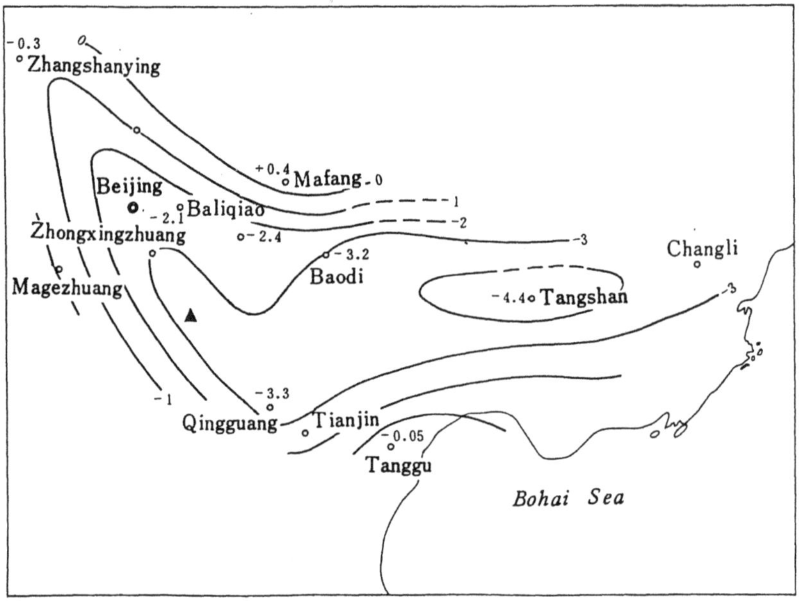

Fig. 5.5. Contours of apparent resistivity anomalies($\Delta\rho_s / \rho_s$) in the Beijing –Tianjin–Tangshan region before the Tangshan mainshock

3) Duration of anomalies and magnitude.

The medium–term apparent resistivity anomalies before the above–mentioned five great earthquakes lasted several years, with longer durations for great earthquakes than for other moderate (M6) earthquakes in China. The anomaly duration versus magnitude relation has been studied by Liu Yinxou of the Tianjin Seismological Bureau, using continuous data of 10 years from Baodi Geoelectric Station. Resistivity anomalies were detected at this station before nine earthquakes with magnitudes ranging from 3.5 to 7.8. The relation between magnitude and duration is as follows:

$$\log T = 0.35\,M - 0.15 \text{ (T in days)}. \tag{5.1}$$

Ding Guoyu (1979) gave a statistical relation between the duration of various medium–term anomalies and magnitude as:

$$\log T = 0.38\,M - 0.34. \quad (5.2)$$

This is similar to the formula derived for resistivity anomalies alone.

In addition, our own statistics on the epicentral distances to stations detecting resistivity anomalies before the nine great earthquakes, show a rough proportionality between the magnitude and the epicentral distance.

4) Imminent apparent resistivity precursors.

Imminent variations of apparent resistivity feature large amplitude, fast rate and abrupt occurrence, and can be observed over a wide area. Furthermore, drastic variations before impending earthquakes are very short in time, and often take place several days prior to the mainshock. The mainshocks normally seem to occur near the extremes of large–amplitude drastic resistivity variations. This fact might be useful for imminent earthquake prediction.

5.2 Variations in Electrical Properties of Deep Crustal Media

The electrical properties of deep crustal media, and their variations, were observed by the magnetotelluric deep sounding method. This has great penetrability and is not shielded by the high resistivity layer of the upper crust. The method is being studied internationally for earthquake prediction.

The Lanzhou Geophysical Institute started in 1965 to investigate crustal structure by the magnetotelluric method. They worked in the northern section of the North–South seismic zone, and in other areas, in order to find an approach for predicting earthquakes. From 1975 to 1976, they recorded electrical variations near the source region before the 1976 Songpan earthquake, in the boundary area between Gansu and Sichuan Provinces. Meanwhile, the Institute of Geology, State Seismological Bureau, also carried out magnetotelluric sounding in the Beijing region, obtaining data from six points. At the Fongheyin point, Daxing County, Beijing, two phases of observations were recorded before the Tangshan mainshock, and one after it. The electrical results before the Tangshan and Songpan earthquakes are outlined below.

5.2.1 Distribution of Electrical Properties of Crust and Upper Mantle in Seismic Regions

Figure 5.6 gives vertical profiles of crust and upper mantle electrical properties in the northern segment of the North–South seismic zone.

1) Low resistivity layers of one to several kilometers thickness generally exist at crustal depths of about 20 km. The apparent resistivity is several ohms–m or lower. These intracrustal low resistivity layers are not obvious outside seismic regions. Other sounding work in North China indicates that these low resistivity layers generally exist at depths of about 20 km in earthquake–prone areas. For example, within the top 20 km of the upper crust in the Beijing region, there is a high resistivity layer (including a granitic layer, an old metamorphic rock series, and overlying sedimentary rocks) with a value of several hundred ohms–m.

There is a thin low resistivity layer, about 1–2 km thick, at about 20 km depth, with an apparent resistivity of several ohms–m. Below this lie tens of kilometers of material with a resistivity of several thousand ohms–m. In addition, results from the Xingtai and Haicheng seismic areas, and the Yishu fracture zone in Shandong Province, also demonstrated that such intracrustal low resistivity layers occur extensively in seismic regions at 20 km depths.

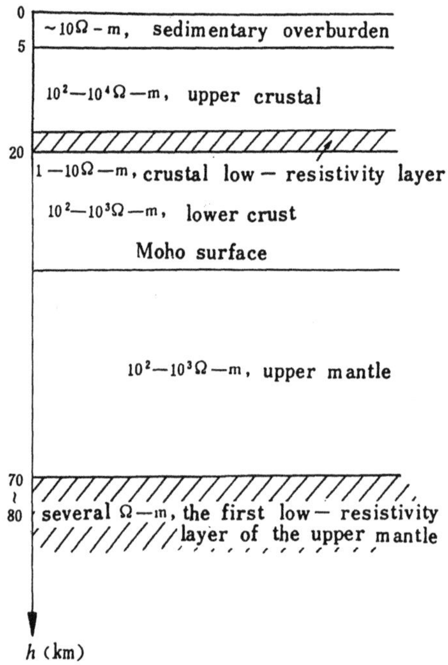

Fig. 5.6. Electric properties of the crust and upper mantle in the north segment of the North–South seismic zone

2) From Fig. 5.6 it is also apparent that the first low resistivity layer in the upper mantle in the seismic region under consideration appears at a depth of 70–80 km, with a resistivity of several ohms–m. This indicates that the first low resistivity layer of the upper mantle within seismic regions is much shallower than in other areas, where it is typically at a depth of 100–180 km. This phenomenon might be one of the physical environmental features for seismogenic preparation.

5.2.2 Anomalous Variations of Deep Crustal Resistivity

1) Anomalous variations of deep crustal apparent resistivity in the Fengheyin area, Beijing, before and after the Tangshan earthquake

Magnetotelluric deep sounding was done three times in Fengheyin area before and after the Tangshan earthquake (November 1975, May 1976, September 1976). The site (Fig. 5.5) has an epicentral distance of 120 km. The observations give consistent apparent resistivity curves, showing that the observations are reliable. Figure 5.7a shows variations of lateral apparent resistivity (ρ_{yx}) before and after the earthquake. The second observations are systematically low, by 30%–40%, relative to the first ones, at periods of less than 200 s, but the post–earthquake measurements are close to the first ones (10%–20% less). The variations of longitudinal apparent resistivity are within the measurement errors. This indicates that there might exist a process of subcrustal apparent resistivity variations, during which the resistivity dropped before the Tangshan earthquake, and recovered rapidly after it. This is basically consistent with the variations of shallow depth apparent resistivity. Figure 5.5 shows that the Fengheyin area is close to the 3% amplitude contour of shallow depth resistivity variations prior to the Tangshan earthquake, and the variation of deep apparent resistivity in the area is an order of magnitude greater than that of shallow resistivity. Besides, Liu Guodong et al. inferred that the drop in deep apparent resistivity may result mainly from variations in the electrical properties of the upper 15 km of the crust, in accordance with the frequency band range of resistivity decrease, and the penetrability of magnetotelluric waves within this band.

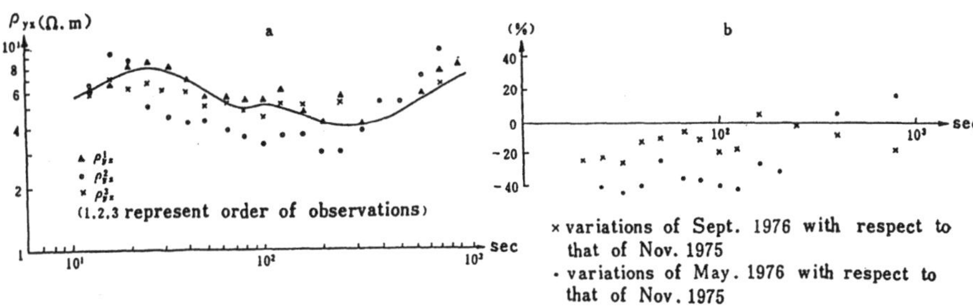

Fig. 5.7. Variations of deep apparent resistivity in the Fengheyin area before and after the Tangshan earthquake; *a* pre–earthquake; *b* comparison of pre–and post–earthquake variations of ρ_{yx}%

2) Variations of deep apparent resistivity near the epicenter before and after the Songpan earthquake

Before and after the Songpan earthquake, the Lanzhou Seismological Institute conducted three magnetotelluric sounding surveys at five points around the epicenter in 1975, 1976 and 1978. The latter two were after the earthquake. Of the five points, Nanping and Zhongzhai are closest to the epicenter, about 50–60 km, as shown in Fig. 5.8a. Figure 5.8b shows the apparent resistivity variations at Nanping. The 1975 value of ρ_{xy} in the mid–period range (60–300 s) was systematically lower than in either 1976 or 1978, with a maximum amplitude of about

50%–70%. This indicates that the apparent resistivity in the depth range of 15–25 km decreased anomalously at Nanping before the Songpan earthquake. The measurements at Zhongzhai in September 1975 and October 1978 were consistent with those at Nanping. As the two measurements at Nanping after the earthquake are similar to each other, and to further measurements done in December 1978 about 2 years after the quake, these can be regarded as the normal value. If this is so, then the deep apparent resistivity near the epicenter must have dropped greatly before the first measurements were made in June 1975. We may extrapolate and infer that the anomalous drop in deep crustal apparent resistivity occurred a year or more before the mainshock. This is similar to the duration of shallow crustal apparent resistivity anomalies.

At Wudu Station, about 90 km north of the Songpan epicenter, anomalous variations of shallow crustal apparent resistivity, decreasing by about 14%, were observed 1 or 2 years before the mainshock. The Wudu station is very close to Nanping and Zhongzhai. Therefore, decreases in apparent crustal resistivity at different crustal depths have been observed by different methods, implying relatively consistent variations of electrical properties of shallow and deep crustal media prior to the earthquake. The deep crustal anomalies are several times greater than those in the shallow crust.

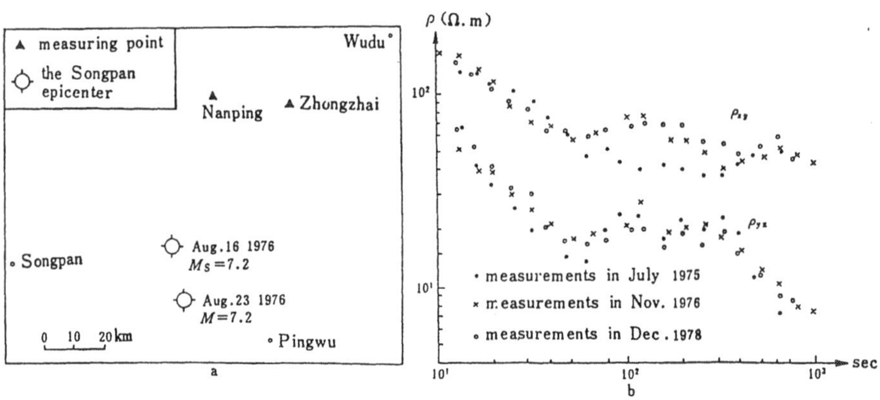

Fig.5.8. Magnetotelluric sounding before and after the Songpan mainshock; *a* location of the points; *b* variations of apparent resistivity at Nanping

In addition, at the Nanping and Zhongzhai points, systematic time–dependent variations were observed in the major axes of inhomogeneous anisotropy of electric properties of the crustal media before and after the Songpan earthquake. This shows that anisotropic variations of medium electric properties occurred before the mainshock when the apparent resistivity dropped.

5.2.3 Possible Mechanisms for the Variations of Deep Crustal Electric Properties

As described above,magnetotelluric sounding studies suggest that anomalous crustal low resistivity layers generally exist in seismic regions. Also, it is found in seismic regions that the first low– resistivity layer in the upper mantle was uplifted, and that for a fairly long period (> 1 year) before the mainshock the apparent resistivity of the deep crust decreased greatly. Such variations of apparent resistivity are possibly associated with the seismogenic process. Laterally inhomogeneous crustal structures, when worked upon by regional dynamic forces, may produce uneven stress distributions. Local stress concentrations may increase nearby boundaries of anomalous low resistivity layers within the mid–crust owing to the inhomogeneity of longitudinal structures. Since the upper part of the low–resistivity layer is more brittle than the lower part, stress increases may cause upwarping fissures at the upper boundary. Since the crustal low–resistivity layers possibly contain much water (this layer is normally explained as the result of lenses of conducting solutions that are highly developed at these depths due to thermo–dehydration of rocks under the high temperatures of the mid crust), the compressive stresses would force water into the fissures. This would cause weakening, thus accelerating the growth of fissures and the penetration of water. Therefore, the resistivity of the medium above the low–resistivity layer decreases greatly. Furthermore, the orientations of the fissures may be controlled by the principal compressive stress under high stress conditions, thus causing the anisotropic variations in the anomalous resistivity decrease.

5.3 Gravity Variations

Pre– and post–earthquake gravity observations are very important for understanding variations in the crustal environment, and for studying the causes and mechanisms of earthquakes. Gravity surveys in Japan before and after the 1964 Niigata earthquake (Fujii 1966), and in the United States before and after the 1964 Alaska earthquake (Barnes 1966), both indicated gravity variations during the earthquake process. Following the Xingtai earthquake, gravity measurements have been successively conducted in North China, Southwest China and Northwest China, a number of continuously recorded gravity stations have been established, and frequent repeated measurements are done along certain fixed gravity profiles (usually called roving gravity measurements). The following will summarize repeated gravity observations obtained near the epicentral areas before and after four great earthquakes (Haicheng, Tangshan, Longling and Songpan). For the Longling earthquake, there are data on only one survey before, and one after the earthquake, and measurements were repeated on only some segments of the profile measured near the Songpan earthquake. Therefore, only a general description is given for these two quakes, whereas the gravity variations will be particularly emphasized for the Haicheng and Tangshan earthquakes.

5.3.1 Pre– and Post–earthquake Gravity Changes

1) Haicheng earthquake.

In June 1972, the Seismological Bureau of Liaoning Province set up a 250–km–long roving gravity profile in a NW–SE direction, from North Town to Zhanghe across the East Liaoning Peninsula. The central segment of the profile passed through Yingkou City and Gaixian County, about 50 km from the M7.3 Haicheng epicenter (Fig. 5.9a). The profile was repeated twice before the earthquake, in November 1972 and May 1973, and many measurements have been taken since, in March 1975, July 1975, and subsequently. The root–mean–square error of measurements between two neighbouring points along the profile was less than 49 microgals. The results of triple measurements made in June 1972 – May 1973 showed that gravity in the southeast section decreased clearly, reaching 352 microgals (Fig. 5.9b). The first survey after the 1975 quake suggested that the gravity field had recovered to its original level of June 1972. The measurements of July 1975 showed a continuing increase, up to 382 microgals, but later measurements showed no notable variations. These observations show large–scale decreases of gravity before earthquakes in areas close to the epicenter, followed by post–earthquake recovery (Chen Yuntai et al. 1980).

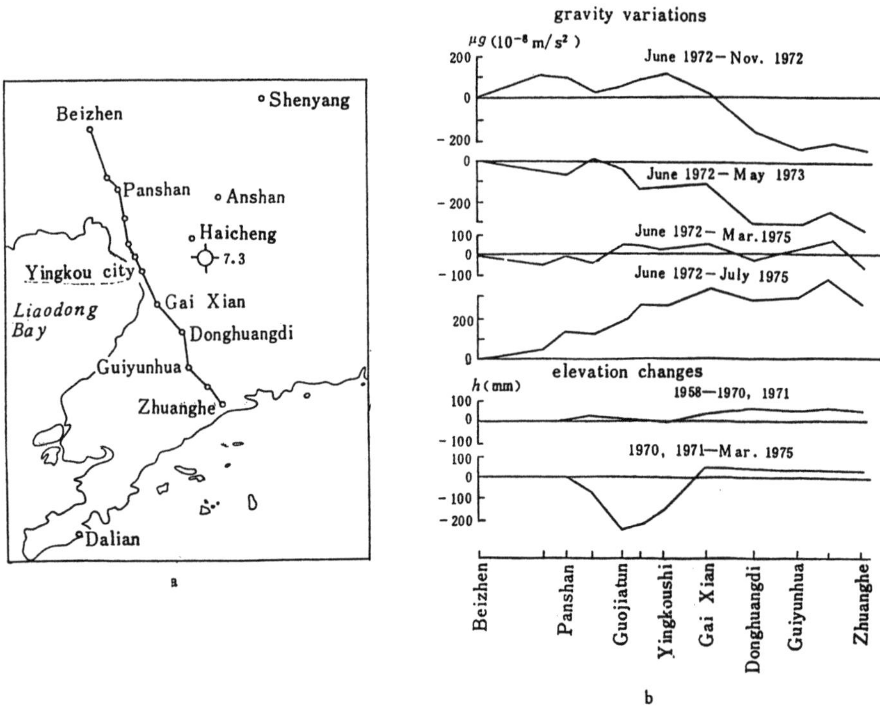

Fig. 5.9. Gravity variations before and after the Haicheng earthquake: *a* North Town–Zhanghe gravity profile; *b* North Town–Zhanghe gravity variations and elevation changes

From Fig. 5.9b it can also be seen that the greatest gravity variations before the earthquake occurred at the Gaixian–Donghuangdi segment, and the differences between the two measurements made in November 1972 and May 1973 amounted to 178 and 188 microgals respectively. This segment is the closest part of the profile to the epicenter (about 50 km). Therefore, pre–earthquake gravity variations are larger at closer epicentral distance. Figure 5.10 shows the total variation with time for the Zhanghe – North Town profile (taking the first measurement of the profile as datum). The results indicate pre–earthquake gravity decreases, and post–earthquake recovery, in the South Liaoning region (Lu Zaoxun 1984).

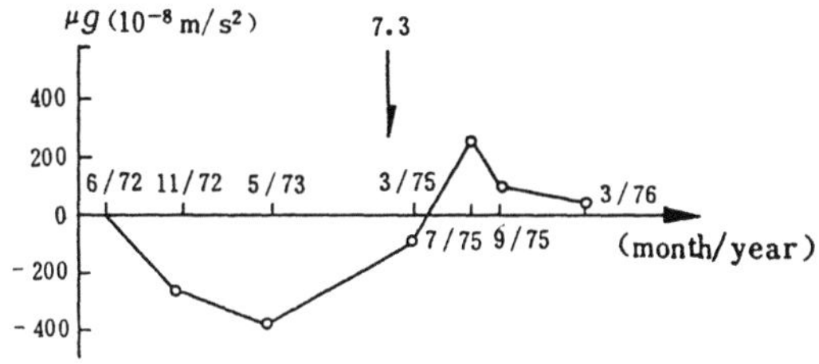

Fig. 5.10. Gravity variations of the entire NorthTown–Zhanghe profile before the Haicheng earthquake

2)Tangshan earthquake.

The Seismological Bureau of Hebei Province started with April 1970 to carry out mobile gravity measurements along lines in the Beijing–Tianjin–Tangshan region. Two lines crossed Tangshan, one being the railway line from Beijing, through Tianjin and Tangshan to Shanhai Pass, the other being the highway from Beijing, through Shanhe, Yutian and Tangshan to Leting (Fig. 5.11a). Each line has its own point in the Tangshan epicentral area, with the former located at Tangshan railway station (Tangshan I), and the latter about 2 km away at Tangshan City (Tangshan II). According to analyses of measurement accuracy by Ma Li et al., the random errors for Tangshan I were +30 gal or so, and +70 gal or so for Tangshan II.

More than 30 measurements were made along the Beijing–Shanhai Pass railway line from 1971 till the Tangshan earthquake, as shown in Fig. 5.11b for Tangshan I, using Beijing as the starting point. The figure shows that the gravity value at Tangshan relative to Beijing underwent a gradual increase since 1971, with a noticeable rise after early 1974; the total change was 130 gal up to the end of 1975. The mean rate of change was 40 gal / a. There was a slight drop in gravity after the end of 1975, and a rapid increase in April 1976. Multiple measurements following the Tangshan earthquake show a gradual decrease in gravity values.

Along the Beijing–Leting highway, the first measurements were made on 24 March 1976, and the second ones in early July, just prior to the earthquake. Taking the first measurements as datum, the gravity variation is shown in Fig. 5.11c. It is seen that the gravity values increased considerably, amounting to 165 gal, along the east segment of the line, near the Tangshan epicenter. The third survey was made the day after the mainshock, and suggested that the gravity values along the entire line had decreased slightly. The fourth measurements on 13 August 1976 showed a continuing decrease in gravity, indicating that the gravity values were recovering progressively following the mainshock.

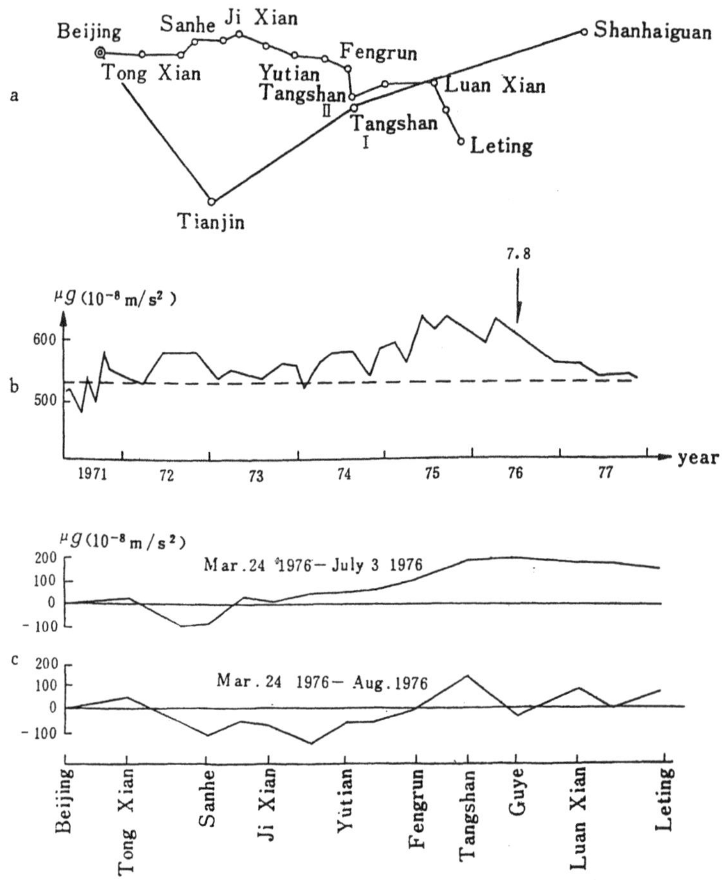

Fig.5.11 Gravity changes in the epicentral area before the Tangshan earthquake; *a* gravity measurement lines (dashed line —— highway, heavy line ——railwary); *b* gravity changesat Tangshan with respect to Beijing; *c* gravity changes along the Beijing through Leting highway

3) Longling and Songpan earthquakes.

In October 1975, before the Longling earthquake, the Seismological Bureau of Yunnan Province made mobile gravity measurements of the closed loop Baoshan–Tengchong–Longling–Baoshan, around the future epicenter. The loop totals 376 km in length, and is measured in ten segments. The survey was repeated in June 1976 after the mainshock. With Baoshan as the datum, the repeated measurements show large scale gravity variations in the area to the west of

Nujiang River, close to the epicenter. Compared with those before the mainshock, the gravity values in the epicentral area decreased by more than 300 gal after the mainshock (Chen Lide 1979).

Before the Songpan earthquake, in October 1974, the Geophysical Prospecting Brigade, State Seismological Bureau, set up a gravity line from Guangyuan, via Wenxian, Nanping, Tazang, Songpan, Jiaochangba, Maowen, Wenchuan, Yinxin and Guan–Xian to Chengdu. The standard measurement error was normally +50 gal. Two months before the earthquake, in June and July 1976, the measurements were repeated along the Chengdu–Jiaochangba segment. The north end of this segment, Maowen–Jiaochangba, is relatively close to the epicenter, at about 70–100 km. Comparison of the measurements taken in late June and mid–July 1976 with those of October 1974 indicates that the gravity varied significantly along the north half of this line, Yinxin–Jiaochangba, which is nearer to the epicenter. In general, the gravity tended to change incrementally towards the epicenter. The largest variations occurred across active faults along the Yinxin–Wenchuan segment (Seismological Bureau of Sichuan Province 1979).

5.3.2 Theoretical Analysis of Gravity Changes Before and After Earth–quakes

The gravity variation at any point on the Earth may originate from three sources: (1) change in elevation of the observation point; (2) density variations in the medium below the observation point; (3) mass migrations below the observation point. As the seismogenic process is possibly accompanied by these factors, anomalous gravity changes may be observed before and after earthquakes.

1) Theoretical extrapolation.

(A) Gravity effects caused by elevation changes and volumetric expansion due to the development, dilatation, and opening and closing of fissures within a seismogenic body

Consider an internally strained seismogenic body as a cylinder of radius, a, and height, H, and assume the side and bottom of the cylinder not to be displaced. During the seismogenic process, the uniform development and expansion of internal fissures will result in a rise, h, of the top of the cylinder. The volume of the cylinder is $V_o = \pi a^2 H$, and the volumetric strain is $= \Delta V / V_o = h / H$ (see Fig. 5.12). Usually, a and H are of the order of tens of kilometers, while h is tens of millimeters, so the volumetric strain is of order 10^{-6}. Let the mass of the cylinder be m_o, and the pre–dilatation density, $\rho_0 = m_0 / V_0$, then the post–dilatation density is given by:

$$\rho = \frac{m_0}{V} = \frac{m_0}{V_0\left(1 + \frac{h}{H}\right)} = \rho_0\left(1 - \frac{h}{H}\right), \tag{5.3}$$

The pre– and post– dilatation density variation is therefore:

$$\Delta\rho = \rho - \rho_0 = -\frac{h}{H}\rho_0. \tag{5.4}$$

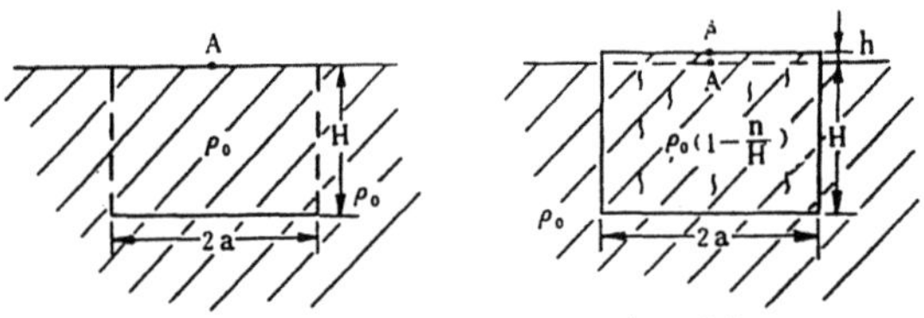

a. before dilatation b. after dilatation

Fig. 5.12. Dilatant deformation of a source body during the seismogenic process

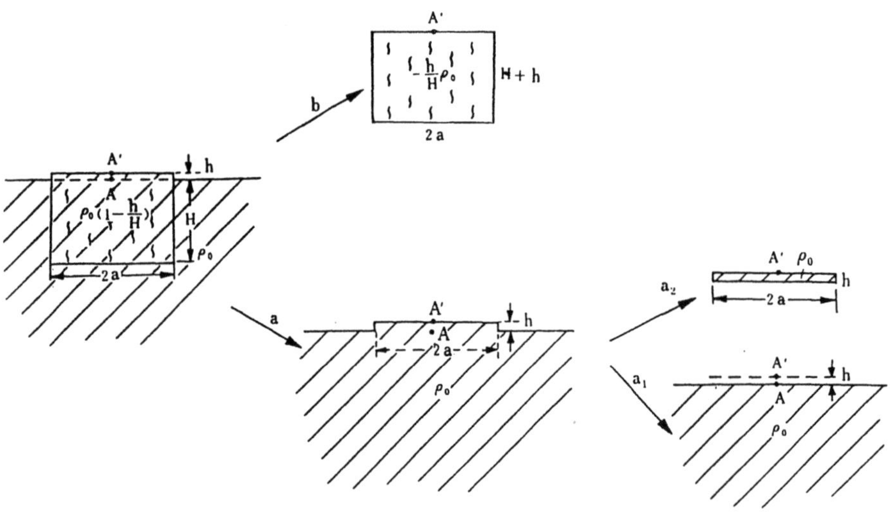

Fig. 5.13. Gravity effects of dilatant deformation of a seismogenic body

As shown in Fig. 5.12, the epicenter, A, is uplifted to A′ after the dilatation. The gravity at A′ can be obtained as the sum of two parts, as shown in Fig. 5.13. Figure 5.13a presents the first part of the uplift h with density unchanged (cylinder with radius a, height h and density ρ_0).Figure 5.13b shows the second part of a cylinder with radius a, height (H+h) and density ($-h\rho_0 / H$). The sum of these gives the gravity at A′. Figure 5.13a can be further divided into $\alpha_1$and α_2, with the former representing free–air effects:

$$\delta g_{a1} = -\frac{8}{3}\pi f \rho_E h, \tag{5.5}$$

where f is the gravitational constant, and $\rho_E = 5.5\ g/cm^3$, giving:

$$\delta g_{a1} = -0.3083h. \tag{5.6}$$

The second term represents the central gravity effects of the top of the cylinder with thickness h, radius a, and density ρ_0:

$$\delta g_{a2} = 2\pi f \rho_0 [h - \sqrt{a^2 + h^2} + a] = 2\pi f\ \rho_0 h, \tag{5.7}$$

where $\rho_0 = 2.67\ g/cm^3$. The gravity value of Fig. 5.13a is therefore as follows:

$$\delta g_a = \delta g_{a1} + \delta g_{a2} = -\frac{8}{3}\pi f \rho_E h + 2\pi f \rho_0 h. \tag{5.8}$$

The gravity effects of Fig. 13b are the central gravity effects of the top of the cylinder with radius a, height (H+h) and density ($-\rho h/H$), as given by:

$$\delta g_b = 2\pi f\left(-\frac{h}{H}\rho_0\right)[(H+h) - \sqrt{a^2 + (H+h)^2} + a]$$
$$= 2\pi f \rho_0\left[1 - \sqrt{1 + \frac{a^2}{H^2}} + \frac{a}{H}\right]h. \tag{5.9}$$

In the above, we have used the approximations $h << H$ and $h << a$. Therefore, the gravity effects, caused by elevation changes and volumetric expansion as produced by the development, expansion, reopening and closing of fissures within the seismogenic body should be the sum of g_b and g_a, namely:

$$\delta g_D = -2\pi f\left[\frac{4}{3}\rho_E - \rho_0\left(\sqrt{1 + \frac{a^2}{H^2}} - \frac{a}{H}\right)\right]h. \tag{5.10}$$

Equation (5.10) shows that this gravity effect, δg_D, lies between the free–air effect ($\delta g_{a1} = -\frac{8}{3}\pi f \rho_E h = -0.308h$), and the Bouguer effect ($\delta g_a = -\frac{8}{3}\pi f \rho_E h + 2\pi f\ \rho_0 h = 0.196h$).The equation agrees with the results of Cheng Yuntai et al. (1980).

(B) Gravity effects due to the migration and filling of deep or distant masses into the original pores or dilatant fissures within a seismogenic body

Consider again the above cylindrical seismogenic body. There are existing pores and fissures within the crustal rocks (the porosity, φ_0, of rocks can vary from $10^{-1}$to 10^{-4}), and new fissures may develop during the seismogenic process. Now, the volumetric change due to dilatant fissures might be $V/V_0 \sim 10^{-6}$, hence the volume of newly developed fissures is less than that of existing fissures. Let the porosity of the seismogenic body be φ, and let a fraction,α, be filled by deep or distant materials with density ρ_F. Then the mass that has migrated into the seismogenic body is m $=\alpha\varphi V\rho_F$. Therefore, the gravity variations at A′ correspond to the effects of a cylinder with radius a, height (H+h) and density $\alpha\varphi\rho_F$:

$$\delta g_m = 2\pi f\ (a\varphi\rho_F)\left[1-\sqrt{1+\frac{a^2}{(H+h)^2}}+\frac{a}{H+h}\right](H+h)$$
$$= 2\pi f\ \alpha\varphi\rho_F(H-\sqrt{H^2+a^2}+a). \tag{5.11}$$

Therefore, the total gravity effects associated with the deformation and mass migration during the seismogenic process are:

$$\delta g = \delta g_D + \delta g_m$$
$$= -2\pi f\left[\frac{4}{3}\rho_E - \rho_0\left(\sqrt{1+\frac{a^2}{H^2}}-\frac{a}{H}\right)\right]h$$
$$+ 2\pi f\ a\varphi\rho_F(H-\sqrt{H^2+a^2}+a). \tag{5.12}$$

2) Analysis of anomalous gravity variations.

The above discussion dealt with the theoretical gravity effects associated with deformation and mass migration in a model of the seismogenic process. The following section will analyze the possible causes of gravity anomalies according to the observations actually made before and after the earthquakes.

Before and after the Haicheng, Tangshan and Longling earthquakes, there were levelling measurements close to the gravity stations, so the gravity effects can be calculated using Eq. (5.10) with the measured elevation changes. For instance, before the Haicheng earthquake, the levelling points near Zhanghe and Chengzituan along the gravity line were uplifted by 60 mm with respect to Yingkou, from 1958 to 1970 / 71. If this value is taken as the approximate deformation during the gravity variations, and if a = H, then the corresponding gravity effects should be about −15 microgal for Zhanghe relative to Yingkou. The observed variation is an order of magnitude larger than this. With the Tangshan earthquake as another example, the levelling points close to the epicenter rose by about 28 mm from 1971 to 1975, before the mainshock. If we take H = 40 km, a = 80 km, then the gravity variation associated with the upheaval should be −8 microgal. In fact, the pre–earthquake gravity variation was more than 100 microgal, again an order of magnitude greater. In addition, after the Longling earthquake, the gravity around the epicenter generally decreased by about 300 microgal compared with that beforehand, while the elevations of some gravity points changed by only tens of millimeters. In summary, the observed gravity variations are much greater than can be explained by vertical deformation alone.

The gravity variations may be attributed mainly to mass migration, since pores and fissures already exist within crustal rocks, channels for mass movements can be formed as the original fissures are perhaps linked up by new fissures developed during the seismogenic process. Assuming that one–tenth of the porosity volume ($\alpha=0.1$) is filled or voided in the course of mass migrations, the resulting gravity effects are great enough. Taking the Tangshan quake as an example, let $\varphi = 5\times 10^{-4}$, $\alpha = 0.1$, a = 80 km, H = 40 km and $\rho_F > \rho_0$, the gravity effect can be obtained from Eq. (5.11):

$$\delta g_m > 171 \text{ microgal.}$$

Similar estimates can be made for the Haicheng and Longling earthquakes.

In addition to the consideration of mass migrations into the fissures of a seismogenic body, there may be upwelling of deep crustal or upper mantle magmatic materials along great, deep fractures that sometimes cut through the crust around source regions. Such material motion could also account for the observed gravity effects. Therefore, gravity variations of several hundred microgal in areas close to the epicenter are principally due to mass migrations and material movements.

5.4 Variations of the Geomagnetic Field

Since the 1966 Xingtai earthquake studies on the relationship between geomagnetic changes and earthquake occurrences have been extensive. At first, the main studies were continuous observations of the magnetic vertical component, and examination of changes in the daily variation patterns for anomalies precursory to earthquakes. This is generally called variable magnetic field analysis. Since 1970, proton precession magnetometers have been applied gradually to observations of earthquake precursors. So it has become possible that further research on the variational regularities of basic magnetic fields and their pre–earthquake anomalous changes can be pursued when at the same time studies on seismo–magnetic information in variable magnetic fields are continued.

5.4.1 Relations Between Daily Variations of Magnetic Field and Earthquake Occurrences

The daily changes of the vertical geomagnetic component are mainly constrained by daily solar variations, with greater intensity and smoother variation at night, and minimum intensity with drastic variation around noon. Therefore, the daily patterns of geomagnetic field variation seem to be regular. However, from continuous observations over many years, it is found that the daily variation pattern of the vertical component is frequently distorted. The distortions include the increase or decrease of daily variation amplitude (difference between daily maximum and minimum), phase shifts (changes in occurrence time of minimum), and reversals (the maximum appears around noon). Sometimes, these distortions appear before earthquakes.

1) Analysis of daily range.

According to observations made at Beijing standard seismic station, Institute of Geophysics, SSB, the daily range of the vertical geomagnetic component (Az) is affected by latitude and longitude, seasons and magnetic storms. The daily range tends to be larger in summer than in winter, and larger in low latitudes than in high latitudes. For a given local area, the A_z of various stations are close to each other. For example, for nine stations within about 200 km in the Beijing region, the daily ranges of all stations are within 1.5 gamma, based on many years′ statistics. On the basis of the normal space–time distributions of A_z, possible local magnetic anomalies before earthquakes can be recognized. The anomalies of $A_z$before the 1976 Tangshan, Longling and Songpan earthquakes are summarized in Fig. 5.14 (Beijing standard seismic station, Institute of Geophysics, SSB, 1978). It is seen that 1–2 days before these three mainshocks, there were high values of daily range around the epicenters, about 5 gamma greater than the normal values nearby.

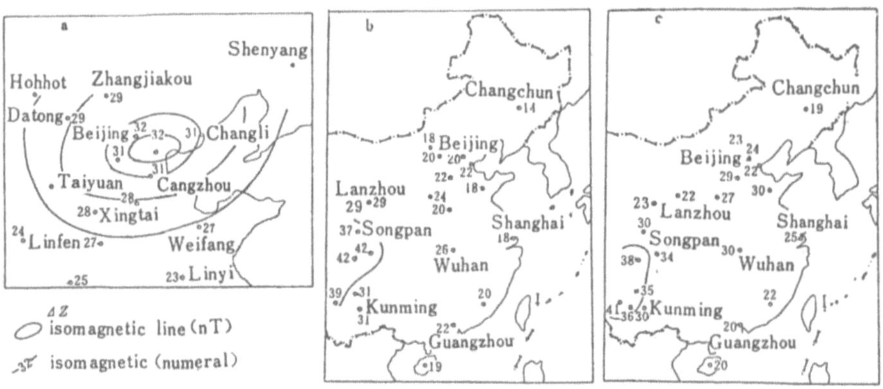

Fig.5.14. Anomalies in daily range before great earthquakes; *a* Tangshan, 26 July 1976; *b* Songpan, 15 August 1976; *c* Longling, 27 May 1976

2) Analysis of low point shift.

The low point shift denotes the change in time of occurrence of the daily minimum. If the time difference between the minimum at a given station and at others nearby is over 3 h, this is called a low point shift anomaly.

There are various cases of pre–earthquake low point shifts. Figure 5.15 shows variations 27 days before the Longling earthquake. The North–South Seismic Zone divides the country into west and east regions, and the appearance time of low points is basically the same within each region, about 11 h in the east, and 14 h in the west. The quake occurred near the dividing line. Similar anomalies appeared before the Songpan, Haicheng and Tangshan earthquakes.

The Seismological Bureau of Sichuan Province (1979) made a statistical analysis of all geomagnetic data from the south–west of the country, for the period of February 1972 to April 1977. They selected appropriate anomaly criteria, and suggested that anomalous distributions of low point shifts can serve as a reference for predicting the locations of strong earthquakes.

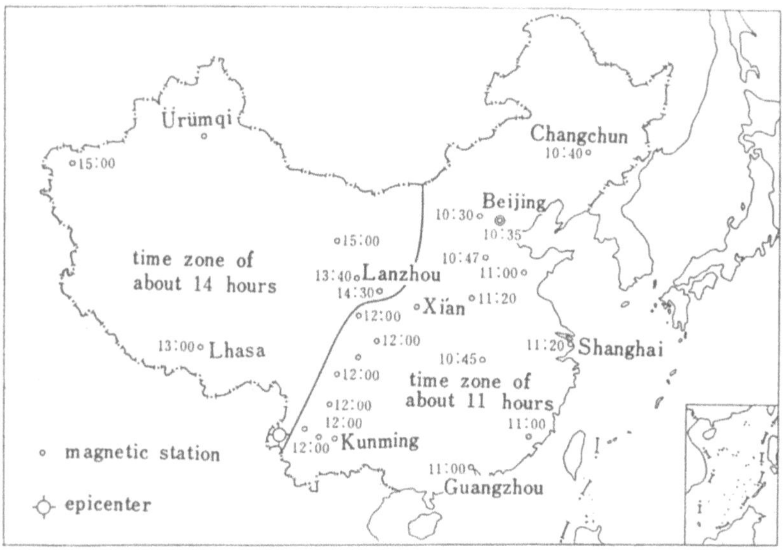

Fig. 5.15. The low point time of the Z component, and the dividing line, before the Longling earthquake (2 May 1976)

Ding Jianhai et al. (1980) split the low point shifts into three types: (1) Nationwide regional anomalies. The low point times at various stations over the nation are divided into two regions, with a difference of more than 2 h between the two. Such anomalies have a fairly high probability of corresponding to strong earthquakes. (2) Large–scale regional anomalies. Within roughly half the country, the low point times are separated into two regions. Such anomalies may correspond to some medium–strong quakes, but with many false alarms. (3) Small–scale regional anomalies. The low point times are divided into two regions over a provincial range. These have no obvious connection with earthquake activity. In addition, Ding Jianhai statistically analyzed low point shift anomalies before 21 earthquakes with $M > 6.8$, and 43 with $M > 5$. He suggested that the possibility for earthquake occurrence seemed larger 27 ± 4 or 41 ± 4 days after the appearance of the anomalies, and that the earthquakes often occurred near the dividing line of sudden changes of low point times.

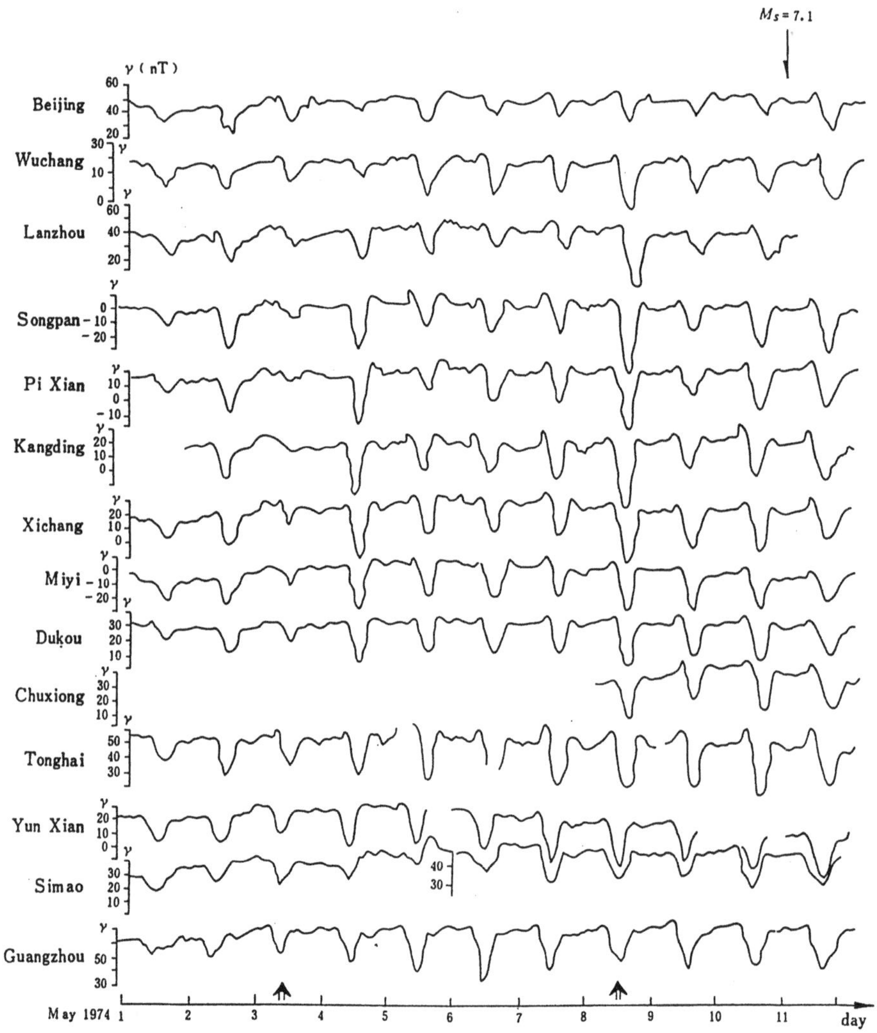

Fig. 5.16. Anomalies in the pattern of daily geomagnetic field variations before the Yongshan earthquake

3) Waveform distortion of daily variations, and reversal of daily variations.

The daily variation pattern is distorted, or even reversed, before some great earthquakes. Before the Yongshan earthquake, the daily ranges at Dukou, Miyi, Xichang, Kangding, Bixian and Songpan stations, which are relatively close to the epicenter, decreased on 3 May 1974, and increased on May 8. Stations outside the epicentral region showed no significant anomalies (Fig. 5.16). Similar distortions also occurred before the Songpan earthquake. In addition, the daily variation was reversed at some stations in the northern part of North China, on 20

January 1975 before the Haicheng earthquake, and such was also the case for the Xingtai earthquake. It should be noted that daily variation distortions are frequently observed during practical observations, but they have less correlation with seismic events.

The preceding paragraphs have simply described some methods that might be used in studying earthquake anomalies in the daily variations of the daily geomagnetic field. For predictions of occurrence times, the multiple method between the time intervals of two magnetic storms was used first, and the idea of the time spanning method was also developed. Then Guo Zengjian et al. (1977) put forward the "times nine" method, which takes 9 days as a multiple beginning with the day of a magnetic storm to predict occurrence times, and by which the appearance times of variable magnetic anomalies (such as low point shifts and distortion of daily variation waveforms) can also be used to speculate on the date of earthquake occurrences.

It should be emphasized that the relationship between magnetic variations and earthquake occurrence is still under research, and that further accumulation of data and in–depth studies are needed. Nevertheless, the research accomplished so far can provide some useful clues, particularly in estimating times of earthquake occurrence.

5.4.2 Anomalies in Long–period Variations of the Geomagnetic Field Before the Tangshan Earthquake

Before the Tangshan earthquake, absolute magnetic measurements were conducted at more than ten geomagnetic stations and points within the Beijing–Tianjin–Tangshan region. As a result, for a fairly long period absolute measurements of the vertical component and the total intensity of the geomagnetic field have been collected.

1) Anomalies detected by the differential analysis method.

Li Cunti et al. (1980) systematically analyzed absolute geomagnetic data from the Beijing–Tianjin–Tangshan region, and calculated monthly averages of mean night values of the vertical component (Z) at each station by selecting data on calm nights. After deriving the monthly intensity at each station, comparisons between stations were made in order to reveal anomalous variations. The Beijing, Changli, Hongshan and Taian stations were selected to form station groups. The variations of Z for each group were obtained with station value minus station value. Figures 17b and c show the time dependence of the vertical components at several stations in the Beijing–Tianjin–Tangshan region, and the differences between some station groups. From Fig. 5.17c it is seen that the variations of the Z components at each station are fairly consistent, and the curves of Z are basically stable without notable anomalies from 1974 to July 1975. After July 1975, the component at Changli station tended to decrease with respect to the outside Hongshan station. This lasted for about 1 year, until the Tangshan mainshock, with more than 10 gammas decrease. This is also the case for Beijing relative to Hongshan, only with a smaller variation. After the mainshock, the component values seemed to increase, but they did not return to their former values; instead, they were relatively constant at their new level.

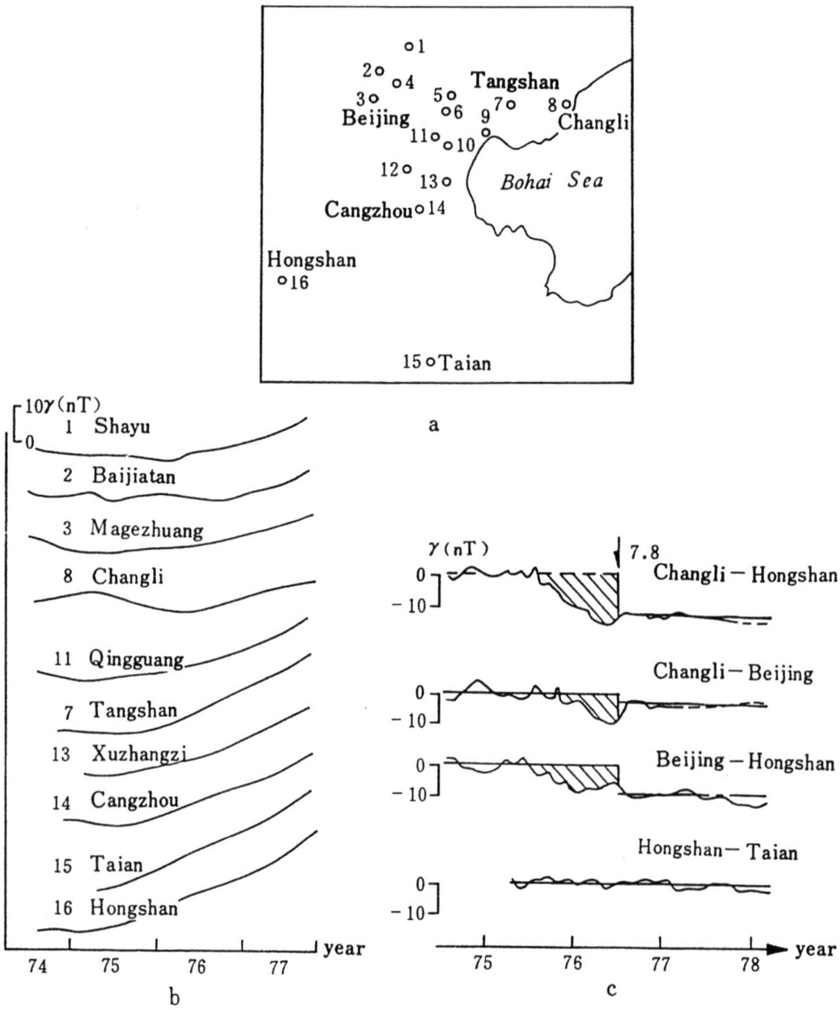

Fig. 5.17. Variations of vertical component (Z) of the geomagnetic field before the Tangshan earthquake; *a* distributions of geomagnetic stations; *b* observations of Z at some stations; *c* variation of Z at the main station groups

2) Anomalies recognized after correction for long–period variations.

The above anomalous variations are obtained by subtracting the observed values at one station from those at another. However, there exist slow, long–period changes in the geomagnetic field itself, with differences in amplitude, phase and rate of variation in different localities. Using the differential method, the long period variations at the different stations cannot be removed. In order to understand the normal distributions of long period variations with latitude and longitude, Lu Zhenye et al. took the annual variation rate of long–period variations (yearly variation amplitude) and the reversal time of long period variations as characteristic quantities to seek for the regularities of spatial distributions of annual variations and reversal times. They found the annual variation rate of long–period changes to be a non–linear function of latitude and longitude (λ,φ), and that it fairly well fitted as a Taylor series. Based on many years' observations at various stations, the coefficients of the Taylor expansion of the annual variation rate of long period variations were obtained by least squares. The distributions of the total intensity (F) and vertical component (Z) of the geomagnetic field at the surface were therefore calculated as a function of and . Using the same method,the variations of reversal times of long period changes in the geomagnetic field were also calculated as functions of λ and φ(reversal time implies a change in direction from increase to decrease, or vice versa). These derived values were used to correct the measurements at various stations before the Tangshan earthquake, thus obtaining the vertical component anomalies given in Table 5.1. It is seen that there are small changes without obvious anomalies at most stations, except for the −11.5 gammas at Changli. Therefore, the corrected geomagnetic anomalies before the Tangshan earthquake occurred principally in the Changli area. There are very few data on absolute measurements made continuously for a long period of time near the epicentral areas of the other great earthquakes, so similar analyses cannot be done for them.

Table 5.1. Anomalies (in gamma) after correction for long period variations

Stations	Hongshan	Wenan	Qingguang	Baodi	Cangzhou	Tangshan	Shayu
Anomalies	0	-2.7	-5.3	-2.9	-2.5	+1.7	-4.3

Stations	Baijiatuan	Ninghe	Majiadian	Changli	Dongditou	Magezhuang
Anomalies	-3.4	-2.8	-2.5	-11.5	+3.0	-4.5

To summarize, anomalous phenomena in the daily variations and in the long–period variations of the geomagnetic field, may appear in the epicentral area before some great earthquakes. As there are numerous factors that can influence geomagnetic field variations, profound research is required on the normal background variation of the geomagnetic field. Only in this way can the possible seismomagnetic information be accurately identified.

CHAPTER 6

Crustal Deformation and Anomalous Variations of Stress

Geodetic work has played an important role in earthquake prediction research. Geodetic results were used in developing the elastic rebound theory, and are now used in studying the dynamics and kinematics of seismic sources, in dividing seismic precursors into stages, and in discussing the stick–slip mechanism and recently developed seismogenic models. As prediction research develops, new requirements are demanded of crustal deformation measurements, such as higher accuracy, more frequent measurement cycle, improved networks and better data processing techniques. However, most of the presently available data are still obtained by conventional geodetic methods.

In this chapter, the geodetic measurements before and after the nine great earthquakes are described to show the characteristic deformations of the latter. The crustal deformation is compared to the geological structures and to the state of stress at the source. Anomalous variations in geodetic networks, short–line levelling, fault–crossing measurements, tilts and tide gauge data are discussed. Since most stress measurements in China are based on the principle of borehole deformation, stress variations before and after the earthquakes are also included here.

6.1 Geodetic Measurements

Following the Xingtai earthquake, the Geodetic Brigade, SSB, and geodetic teams from various Provincial Bureaus have successively reconstructed, resurveyed and expanded the existing geodetic networks in certain critical areas. They have also begun to set up deformation stations and to carry out mobile measurements across faults. Most of the seismic areas were promptly resurveyed following the earthquakes. Hence very valuable, though incomplete, geodetic data have been collected over the last 15 years. These data are collected in the summary reports on the great earthquakes by the Geodetic Brigade, SSB, and geodetic teams from various Provincial Bureaus. Network adjustments and analysis have been carried out on these data, and as a result several important conclusions have been drawn. These constitute the basis for our discussion of the characteristics of the nine great earthquakes.

Most of the network data on the nine earthquakes have been processed by free–network adjustment. In order to remove the influence of starting errors, some of the data have been corrected by the affine conversion method, following the free–network adjustment. When there are repeated triangulations or trilaterations, the average strains are generally calculated by the plane strain analysis of Jaeger (1956) and Frank (1966). Levelling measurements were not adjusted as only one or two lines passed through the epicentral areas. Rather, differences between surveys were plotted as a profile of elevation changes. A running average line is defined graphically to fit the general tendency of the profile, then any large deviations are assumed to represent local crustal deformation.

6.1.1 Xingtai Earthquake

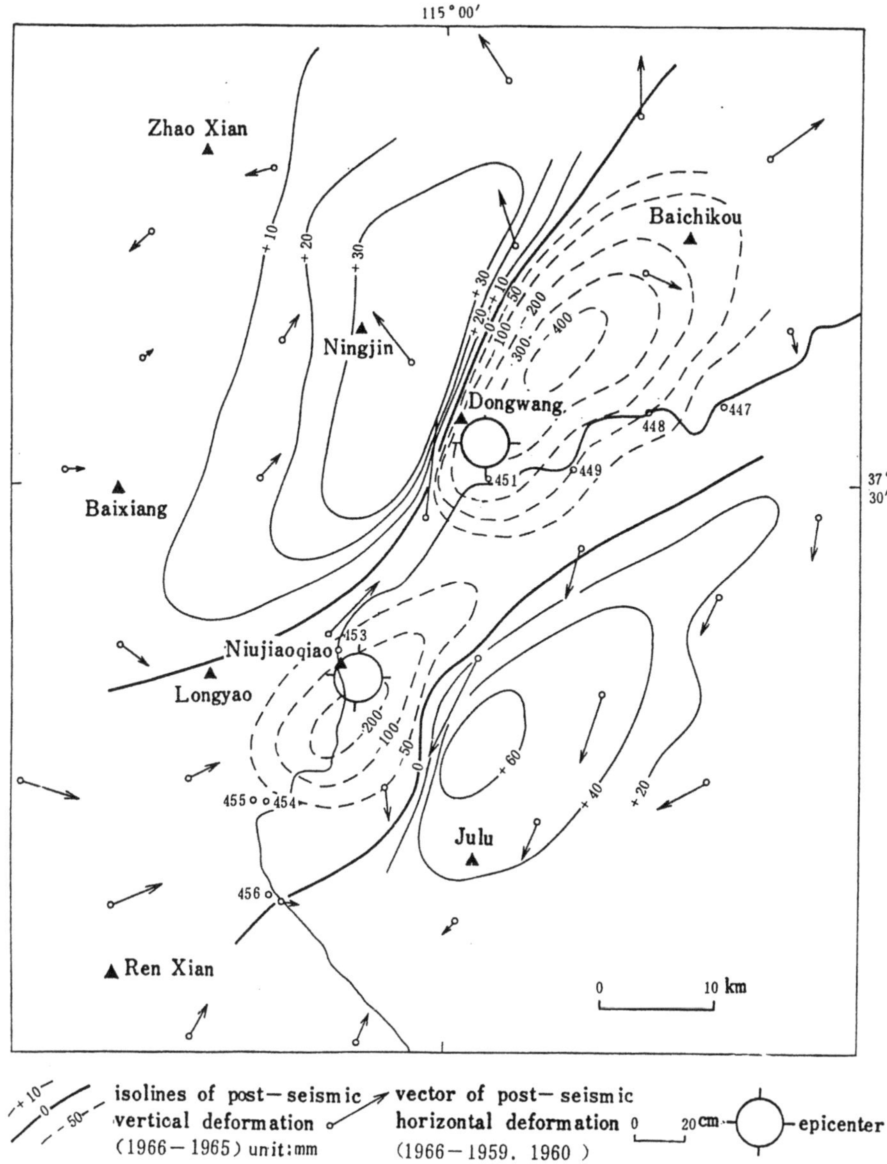

Fig. 6.1. Surface deformation before and after the Xingtai earthquake

1) Levelling measurements.

Levelling was conducted in 1920, 1950 and 1965 before the earthquake, and in 1966 following it. There was only one repeated line along Fuyanghe before the quake (Fig. 6.1). For this line, the 2 standard deviation of elevation differences corresponding to the same benchmarks in different time is $\pm 7.4\sqrt{R}$ (in millimeters, where R is the length of the line in kilometers). The elevation changes at various points along the line show that the surface is higher in the south and lower in the north, and that it tilts northeastwardly, in agreement with the trend of the relief. The locally depressed area (point No. 499, −210 mm) and locally uplifted area (point No. 456, +140 mm) have been formed in the past 35 years. The meizoseismal area was depressed greatly after the mainshock, and there were two subdued areas in both north and south, with both sides uplifting slightly. The strike of the subdued zone was basically identical to that of the Shulu fault depression (Fig. 6.1).

2) Horizontal deformation measurements.

Triangulation measurements were made once before (1959–1960) and once after (1966) the Xingtai earthquake. The first measurement was second order, but the net was remeasured with first order accuracy. The data were processed by free–network adjustment, and modified by affine conversion, to obtain right–lateral displacement vectors before and after the earthquake with a maximum offset of 0.8m or so (Fig. 6.1).

6.1.2 Tonghai Earthquake

1) Levelling measurements.

There is a levelling route along the Qujiang fracture zone through the epicentral area (Fig. 6.2a). Measurements were made before the mainshock in 1958–59 and 1960, and following it in 1970 and 1972. The measurement accuracy in 1958–59 was only third order. Relative to 1958, the elevation difference became gradually large from Daliyuan in the northwest, to the southeast of the area, before the earthquake occurred in 1969. There was also an area of 60 mm relative subsidence between Gaoda and Guanyi. During and after the mainshock (1970–69 and 1972–69), the Gaoda, Quxi and Guanyi areas continued to descend strongly, with a maximum amplitude of −230 mm, while the north side of Qujiang fault uplifted with a peak value of some +60 mm.

2) Horizontal deformation measurements.

The only results from the epicentral area are from a second order "filling in" triangulation net that was measured for a topographic map in 1958. Thirty–six sides were remeasured in 1970, after the mainshock, with an actual precision of side length of 10–20 cm. The data were processed by free network adjustment; the vector displacements of various points are plotted in Fig. 6.2a, and show right–lateral dislocation with relative motion of 2 m or so.

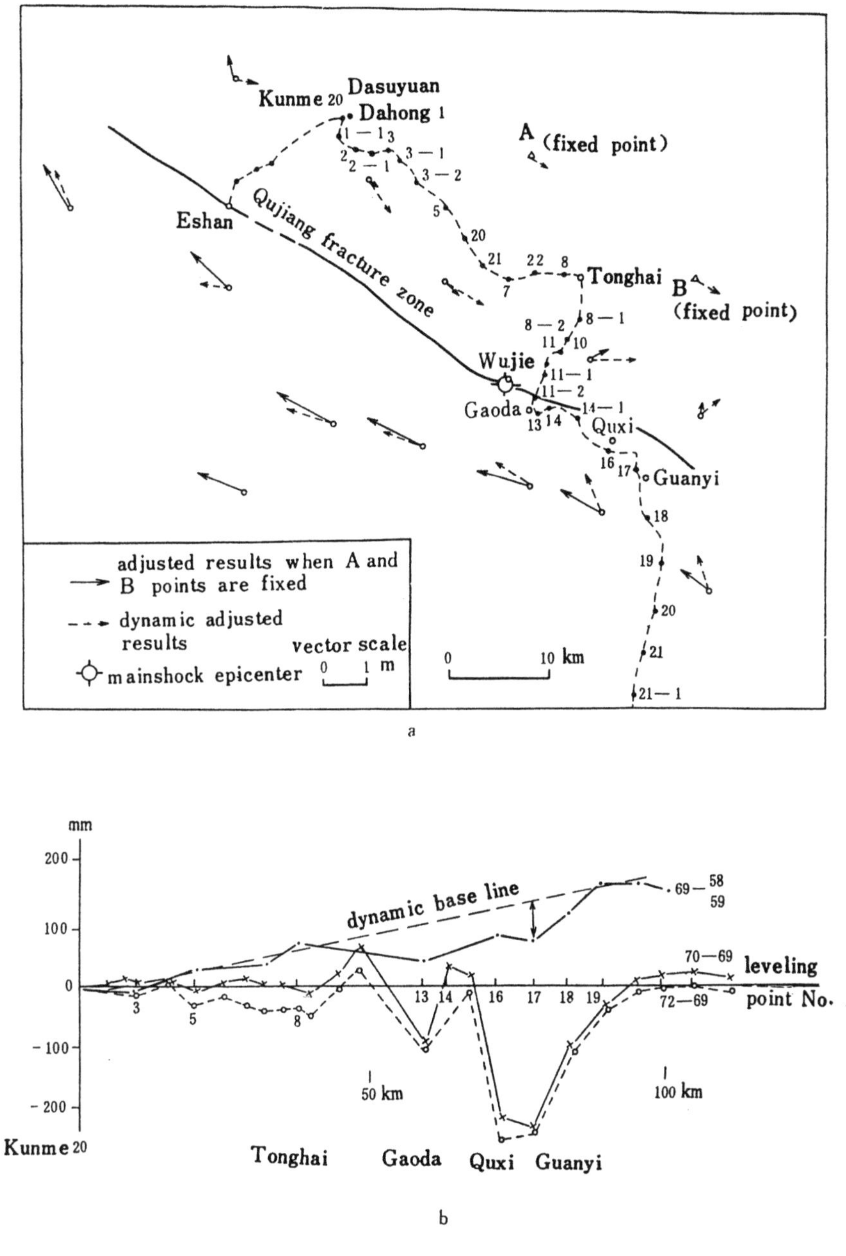

Fig. 6.2. Deformation measurements before and after the Tonghai earthquake; *a* levelling routes in the epicentral area, and horizontal displacement diagrams before and after the mainshock ;*b* vertical deformation profiles before and after the mainshock

6.1.3 Luhuo Earthquake

No horizontal deformation measurements were made for this earthquake. Second order levelling was available for one measurement in 1960 in the epicentral area, and the net was remeasured with first order accuracy after the mainshock. The results suggest that there were subsidence centers caused by the earthquake near Xialatuo and Zhuwo, both areas of intensity–X. From Zhuwo southward, the subsidence amounted to 200–300 mm, and overlapped the effects of the 1967 M6.7 earthquake; and Xialatuo sank nearly 200 mm. The coseismic surface ruptures showed a maximum horizontal displacement of about 3.6 m, with typical strike–slip features. The ratio of horizontal to vertical dislocation was 12:1.

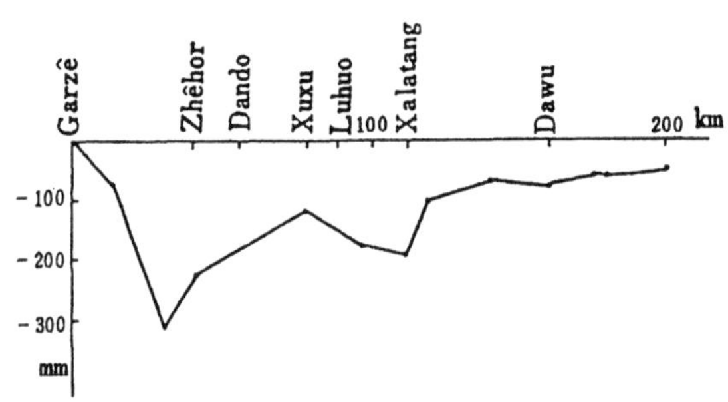

Fig. 6.3. vertical deformation before and after the Luhuo earthquake (1973–1960)

6.1.4 Yongshan Earthquake

1) Levelling measurements.

Within the intensity–VI area, about 50 km south of the epicentral area, there was a second–order levelling route along Daguanhe (Huanggexi–Yanjin line) (Fig. 6.4). There are data on one measurement before (1958) and one after (1974) the mainshock. Comparisons between the pre– and post–earthquake measurements show that the area between point No. 11 and point No. 4 uplifted by 8 mm, while the surface to the west of point No. 4 sank by 18 mm.

2) Horizontal deformation measurements.

The first triangulation measurements were made in 1958, and seven sides were resurveyed with a microwave rangefinder after the mainshock. Comparisons between the two show that various sides were shortened after the mainshock, but that the nearly north–south side was less shortened, while the nearly east–west side contracted significantly with a maximum of about 1.15 m. The orientation of the principal compressional strain, estimated from the linear strain of each side, was dominantly east–west, in agreement with fault compression during the earthquake (Fig. 6.4).

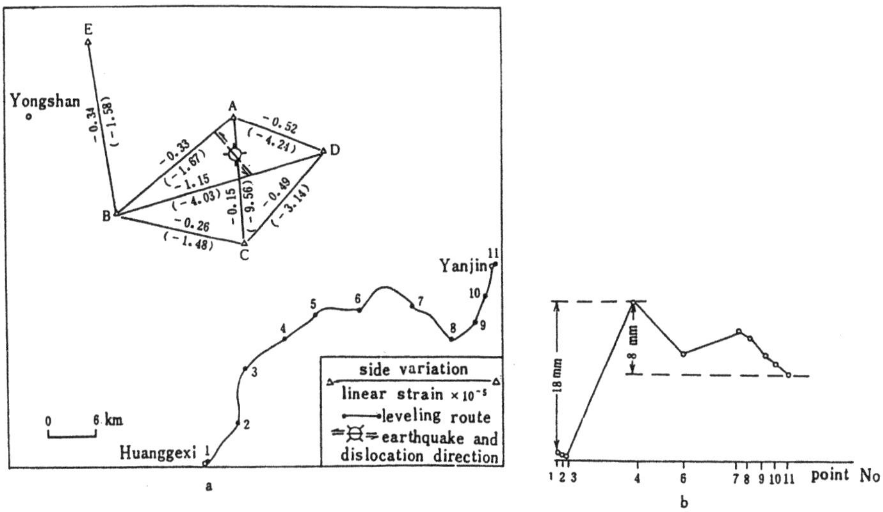

Fig. 6.4. Vertical deformation measurements before and after the Yongshan earthquake; *a* distribution of net and routes in the epicentral area; *b* vertical deformation before and after the mainshock (1974–1958)

6.1.5 Haicheng Earthquake

1) Levelling measurements.

National first order levelling loops have been established in Liaodong Peninsula. Before the earthquake, overall measurements were carried out in 1957–1958 and in 1970, and in 1973 just the Goubanzi–Panshan–Yingkou line was remeasured (Fig. 6.5a). The triangular loop of Haicheng–Gaixian–Xiuyan–Haicheng close to the epicenter was remeasured after the mainshock. The pre–earthquake deformation has suggested that the earthquake occurred just at the saddle–shaped part between the Panshan upheaval and the Liaodong Peninsula uplift (Fig. 6.5a), and that the saddle part subsided after the mainshock. The major axis of the sinking area was directed NW, with two subsidence centers, one of −140 mm near the epicenter, and the other at the lower reach of Liaohe Plain, which sank 260 mm at the mouth of the lower Liaohe due to sand liquefaction (Fig. 6.5b).

2) Horizontal deformation measurements.

Second order triangulation measurements were made in the epicentral area in 1958, and some sides were remeasured by microwave rangefinder and laser rangefinder, with a measurement error on each side of less than ± 4 cm. The results reveal that those lines that cross the NW earthquake rupture zone were elongated, thus indicating tensional faulting.

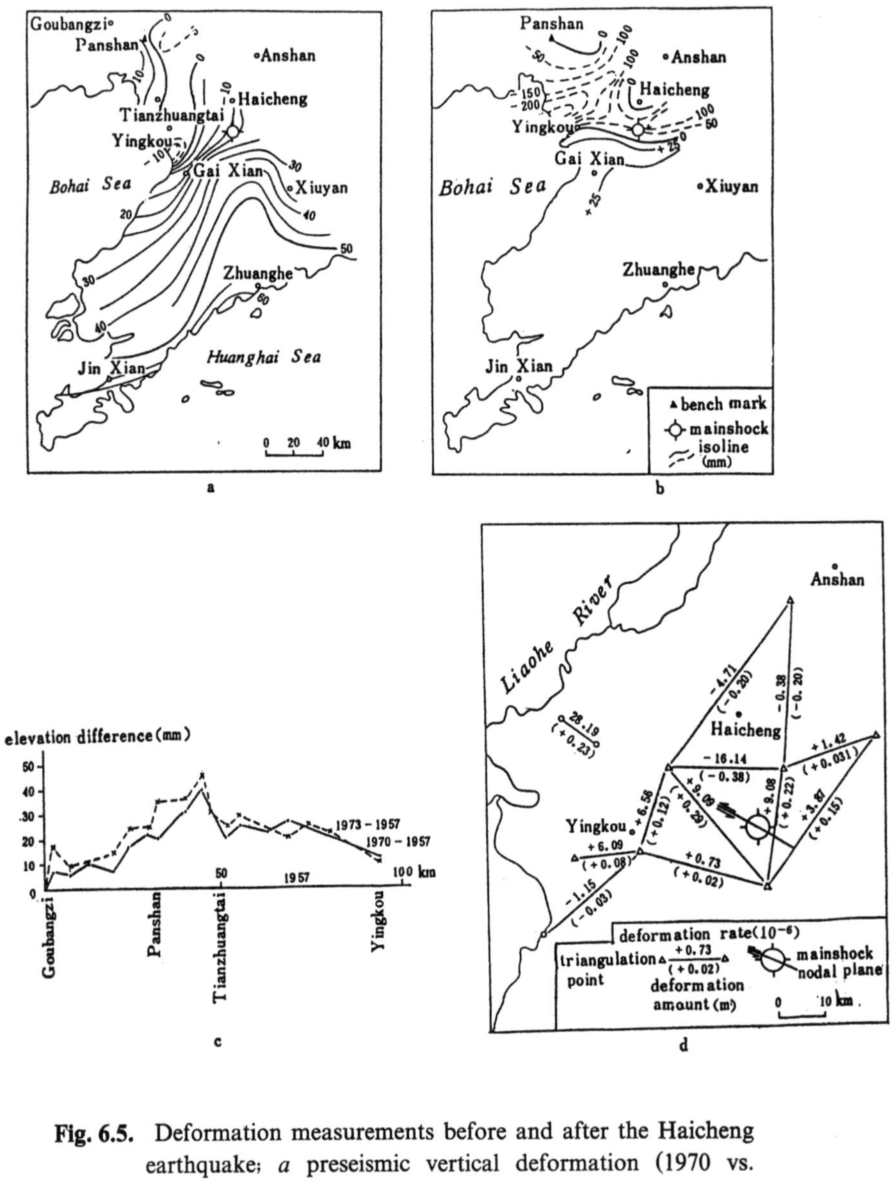

Fig. 6.5. Deformation measurements before and after the Haicheng earthquake; *a* preseismic vertical deformation (1970 vs. 1957–1958); *b* post–seismic vertical deformation (1975 vs. 1970); *c* vertical deformation of the Goubangzi–Panshan–Yingkou region; *d* horizontal deformation before and after the earthquake

6.1.6 Longling Earthquake

1) Levelling measurements.

There were two routes in the epicentral area from which repeat levelling data were obtained. One route ran from Tengchong to Luxi, and was measured in 1958 (2nd order) and 1973 (1st order) prior to the mainshock. The segment from Tengmang (point 8) to Manggeng (point 1) was remeasured after the mainshock with first order accuracy. The other route ran from Daguanshi to Longling. It was first measured in 1973, and the Dalong segment (points 6 to 27) was resurveyed after the mainshock with first order accuracy (Fig. 6.6a). The various pre– and post–seismic data are shown in Fig. 6.6b. The figure shows that the Tengchong–Longling segment was uplifted in 1973, relative to 1958, by an average of 70 mm. It continued to rise after the mainshock, by an average of 20–30 mm compared to 1973. However, point 13 in the Longling Basin and point 18 in the Luxi Basin appeared to sink relatively by 40 mm. The comparisons between the pre– and post–seismic data for the Dalong (point 4) to Longling route indicate that the area to the west of point 23 was uplifted by +40 mm on average, while the area to the east of it, centered at point 22 in the Zhenan Basin, sank by about 170 mm.

2) Horizontal deformation measurements.

The epicentral area contains some first order and second order triangulation; these nets were surveyed separately in 1956 and 1959. The root–mean–square error of the weakest point after adjustment was 0.12 m. Twelve sides were remeasured by microwave rangefinder after the earthquake. The results show that the sides in a NE or nearly NS orientation were shortened, while those in a NW or nearly EW direction were less shortened or elongated. The sides of the entire net were, in general, shortened. Plane strain estimates show that the principal compressive strain axis was oriented NNE.

6.1.7 Tangshan Earthquake

1) Levelling measurements.

There are about 140 levelling points in the Tangshan area and its vicinity, with rather abundant observation data. Parts of these routes were surveyed in 1954, 1959 and 1967–1975. There were 11 sets of observations before the mainshock, and the points were resurveyed in 1976 and 1979 after the quake. All the data obtained in previous years have been processed by free–network adjustment, giving standard errors in elevation of $\pm$ 8 to $\pm$ 12 mm. The data near Changli, the east part of the observation area, have shown less annual variations in the past (C and D in Fig. 6.7a), but the area to the southwest has tended to sink gradually as it is close to an industrial water pumping area. The levelling points in the epicentral area (B in Fig. 6.7a) varied slowly in 1954–1967, and uplifted significantly in 1968 and 1969. The Luanxian and other points were also uplifted (B in Fig. 6.7a). An upheaval center was formed between Fengnan and Ninghe, with a maximum amplitude of 50 mm (Fig. 6.7b), but it began to subside from 1970 on. The range and the appearance time of such anomalous uplift in the epicentral area seems to have been associated with the M7.8 Tangshan earthquake, and was possibly an early indication of the seismogenic process of this earthquake.

The differential motions on either side of the earthquake–producing fault were clear–cut after the mainshock, with the north side up and the south side down. Three subsidence centers were produced in the Luanxian, Tangshan and Ninghe areas (Fig. 6.7c), with level changes of −1000 mm, −740 mm and −1500 mm, respectively (zhang zusheng 1984).

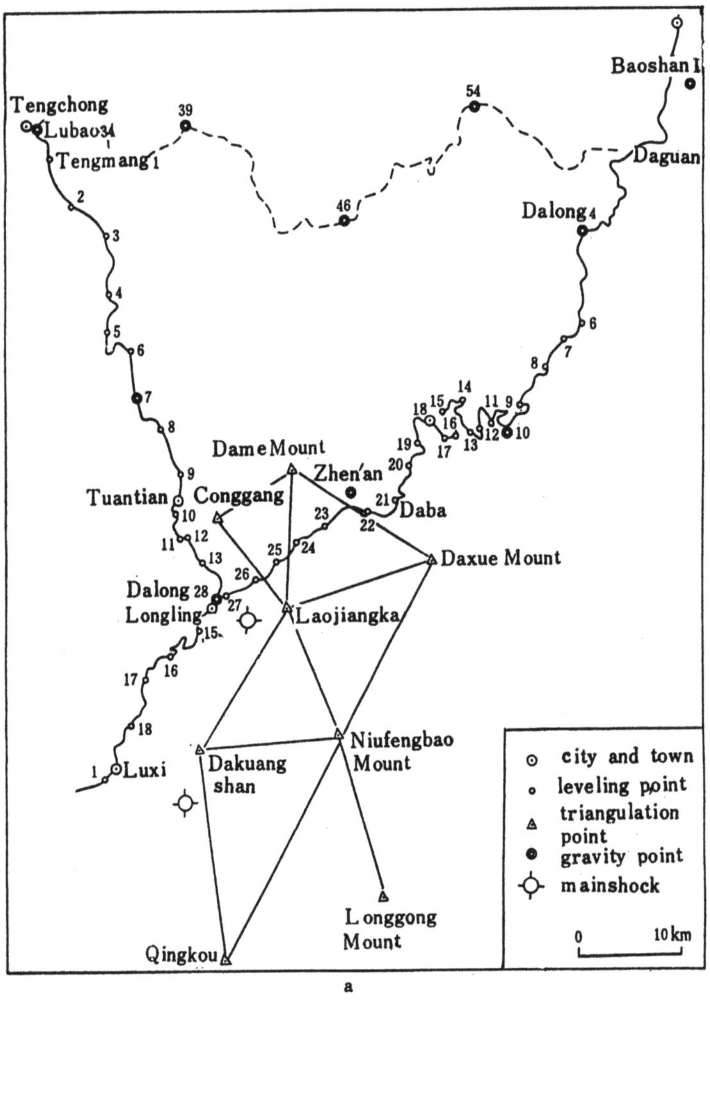

a

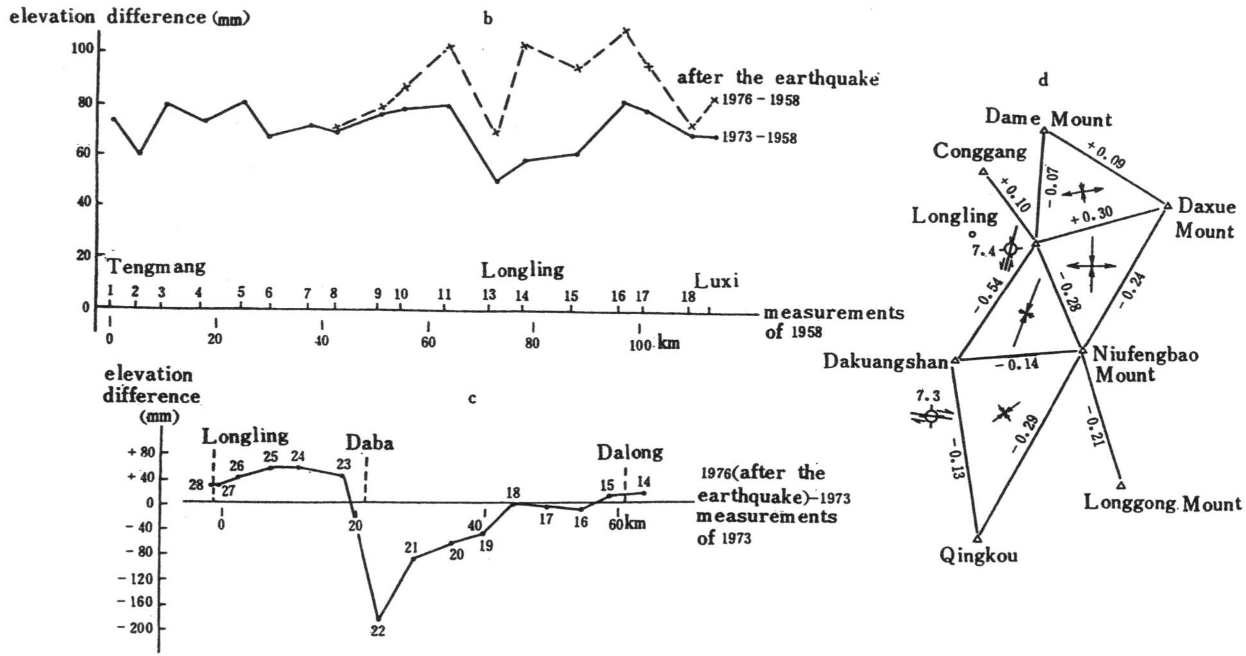

Fig. 6.6. Deformation measurements before and after the Longling earthquake; *a* Distrubution of net and routes in the seismic area; *b* profile of elevation changes along the Tengchong–Luxi line; *c* profile of elevation changes along the Longling–Dalong line; *d* line–length changes and horizontal strain before and after the earthquake

2) Horizontal deformation measurements.

There have been four repeats of horizontal deformation measurements in the Tangshan area (1954–1960, 1971, 1976 and 1977).The first and fourth were triangulation surveys, and the relative root mean square (rms) error in length of the weakest side was about 10^{-5}after adjustment. The other measurements were trilateration, and had a relative rms error of about $2–3 \times 10^{-6}$. As the accuracies of the pre–seismic surveys were low, the variations of the side lengths were mostly within the error. The coseismic deformation suggested that the various points in the epicentral area were shifted in a clockwise sense, with a relative displacement of 2.5 m between the two sides (Fig. 6.7c).

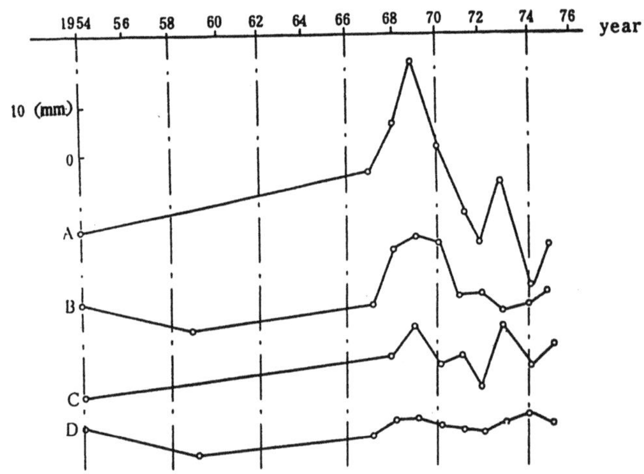

a

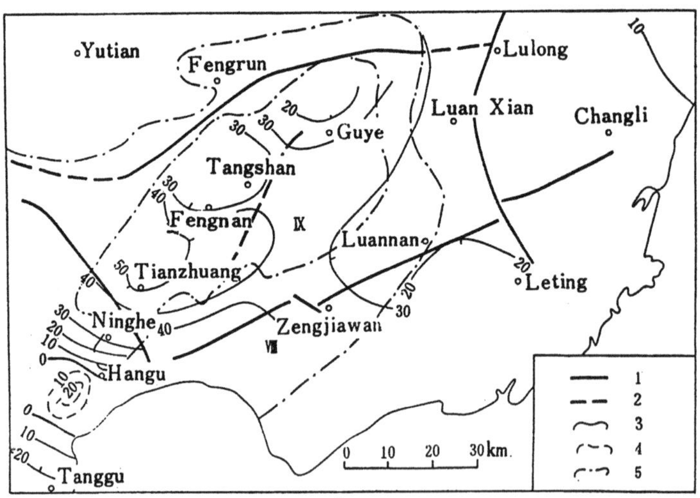

b

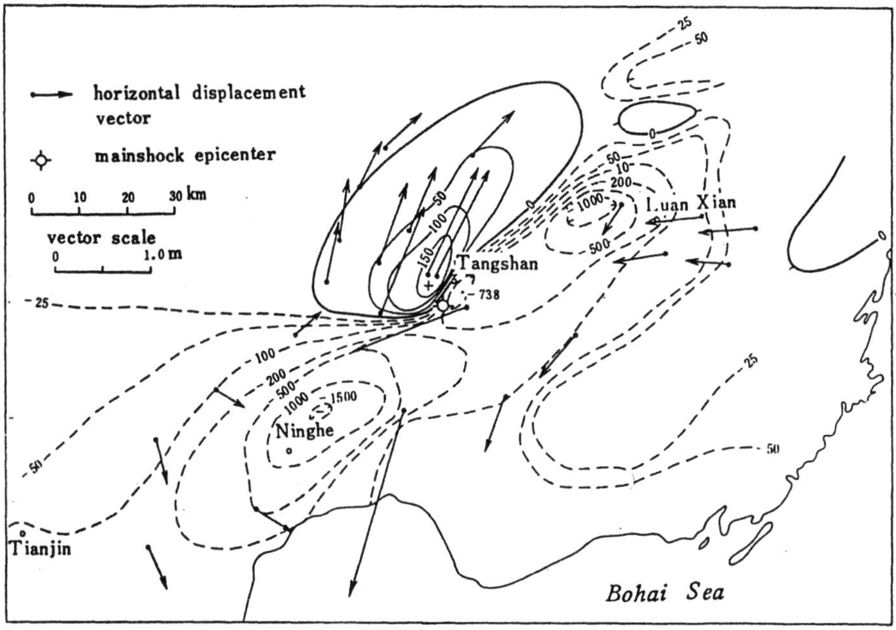

c

Fig. 6.7. Deformation measurements before and after the Tangshan earthquake; *a* vertical deformation of the epicentral area before the earthquake–time dependent elevation changes; *b* contour map of vertical deformation; *1* fault; *2* speculative fault; *3* uplift; *4* subsidence (in mm);*5* isoseismal line; *c* pre– and post–seismic deformation diagram (in mm)

6.1.8 Songpan Earthquake

1) Levelling measurements.

The Nanping–Songpan–Wenchuan line runs close to the epicentral area. There were five measurements along the Songpan–Nanping segment, with four before the earthquake (1960, July–September 1975, September–November 1975, and June – mid–August 1976), and one after it (1976). There were only two measurements along the Songpan–Wenchuan segment before the earthquake (1958–1960 and 1975), and none afterwards.

According to the analysis of the Seismological Bureau of Sichuan Province, the elevation changes from the 1960's to 1975 along the Songpan–Wenchuan segment were tiny, showing no significant differential motions. Along the 150 km line from Songpan to Nanping, with point No. 60 as the base point, the results of the first measurements in 1975 (July–September) relative to 1960, indicated that point No. 4 was a center of subsidence and point No. 10 a center of uplift, with amplitudes of −135 mm and +177 mm, respectively (Fig. 6.8). The subsidence close to points No. 1–3 and at point No. 4 near the epicentral area of the 1960 M6.7 Zhangla–Songpan earthquake, might include the effects of this earthquake. The second mesurements in 1975 showed that all points had uplifted significantly, with a maximum of 15 mm within 2 months. The measurements made 2 months before the 1976 quake demonstrated relative subsidence at various points, i.e. reverse motion as compared to the second 1975 results.

After the earthquake, pionts No. 4–6 were uplifted distinctly, with a maximum amplitude of 10 mm, and various points to the east of point No. 12 subsided, with a maximum of 20 mm.

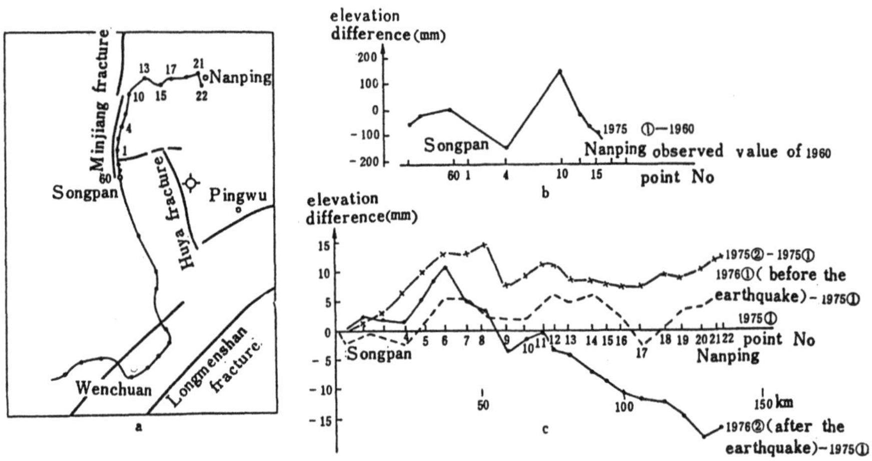

Fig. 6.8. Levelling routes and elevation changes in the Songpan earthquake area; *a* levelling routes; *b* and *c* elevation profiles

As the Bohai earthquake originated under the sea, no deformation data were obtained from the epicentral area. However, the crust along the coastline of the Bohai Sea has deformed considerably during at least the last 10 years.

The deformation characteristics in the epicentral area before and after the above earthquakes can be summarized as follows:

1) Heredity of tectonic activity.

Figure 6.9 compares the vertical deformation following the above earthquakes with the tendency of neotectonic activity. For earthquakes occurring in the Quaternary grabens and basin margins, points within the grabens and basins subside markedly during the quakes, while points outside are relatively uplifted. Examples are the Xingtai, Tonghai and Longling earthquakes (Figs. 6.1, 6.2b and 6.6). For some other earthquakes that occurred at the boundaries between neotectonic uplift and subsidence, or at the large fracture zones, the coseismic deformations retained the overall tendency of the neotectonic activity. For example, the crustal deformations of the Haicheng, Tangshan and Songpan earthquakes kept uplifting (or subsiding) at the points in a narrow zone on either side of the earthquake–producing faults. Meanwhile, for some earthquakes, such as Xingtai, Tonghai, Tangshan and Songpan, the pre–seismic anomalous deformation also showed hereditary characteristics. These data demonstrate that the directions of the pre– and post–seismic crustal deformations are the same as those of the recent neotectonic motions.

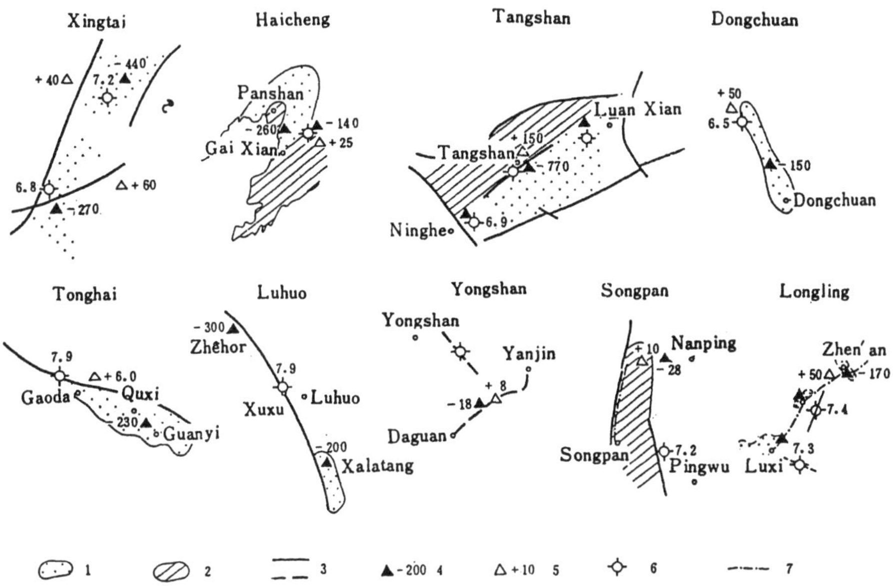

Fig. 6.9. Diagram of vertical deformation before and after some earthquakes; *1* basin or long descending area; *2* long uplifting area; 3 fault; *4* subsidence center with value in mm; *5* uplift area with value in mm; *6* epicenter; *7* levelling route

2) Coseismic subsidence larger than uplift.

As shown in Fig. 6.9, the subsidence associated with the earthquakes is normally larger than the uplift. This might result from the quakes occurring in basins or at basin margins, together with gravity readjustments during the quakes. As mentioned in Chap. 3, about 70% of historical strong earthquakes in the Yunnan and Sichuan regions are related to recent fault–depressed basins, and the relationship is even more pronounced in North China. Such large subsidence in epicentral areas may also be due to the predominant horizontal dislocation and dip–slip during the earthquakes, as well as to the uplift and subsidence areas at the ends of strike–slip faults. Examples are the Xingtai, Haicheng and Tangshan quakes. Focal mechanism solutions and coseismic ground fissuring demonstrate that all these earthquakes were strike–slip with a minor dip–slip component. However, the post–earthquake deformation measurements indicate that vertical subsidence was considerable, which might result in a gravity readjustment, after horizontal dislocations occurred. For example, two dip–slip earthquakes took place to the south after the Tangshan quake. Such epicentral subsidence may be due to the fact that a large quantity of groundwater was erupted over the surface during the earthquake, thus providing the space for local large– amplitude gravitational sinking, since local arching and dislocations would occur in the boundaries between various layers during the seismogenic process (see Chap. 11).

3) Phasic crustal deformation.

Table 6.1. Pre–seismic deformation rates and lead times

earthquake and point No.	date for twice survey	deformation amount	rate	lead time
Xingtai, No.449 point	1955–1920	-210 (mm)	-5(mm/year)	29 (year)
	1965–1955	+ 10	1	5
Tonghai, No.17 point	1969–1959	- 60	-6	6
Haicheng, Tianzhuang point	1970–1958	+ 40	3.3	11
	1973–1970	+ 6	2	2.5
Longling, No.11 point	1973–1958	+ 80	5.3	10
Tangshan epicenter	1967–1954	+ 10	0.7	16
	1969–1967	24	12	8
	1975–1969	- 50	-8	3
Songpon, No.10 point	1975–1960	+177	12	9
	1975–1935	- 10	5/month	1
	1976–1975	- 8	1/month	0.5
Niigati, Japan	1935–1899	+ 40	1	46
	1954–1935	+120	6	19
	1960–1954	+ 50	8	6
	1961–1960	- 10	-1	2.5

C. Tsuboi reported in 1933 on the existence of various phases of deformation before, during and after several Japanese earthquakes, and proposed the basic idea of the development of deformation in distinct phases. Since then, many people worldwide have tried to verify these ideas. Although it is difficult to find examples of quakes that have perfect data on several phases, the basic model has been proposed as follows. Mesherikov (1968) proposed that deformation before large earthquakes might be separated into alpha and beta phases. From the point of view of earthquake predicton, it is important to detect and evaluate the beta stage, because it is difficult to distinguish the low rate of the alpha phase from aseismic crustal deformation. On the basis of the above limited data, the pre–seismic deformation rates and lead–times are tabulated in Table 6.1. The lead time is calculated as the median year between the two measurement periods. From Table 6.1 it is clear that the deformation rates vary greatly during different periods before each earthquake, but that the general tendency is for relatively high deformation rates 6–10 years before $M > 7$ quakes, followed by even higher or reversed rates 6 months to a year before some earthquakes.

4) Regionalization of deformation characteristics.

a. As mentioned above, the coseismic vertical subsidence was greater than the uplift. The ratio of maximum subsidence to maximum uplift in North China was 5:1 to 7:1, while the ratio was 1.5:1 to 4:1 in Southwest China (Table 6.2). The significant vertical deformation areas were close to the mainshock epicenters in North China, while in Southwest China they were mostly at the ends of the epicentral zones (Fig. 6.9).

b. The ratio of horizontal to vertical deformation during earthquakes in North China was 2:1 to 3:1, while in Southwest China it was 9:1 to 12:1 (Table 6.2). Hence, the earthquakes in the southwest are more typically strike–slip than those in the north.

c. The seismic faults in North China are predominantly characterized by strike–slip normal faulting, while those in Southwest China are mostly strike–slip thrust faulting. Figure 6.10 shows horizontal tensile offsets of 0.3 m and 0.5 m, respectively, for the Xingtai and Tangshan quakes, based on the projection of horizontal displacement on either side of the fault on the direction perpendicular to fault strike (Figs. 6.1 and 6.7). The Tonghai quake in Southwest China did not show any tensional activity. From length measurements of lines crossing seismic faults, the lines in the Haicheng area were lengthened, while similar lines in the Yongshan and Longling areas contracted.

Furthermore, Fig. 6.10 shows that the attenuation rate of horizontal displacement for the Xingtai, Tangshan and Tonghai earthquakes (i.e., the average rate of decrease of displacement with distance from the fault, shown by the slope in the figure), were $(1-3)\times 10^{-5}$, $(4-5)\times 10^{-5}$ and 3×10^{-5}, respectively, while the attenuation rates of vertical displacement were within $(1-2.5)\times 10^{-5}$. However, these figures are much smaller than those for earthquakes at plate boundaries. For the 1906 San Francisco earthquake in the U.S.A., the 1923 Kanto quake in Japan and the 1964 Alaska quake in the U.S.A., the attenuation rates of horizontal displacement were 1.7×10^{-4}, 7.5×10^{-5} and 8.4×10^{-5} respectively. Again, for the 1954 Mikawa earthquake in Japan and the 1968 Inangahua earthquake in New Zealand, the attenuation rates of horizontal displacements were 1.3×10^{-4} and 2.4×10^{-4}, respectively. Hence, there are obvious discrepancies in attenuation rates between intraplate and interplate earthquakes, with the interplate attenuations faster than the intraplate ones. It is thus clear that in this regard there may be regional characteristics for different types of earthquakes.

Table 6.2. Comparison of coseismic deformation parameters for some great earthquakes

item \ earthquake	Xingtai	Haicheng	Tangshan	Tonghai	Luhuo	Yongshan	Longling	Songpan
maximum subsidence amount (cm)	-44	-14	-74	-23	-30	-1.8	-17	-2.8
maximum uplift amount (cm)	6	2.5	15	5		0.8	5	1
ratio of subsidence to uplift	7:1	5.5:1	5:1	3.8:1		1.5:1	3.2:1	2.8:1
maximum horizontal displacement (cm)	80	40	200	200	360			
ratio of horizontal to vertical displacements	2:1	3:1	2:1	9:1	12:1			
attenuation rate of vertical displacement (10^{-5})	2–2.5		0.8–2					
attenuation rate of horizontal displacement (10^{-5})	-0.8– 3		4 – 5	3				
orientation of fault dislocation	right-lateral	left-lateral	right-lateral	right-lateral	left-lateral	right-lateral		
faulting	strike-slip normal	strike-slip normal	strike-slip normal	strike-slip thrust	strike-slip thrust	strike-slip thrust		
fault extension(m) or side strain +,extension; -,contraction	0.3	1×10^{-5}	0.5			-4×10^{-5}		

The above discussion was based mostly on vertical deformation data, as the horizontal data seldom showed pre–seismic anomalies. On the one hand, this might be because the evident phenomena of preseismic strain accumulation and fault creep are confined to within 20–30 km of faults. Therefore, observation points must be concentrated densely on either side of a fault in order to detect pre–seismic anomalies. Levelling measurements basically satisfy this requirement, as the interval between points is generally 3–4 km, whereas the distance between horizontal deformation points is generally about 20 km. On the other hand, as the accuracy of horizontal deformation measurements is low and the measurement interval long at present, significant anomalous signals could hardly be detected in any case.

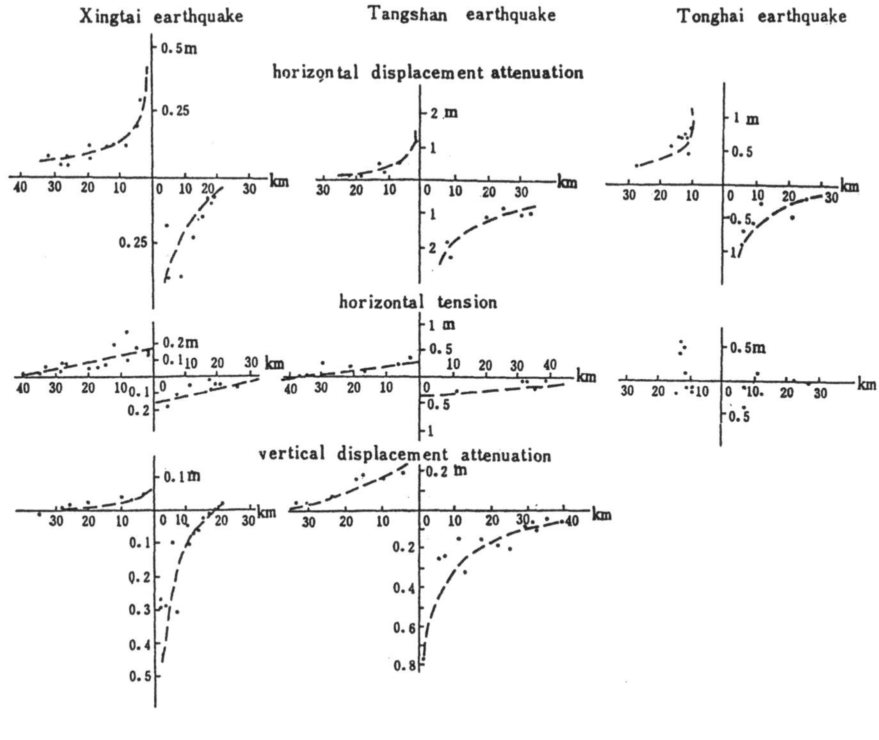

Fig.6.10. Attenuation of displacement caused by earthquakes as a function of distance from the fault

6.2 Across–fault Measurements, Tilt and Sea–level Observations

6.2.1 Across–fault Measurements

Anomalous variations of across–fault deformation measurements have been observed to different extents before some of the nine great earthquakes. The anomalies are observed at certain stations at 100–200 km from the epicenter. Because of the different geological structures at different stations, the anomalous precursory phenomena are not displayed at all stations.

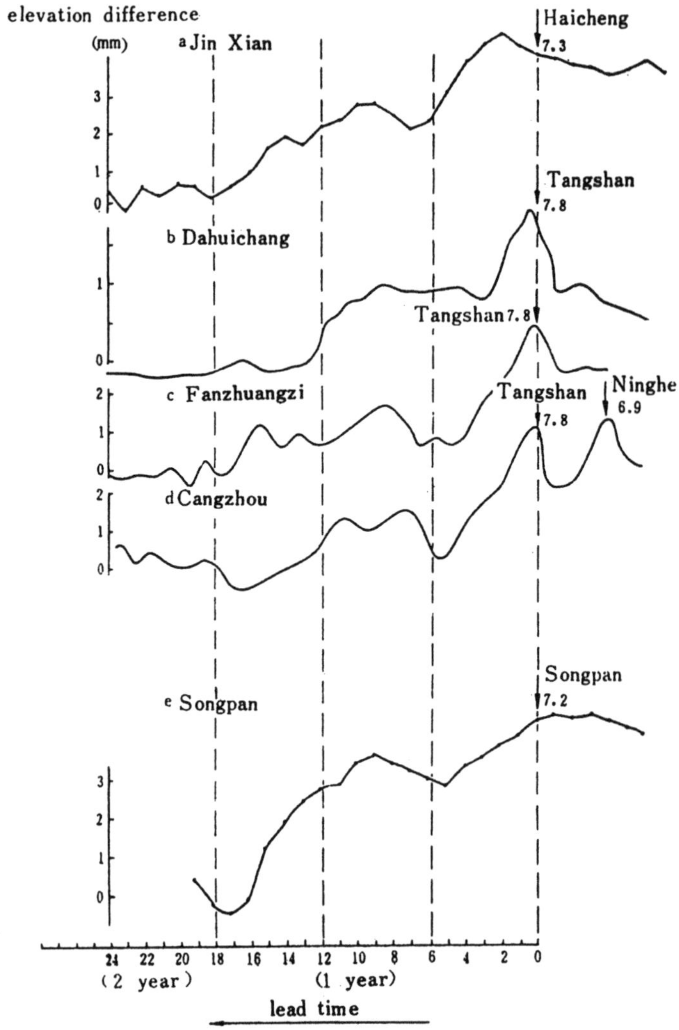

Fig. 6.11. Curves of elevation changes for short levelling routes across faults

The variation of short levelling data from Jinxian station before the Haicheng earthquake is shown in Fig. 6.11a. The station is situated 200 km southwest of the epicenter, and the measuring line, 560 m in length, is across the NE Jinzhou fracture zone. Eighteen months before the mainshock, the height difference had not changed very much, but then the deformation rate accelerated, showing relative subsidence of the west side of the fault. Six months before the mainshock, the deformation rate further accelerated, and the mainshock occurred 2 months after the deformation reversed. The maximum relative dislocation between the fault walls amouted to 4.3 mm.

Figures 6.11b, c and d show the height difference changes of short levelling lines at Dahuichang deformation station across the Babaoshan fault, Beijing (Δ= 165 km), and Cangzhou station across the Cangzhou fault (Δ=190 km). The anomalies at these stations appeared nearly 18 months before the earthquake, and the mainshock occurred as the anomalies reached their peak and slightly reversed. The anomaly magnitudes were 2.0, 4.2 and 3.7 mm, respectively. In addition, increasing creep of certain faults within an epicentral distance of 110 km was also observed 2–3 months before the mainshock. This will be discussed in Sect. 4 of this chapter.

Figure 6.11e shows the short levelling data recorded at Songpan station, about 50 km from the epicenter, before the Songpan earthquake. The levelling route crossed the NS Minjiang River fault (parallel to the earthquake–producing fault) and was about 268 m in length. The height difference curves show that the difference increased clearly 1 year before the mainshock, turned down 6 months before the mainshock, then kept rising until the mainshock.

The above earthquake cases feature the following characteristics:

a. For earthquakes of M $>$7, the anomalies were observed 18 months before the mainshocks, with their shape showing a process: ongoing development– turning down – speedy development – reversing (or non–reversing) – earthquake occurrence.

b. Though the epicentral distances were different, the shapes of the anomalies were alike. Stations showing anomalies were not necessarily located on the earthquake–generating faults. There was a migration of the anomalous points outward into the nearby epicentral area within 3 months before the Tangshan earthquake (see the following Sect. 6.4).

6.2.2 Tilt Observations

There were tiltmeters installed in the vicinity of the nine earthquakes, but very few convincing anomalies were registered, mostly due to poor observation conditions and to instrumental instability. Fairly good results have been achieved only before a few earthquakes.

1) Tilt anomalies at Kangding station before the Luhuo earthquake.

The Kangding station is located at the southwest end of the Luhuo earthquake fault, at an epicentral distance of about 220 km. The instruments were installed in a tunnel, and were capable of recording the regular changes of earth tides. Analysis done by the Seismological Brigade of Chengdu City, State Seismological Bureau, showed that distinct anomalies were observed at this station before the Luhuo earthquake. Beginning with 30 January 1973, the instrument was malfunctioning, and the operator adjusted it on January 31. Normal operation was restored on February 1, and starting with February 5, the recorded earth tide waveform was distorted. The residual obtained by subtracting normal waveform values from the distorted waveform was regarded as a precursor to the mainshock.

2) Tilt anomalies at Shengyang station before the Haicheng earthquake.

The Shengyang station is situated in the east part of Shengyang City, at an epicentral distance of 140 km; here the instrument is also installed in a tunnel. After June 1974, the two components of tilt recorded at the station showed a clear linear relation, and statistical results indicate that two components changed simultaneously. However, from the end of 1974 the recordings showed significant non–simultaneous anomalies, with the direction of tilt varying clearly, and pointing to the mainshock epicenter.

3) Tilt anomalies at Xiangshan (Fragrance Hill) station before the Tangshan earthquake.

According to the Working Team of the Tangshan earthquake (1977), a sizeable amplitude of eastward tilt was observed by the bubble tiltmeter at this station in Beijing, 4 days before the Tangshan mainshock. Fluctuations with 1 min and half–hour periods were also observed 5 hours before the mainshock.

6.2.3 Sea Level Observations

A fairly long period of observation data are available from the Yingkou, Huludao, Qinhuangdao, Tanggu and Longkou tide gauge stations along the Bohai coast. The Seismological Section, Information Institute, State Oceanographic Bureau, used the comparison method among neighbouring stations to reduce or eliminate the effects caused by meteorological, maritime and long–period astronomical factors, so that the sea–level variations before the earthquakes could be analyzed. The differences between the annual average values and mean values of multiple years (that is, yearly average differences) were first derived for various stations, then subtractions were made among neighbouring stations. The Qinhuangdao station was taken as datum because its tidal recordings had changed only very slightly over many years. Relative sea–level variations were so obtained at Yingkou and Tanggu stations before the Haicheng and Tangshan earthquakes. The main results are described below.

1) Haicheng earthquake.

Yingkou station is about 50 km from the epicenter, and the sea–level recorded there tended to fall slowly from 1955 to 1968, about 0.4 cm every year on average. But it began to rise clearly in 1970, at a rate of 0.9 cm / yr, amounting to an accumulation of 4.7 cm over 6 years. This was basically the same as the obvious subsidence of sea–level from 1970 given by levellng data. Also, the sea–level kept rising after the mainshock (while the ground subsided). The duration of significant sea level change before the mainshock was 6 years.

2) Tangshan earthquake.

Tanggu tide gauge station is 75 km from the epicenter, and the sea– level changed very little from 1950 to 1970, exhibiting a slow rise of 0.1 cm / yr. For 5 years after 1970, changes were very clear, totalling about 7 cm, at a rate of 1.4 cm / yr. This was also basically in agreement with the Tanggu surface subsidence of many years, which was somewhat accelerated after 1970.

3) Bohai earthquake.

Longkou station is 80 km from the epicenter; it began operation in 1961. The main tendency during 8 years before the mainshock was sea– level subsidence. It was notable that this reached a minimum 1 year before the mainshock, and that the sea–level tended to rise after the mainshock, in agreement with geodetic measurements conducted in Shandong peninsula after 1970.

In addition to the sea–level variations mentioned above, anomalous tidal waveforms were also recorded 1 or 2 days before the earthquakes. For instance, from 1500 hours on 3 February 1975, just 1 day before the Haicheng earthquake, an anomalous waveform appeared intermittently in the recording curve. The curve looked relatively smooth from 0000 to 0900 hours on February 4. Then, from 0900 to 1835 hours, an irregular zig–zag waveform appeared successively, with an amplitude of 1–5 cm and a period of 10–15 min. Finally, such anomalies died off about 1 h before the mainshock.

Again, before the Tangshan earthquake, short–period fluctuations overlapping and distorting the tidal curve were observed at Tanggu station from 1200 on July 24, just 4 days before the mainshock.

The above facts imply that sea–level observations may provide data on the long–term variations and shorter period rapid changes of crustal deformation along the coast. Tide gauges can also record the distorted tidal signals 1 to 4 days before the impending earthquakes. Such work not only tends to compensate for the inadequacies in long–period geodetic observations, but also provides clues for imminent earthquake prediction.

6.3 Inversion of Focal Parameters from Crustal Deformation Data

Since the dislocation theory of semi–infinite elastic media was suggested by Steketee (1958), the quantitative relation between pre– and post–seismic crustal deformation and the stress state at the hypocenter has been established by the efforts of many researchers. Both in China and abroad, numerous results have been published concerning the inversion of deformation data for mechanical properties at the hypocenter. These include the results of the Xingtai, Tonghai and Tangshan earthquakes.

By using deformation data obtained before and after the Xingtai earthquake, Chen Yuntai et al.(1975) inverted the focal parameters of this earthquake, proposing a compound fault model. This consisted of six rectangular faults, with three narrower faults of smaller dip exposed at the surface, and three wider faults of larger dip below. The three exposed faults all dipped 45° with a width of 15 km, while the lower three all dipped 82° with a width of 30 km. The inversion results showed that the M6.8 and 7.2 quakes in the sequence were due mainly to slip on the upper faults, while the aftershocks resulted from slip on the lower faults. The fault with the greatest slip had a strike–slip dislocation of −134 cm (right–lateral) and a dip–slip dislocation of +50 cm (normal faulting).

For the Tonghai earthquake, the unified method of combining triangulation net adjustments and fault parameters was used to derive the solutions. The results showed that the fault strike was N60° W, dipping 84° SW; the fault length was

52 km, width 22 km, and it was exposed at the surface. The average dislocations along strike and dip were −2.24 m and +0.14 m, respectively, showing right–lateral normal faulting. Most of the results discussed above were roughly coincident with those of macroscopic and P–wave first motion analyses, but the dip of the fault plane contradicted the macroscopic and seismological results; hence the faulting mode differed from other results. While discussing the relation between the faulting of this earthquake and its other aspects, the NE fault plane dip was assumed, resulting in right −lateral thrust faulting.

The inversion for the Tangshan earthquake indicates that the fault was generally a right–lateral normal fault with strike N49 ° E, dip 76 ° to the SE, length 84 km and width 34 km. The dislocations were −459 cm and +50 cm, respectively, in the strike–slip and dip–slip directions, and the seismic moment was 4.3×10^{27}dyne cm. Chen Yuntai et al. found that the inverted fault strike, dislocation magnitude and seismic moment differed very much from the seismological results, and suggested that the rupture direction changed during the earthquake, and that fault creep might have occurred before and after the mainshock.

In general, fault parameters derived from deformation data were roughly in agreement with the macroscopic and seismological results (Table 2.3). The following two points can be made.

1) Inversion studies were done only on the above three earthquakes, because of insufficient data for the others. However, the stress state at the source extrapolated from the plane strain results and qualitative analyses were in general conformity with seismological and other data. In general, it is shown that the coseismic deformation resulted from fault dislocations at hypocenters.

2) For the above three earthquakes, there was poor agreement between the vertical deformation data and the inversion results. In Table 2.3 the dip dislocations deduced for the Xingtai, Tonghai and Tangshan earthquakes are 50, 14 and 50 cm, respectively, while their maximum observed vertical deformations are 44, 23 and 74 cm. Furthermore, the locations and extents of the maximum deformation are also different from the inversion results. Such phenomena demonstrate that inhomogeneity of the medium and fault plane irregularity should also be considered in addition to the primary effects of horizontal dislocation and the secondary dip–slip movements.

6.4 Fault Creep

6.4.1 Modes of Fault Creep

Before the 1960's, two types of fault movement, namely creep and sudden dislocation, were already recognized. However, little knowledge has been acquired on details of creep and it is simply described as a kind of slow motion, whereas sudden dislocation of the fault (i.e., an earthquake) is considered as capable of exciting elastic waves and producing finite stress drop. Nevertheless, the relation between fault creep and sudden dislocation is not clear. Since fault creep was observed at a winery near Hollister, California, in 1960, extensive observational, experimental and theoretical research has developed, and a number of significant results have been achieved. With improvements in experimental rock mechanics and developments of field observation work, a greater understanding of fault creep has been obtained. The following four creep modes can be recognized.

1) Stable creep, which proceeds at a relatively uniform and slow rate, with roughly continuous displacement, similar to the stable sliding in experiments. Field observations show sliding rates of several centimeters per year.

2) Intermittent creep, which proceeds at a non–uniform rate, with a certain amount of slip accomplished in a short period, followed by a long cessation of activity, and so on. The San Andreas fault observations show no more than several millimeters of creep on each occassion.

3) Precursory creep, similar to the stable sliding before stick– slip during rock mechanics experiments. This occurs at a fast rate ,from several years to several minutes before earthquakes. The creep increased markedly, as recorded by creepmeters at Hollister, about 800 days before the M5.6 and 5.5 Hollister earthquakes of 9 April 1961.

4) Post–seismic creep, which occurs at the earthquake producing fault, with a creep rate logarithmically decreasing with time, and lasting from several years to several decades after the earthquakes.

There are two methods for researching faulting, the dynamic method which assumes that if the force acting is known, then the movement of the medium in the vicinity of the fault can be derived; and the kinematic method which assumes that if the dislocation plane is known, the faulting can be estimated. The results obtained by these two methods are similar, namely, if a source region is assumed to be acted on by a force, it can be said that there exist displacement discontinuities – dislocation planes. Since it is assumed that the acting force is known, there used to be much controversy over a single–couple or a double–couple source. That was why a dislocation model was used in place of the force source model in the late 1950′s. Dislocation theory allows detailed description of the fault plane from surface displacement observations. Although this method approximates the medium as fully elastic, it is much better than the earlier dynamic method. Dislocation theory can be used to interpret continuous creepmeter observations of surface sliding, and also to interpret periodically measured dense geodetic networks, in terms of subterranean fault creep. Creepmeter networks are not yet established in China, but the presently available geodetic data do indicate some pre– and post–seismic fault creep phenomena.

6.4.2 Pre– and Post–seismic Fault Creep

1) Pre–seismic fault creep.

Luo Zhouli (1980) discussed the stress field of a source under the action of a horizontal single force, and compared the theoretical vertical displacement field under plane stress. He thus derived the modes of stress action for the Xingtai, Tonghai, Luhuo and Songpan earthquakes (Fig. 6.12). Though his results emphasize the modes of stress acting on a pre–seismic source, we may also understand that there exist dislocation planes (creeping faults) within a source area before an earthquake. According to dislocation theory, similar surface displacement can be given by different sizes and depths of dislocations. Since few data were obtained before the earthquakes, it was difficult to derive quantitative fault creep parameters. However, qualitatively speaking, creep phenomena may have occurred at the earthquake–producing faults before the Xingtai, Tongtai, Songpan and Luhuo earthquakes.

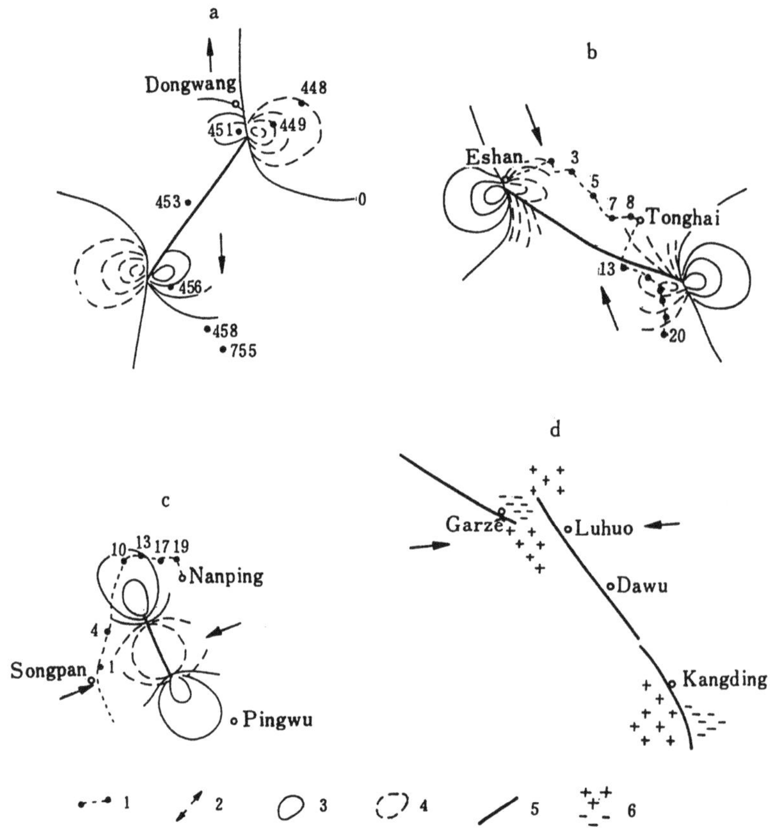

Fig. 6.12. Theoretical pre–seismic vertical deformation field; *a* Xingtai earthquake; *b* Tonghai; *c* Songpan; *d* Luhuo; *1* levelling points and routes; *2* orientation of tectonic stresses; *3* area of uplift; *4* area of subsidence; *5* fault; *6* + and – indicate uplift and subsidence, respectively

In addition to the earthquakes shown in Fig. 6.12, levelling measurements across the Tangshan fault before and after the Tangshan earthquake show elevation changes that signify fault creep, as well as strain accumulation and elastic rebound. These results can be considered as fault creep superimposed on secular strain accumulation. The vertical displacements at various points within the epicenter exhibit symmetric distribution patterns. Zhang Yingzhen (1981) used the uniform strain accumulation and elastic dislocation theory to invert the fault creep parameters before the Tangshan earthquake, obtaining theoretical values of displacement at various points within the epicentral region (Fig. 6.13). The results showed that the creeping fault was at a depth of 2–8 km, with a strike of N 47 ° E, dipping 87 ° SE. The length of the creeping section increased slowly to 8 km from 1969 to 1975, with strike–slip and dip–slip dislocations amounting to −104 cm and +8 cm, respectively. The mean strike–slip and dip–slip velocities were 18.6 cm / yr and 1.4 cm / yr, and the accumulation rate was 0.9×10^{-7} / yr. The parameters of the creeping fault were very close to those for the seismic fault derived for post–seismic deformation and seismic waves, thus indicating that the earthquake may have been the outcome of further development of the creeping fault.

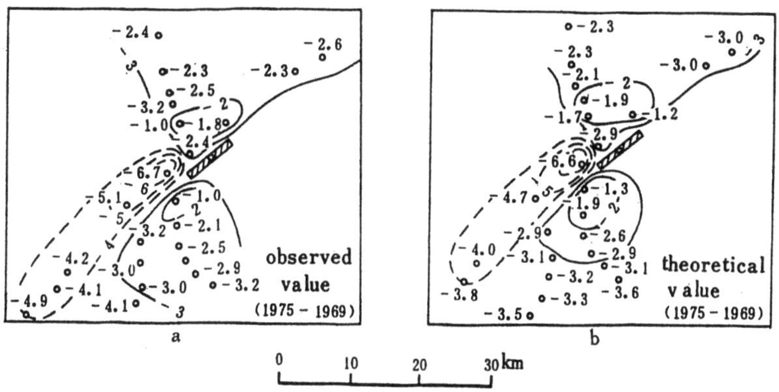

Fig. 6.13. Contour map of vertical displacements near the creeping fault at Tangshan. *Solid line* indicates uplift, *dashed line* subsidence, *shadowed rectangle* the creeping fault

At the time when the Tangshan fault was creeping, the Ji Canal fault near Baodi also crept. Inversion of surface data showed that the strike of the creeping fault was in agreement with the NW band of M > 3 earthquakes in the Ninghe–Baodi–Shanhe region that followed the Tangshan quake.

Several months before the Tangshan earthquake, rapid fault creep was observed at the Zhangdaokou station and at the Ninghe mobile observation point across the NE fault southwest of the epicenter. The Ninghe point changed by 7 mm within 9 months, with a creep rate much faster than the above average rate, several years before the earthquake (Fig. 6.14).

2) Post–seismic creep.

Figure 6.15 gives the time–dependent attenuation of height changes near the epicenters after the Xingtai, Tonghai and Tangshan earthquakes. The observation points were close to the earthquake–generating faults, hence mirroring the post–seismic faulting. After the Tangshan earthquake, in addition to the continuous creep observed at the Ninghe and Zhangdaokou points, instruments were installed in Zhaoguchuan mine in the meizoseismal region for monitoring the behaviour of the 5th fault striking about N45 ° E. Relative displacements of the fault walls were a vertical shortening of 1.9 mm and an oblique extension of 5.5 mm between February and October 1977, implying that horizontal activity still dominated after the earthquake.

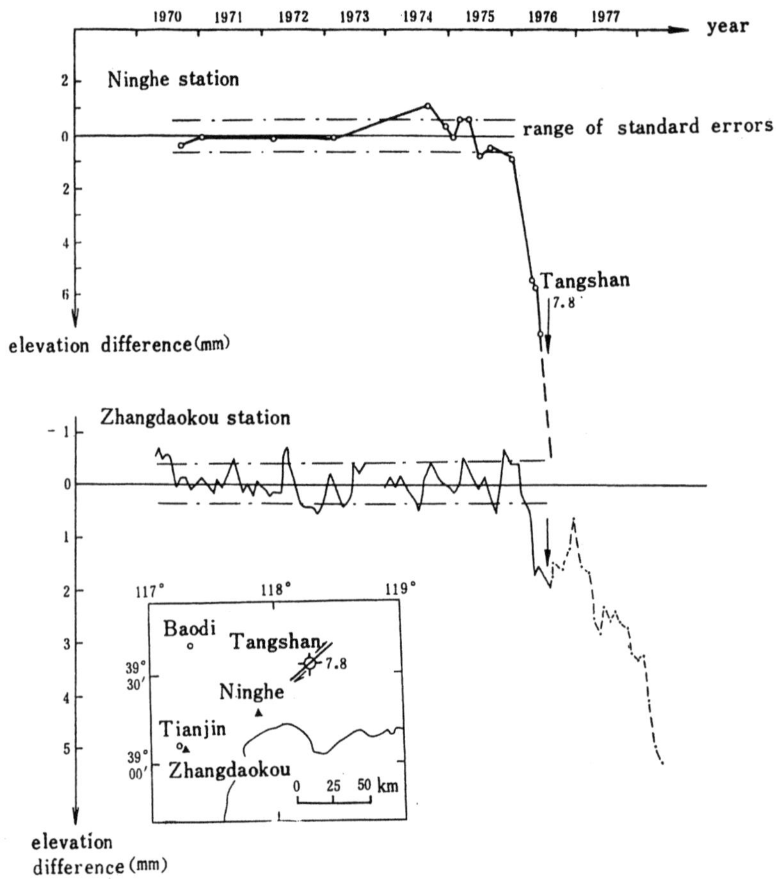

Fig. 6.14. Fault creep at the Ninghe point and the Zhangdaokou station before the Tangshan earthquake

Figure 6.15 shows that post–seismic fault creep lasted for 4–5 years, and that the creep rate attenuated gradually.

The preseismic fault creep rate seemed to accelerate gradually, which was verified by laboratory results. Therefore, one of the key issues for earthquake prediction research is to monitor variations in fault creep rate.

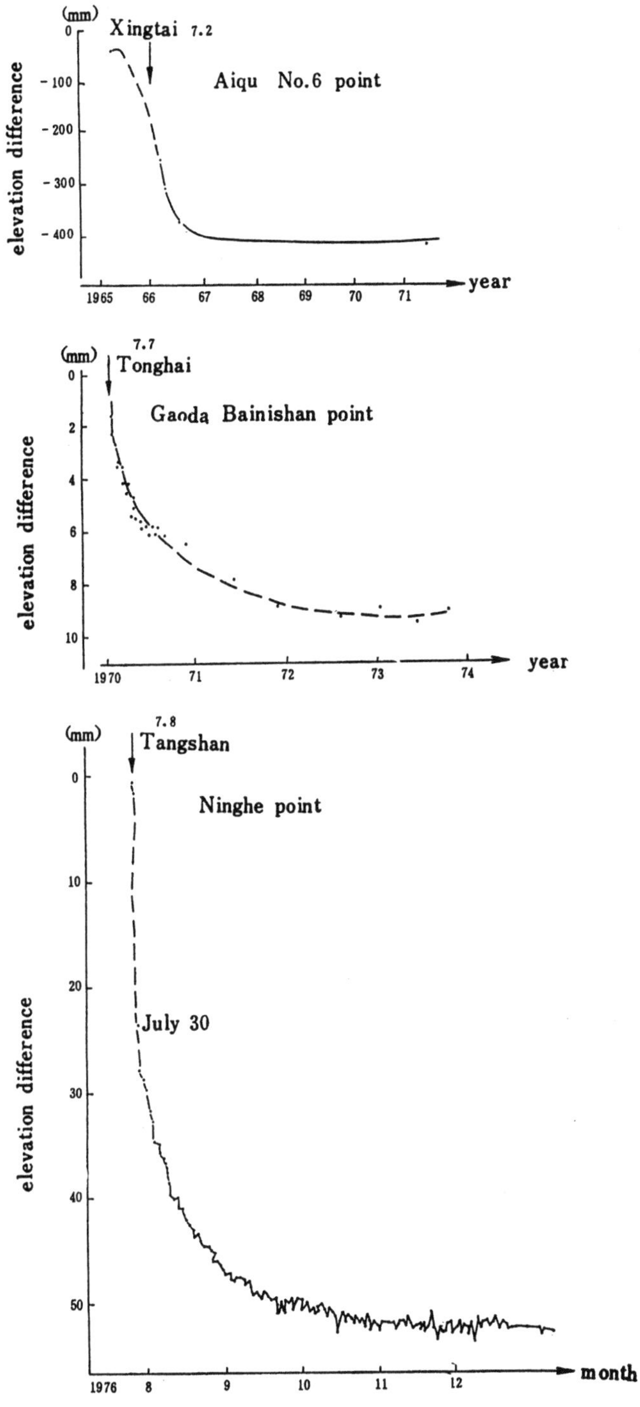

Fig. 6.15. Fault creep after the Xingtai, Tonghai and Tangshan earthquakes

6.5 Anomalous Earth Stress Variations

After the Xingtai earthquake, stations for measuring stress variations were set up in the major seismic regions of China. Borehole piezomagnetic stress gauges and vibrating–wire strain gauges were mainly used, the former at most of the stations. This method consists in putting the piezomagnetic stressmeter into a borehole normal to the surface, and measuring radial displacement of the borehole to derive stress variations. This method is usually called "measurement of relative stress by the electro–induction method". At present, the depths of boreholes range from tens of meters to a hundred meter or more.

Stress anomalies were measured at some stations within 50–300 km epicentral distance of six of the nine earthquakes (excepting the Xingtai, Bohai and Yongshan quakes).

6.5.1 Luhuo Earthquake

Before this earthquake, stress anomalies were detected at the Luding station situated at the south end of the Xianshuihe fracture zone (Fig. 6.16). From mid–May 1972, the values measured by the N70 ° W and N50 ° E elements gradually decreased (increase of compressive stress), while the values measured by the N10 ° W element increased (decrease of compressive stress). The M7.9 Luhuo earthquake occurred just when the values returned to their normal baseline value; the duration of the anomalies was about 8 months. The amplitudes measured by the three elements suggest that the principal compressive stress axis was roughly EW, and the principal tensional stress axis roughly NS, for the additional stresses measured at this station. This is in basic agreement with the mechanical state of the source region derived from seismological data and macroscopic investigations of the earthquake fault.

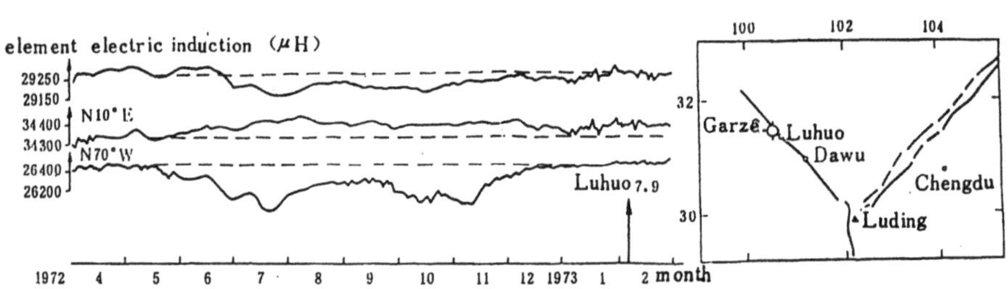

Fig. 6.16. Stress changes measured by the electro–induction method at Luding station before the Luhuo earthquake

6.5.2 Haicheng Earthquake

Analyses by the Working Group of the Haicheng earthquake showed that fairly large amplitude stress anomalies were observed at the Jinzhou and Shengyang stations before the earthquake. Based on the observed values, the maximum and minimum principal compressive stresses were calculated. At the Shengyang and Jinzhou stations, they were 5.4 kg / $cm^2$and 3.9 kg / cm^2, and the corresponding azimuths were N70 ° E and N50 ° W.

6.5.3 Longling Earthquake

Analyses by the Earth Stress Research Group of the Geological Brigade, Seismological Bureau of Yunnan Province, indicate that electro–induction anomalies were measured at Xiaguan station, about 150 km from the epicenter, before the Longling earthquake. From the values measured by three elements, the maximum principal compressive stress was 13.26 kg / cm^2in azimuth N36 ° E.

6.5.4 Songpan Earthquake

Analyses by the Seismological Bureau of Sichuan Province (1979) indicate a large amplitude stress decrease after September 1975 at the Songpan stress measuring station, about 50 km from the epicenter, with the most significant variations in the EW element. The Songpan earthquake happened when the stress values reached minimum (compressive stress maximum). From observations of three elements, it can be inferred that the orientation of compressive stress at this station was nearly EW before the mainshock.

6.5.5 Tangshan Earthquake

Analyses by Huang Xiangning and others suggest that the potential variations of some anomalies recorded at certain stations in North China lasted for about a year before the Tangshan earthquake. The variations were similar in character to some of the earthquakes discussed above. In addition, at the Tangshan stress measuring station near the epicenter, notable stress fluctuations were observed 6 months before the earthqauke, and sudden changes were measured shortly before the mainshock.

Potential (long–term) variations of electro–induction stress values have occurred before the earthquakes discussed above. However, the characteristic shapes of anomalies and their relations to the earthquake occurrences were different. Some earthquakes took place at the peak of the stress–variation (e.g., Songpan); some showed stress increase – recovery – earthquake occurrence (e.g., Luhuo); while others showed stress increase – recovery – increase – earthquake occurrence (e.g., Longling). The diverse character of these anomalies can be partly explained by the effects of annual changes, as well as by synchronous changes in the stressed elements and the suspension elements of the observation system. In summary, the relationship between electro–induction stress values and earth–

quake occurrence is rather intricate.

Several years' experimental observations show that present borehole vibrating–wire strainmeters have a sensitivity of 10^{-9}, and these can be used to record earth tide signals accurately.

At the Xingtai, Zhengzhou and Luoyang stations, anomalous variations of "strain pulse" type were recorded before the Tangshan and Songpan earthquakes, as well as before local earthquakes of M 3–5. The anomaly amplitudes amounted to $10^{-7}$and they were observed several days to 20 days before the earthquakes.

In conclusion, the above analyses of crustal deformation and stress imply that certain common characteristics do exist for these earthquakes, namely: (1) 6–10 years beforehand, a high deformation rate appears within the epicentral area, and then the rate decreases; 1–2 years before the quake, the deformation rate is heightened or reversed; (2) Before and after some earthquakes, the mainshock fault appears to creep slowly, and remarkable faulting is observed outside the epicentral area 1–2 years before the quake; (3) Tidal strains and tilts are distorted prior to the impending earthquake; (4) The ground deformation associated with the earthquake is compatible with the neotectonic activity. The stress state at the source inferred from the ground deformation is in general agreement with the seismological data. For vertical deformation during the earthquake, subsidence is larger than uplift; (5) The crustal deformations associated with the earthquakes were regionally coherent, with the earthquakes in North China having strike–slip normal faulting with a tensional component, and those in Southwest China having strike–slip thrust faulting with a compressive component. In addition, there were obvious differences in the ratio between horizontal and vertical deformation, and between subsidence and uplift, in the two regions.

Furthermore, 1–2 years before the earthquakes, the electro–induction stress values tended to decrease (indicating a compressive stress increase), and sudden changes in stress values were observed at some stations before the impending earthquakes.

CHAPTER 7

Abnormal Changes of Ground Water (Gas)

In line with the abundant 3000–year historical record in China and the actual observations obtained in earthquake–stricken areas in the last 10 years or so, many anomalous phenomena in the natural world before and after large earthquakes appear to be correlated with dynamic anomalies of ground water (gas).

Variations of ground water (gas) before large earthquakes were noticed in ancient China. Therefore, the relation between subterranean water anomalies and earthquake occurrences remains a traditional research subject. Early in 1663, the "Longde County Annals" noted that "there are about six kinds of earthquake precursor". The report described ground–water anomalies before earthquakes, such as "well water suddenly becomes muddy, like prepared Chinese ink, and soils and slags float; pools and swamps bubble like boiling tea without any reason; these phenomena must be followed by earthquakes".

There were about 430 ground–water (gas) anomalies observed in the epicentral area and its vicinity before the 1966 Xingtai earthquake. The relations between these anomalies and the earthquakes were studied so as to predict strong aftershocks, and initial successes were achieved. In 1966, following the concern shown by Premier Zhou Enlai himself, the ground–water observation network in the Beijing–Tianjin–Tangshan region was organized under the direction of Professor Li Siguang. Observations have been carried out in earthquake risk areas and other key monitoring areas throughout the country. They have been conducted by a combination of professionals and amateurs, using an integration of fixed observation points with temporary monitoring wells, and a combination of instrumental recordings with macroscopic investigations. For ground–water observations, Red Banner–1, HCJ–1 and SW–40 type water autographic recorders, with 0.5cm accuracy, have been installed in professional deep wells, and simple tools like measuring ropes have been used in amateur observation wells. For the chemical composition of ground water (including water radon content) the 128 type automatic radon meter, FD–125 type radon thorium analyzer, FD–128 scintillation counter, new 105K type static electrometer and FD–105 type static electrometers were generally employed in professional wells, while 105 type static electrometers were used in amatear wells with some 103 type gas emitter and other equipment used at a few points. For observations of water temperature, the highly accurate Continuous Temperature Monitor (with 0.01℃ accuracy) and the semiconductor water thermometer have been deployed at only a few points, and the maximum / minimum thermometer or medical thermometer (with 0.5 ° C accuracy) have generally been used at amateur wells.

Observations of about 60 earthquake cases have been accumulated in the past 15 years. For the nine earthquakes with M over 7, there are about 60 systematic observations on water–level changes; about 40 radon concentration observations; and about 10 water temperature measurements. In addition, macroscopic observations on ground–water level changes were conducted by amateurs within the epicentral areas and their vicinities before and after six of the earthquakes, thus providing the characteristic patterns of areal distribution of shallow ground–water level variations before and after great earthquakes.

7.1 Process of Anomalous Changes of Ground Water (Gas) Before and After Large Earthquakes

7.1.1 The Dynamic Trend of Ground Water (Gas) and the Factors Affecting It

The dynamic trend of ground water (gas) denotes the variation in time and space of ground water level, water quantity, water temperature, chemical composition and physical nature of ground–water, because of multiple factors. These factors include natural ones (generally meteorology, biology, astronomy, geology and so on) and man–made ones (generally exploration and irrigation of ground water).

The dynamic trend of groundwater has been studied at shallow depths. Hence, traditionally, the meteorological, hydrological, biological and man–made activities have been thought to be the main factors influencing the ground–water trend. Then, as the observation depth was increased, and the duration and accuracy of recordings extended, it was found that the dynamic trends of ground water were still unstable in the case of deep wells with sufficiently good sealing to exclude almost completely the meteorological (except air pressure), hydrological and man–made effects. This strongly suggests an astronomical or geological origin. The tidal variations of ground water associated with lunar and solar gravitation were observed at the end of the nineteenth century, and the influences of sunspots and magnetic storms on ground–water changes were observed in the mid–twentieth century. Also, the anomalous zones of ground water trend caused by crustal movements were noticed during petroleum exploration. Observations along the San Andreas fault in the United States have indicated that water–level changes are related to fault creep. Therefore, the dynamic trend of ground water can relate to the geodynamic process, which involves plate motion, upwelling of mantle material, volumetric strain of the crust, and even details of fault movement and the seismogenic process of an earthquake.

This chapter will begin with the dynamic trends of ground–water level, hydrochemistry (radon) and water temperature, then discuss the process of tectonic movement which illustrates the preparation and occurrence of an earthquake, and the post–earthquake readjustments.

7.1.2 Variation of Ground–water Level

1) Tangshan earthquake.

The Tangshan earthquake occurred within the Beijing–Tianjin–Tangshan ground–water observation network. Therefore, an unprecedented amount of data have been collected. However, as the quake took place in a heavily populated and industrialized city, the seismic information was not recognized before the mainshock owing to the intense disturbances.

It has been found through post–earthquake analysis (Fig. 7.1a) that:

(1) The ground–water level began to decrease slowly, measured on many years' basically stable background, 3–4 years (1972–1973) before the mainshock. Anomalies occurred throughout the Tangshan depression, with amplitudes about several meters.

(2) 3–4 months before the mainshock, the decrease of water–level speeded up simultaneously in several wells, and generally became the lowest in history in the Tangshan, Ninghe and Fengnan areas. A large number of pumping wells stopped producing (for example the Commodity Corporation in New Tangshan urban area, Xinjuntu Chemical Fertilizer Plant, 422 Cement Plant and Tangshan Steel Plant) and numerous factories and enterprises partly ceased operation.

(3) Several days and even several hours before the mainshock, the water levels in the epicentral and adjacent areas started to rise drastically, some wells experienced sudden changes in water–level, and some wells even became artesian from 4min to 3h before the mainshock. Some autographic recordings and densely sampled readings have verified this phenomenon. Three hours before the M7.1 aftershock on the day of the mainshock, the water level in Biaokou well ($\Delta = 125$ km) increased rapidly by about 8cm relative to its long–term decreasing background. It again rose by about 25cm, 4h before the M6.9 aftershock on December 15, hence showing a distinct change in direction just before the earthquake ("imminent rebound").

(4) The water–level continued to rise substantially after the mainshock, and even upwelled and gushed out automatically.

Some anomalous ground–water changes also appeared before and after the mainshock in areas away from Tangshan. For example, the water–level in the well at Luannan meteorological station kept rising drastically, completely swamping the normal annual variation, from mid–June 1976 on; discontinuous pulse–like variations occurred just prior to the mainshock; and the baseline of normal annual variations recovered after the mainshock. In some wells near the Babaoshan fault, Beijing, the water–levels tended to suddenly turn from rising to staying stable or decreasing, and then dropped drastically after the mainshock. At the Panshan well, located at the north end of the Tancheng–Yingkou fracture zone, the water–level dropped steadily for several months before the mainshock, turned upwards during the earthquake, and rose drastically after it.

Though there are few observations of epicentral ground–water variations for other earthquakes, a brief description of these follows (Fig. 7.1).

2) Xingtai earthquake.

Results obtained by the macroscopic inspection groups from the Chinese Academy of Sciences , the Ministry of Geology and the Ministry of Petroleum, show that the water–level in the Xingtai area kept falling for over a year before the mainshock. From 1956 to 1963, the local water depth of the first aquifer was

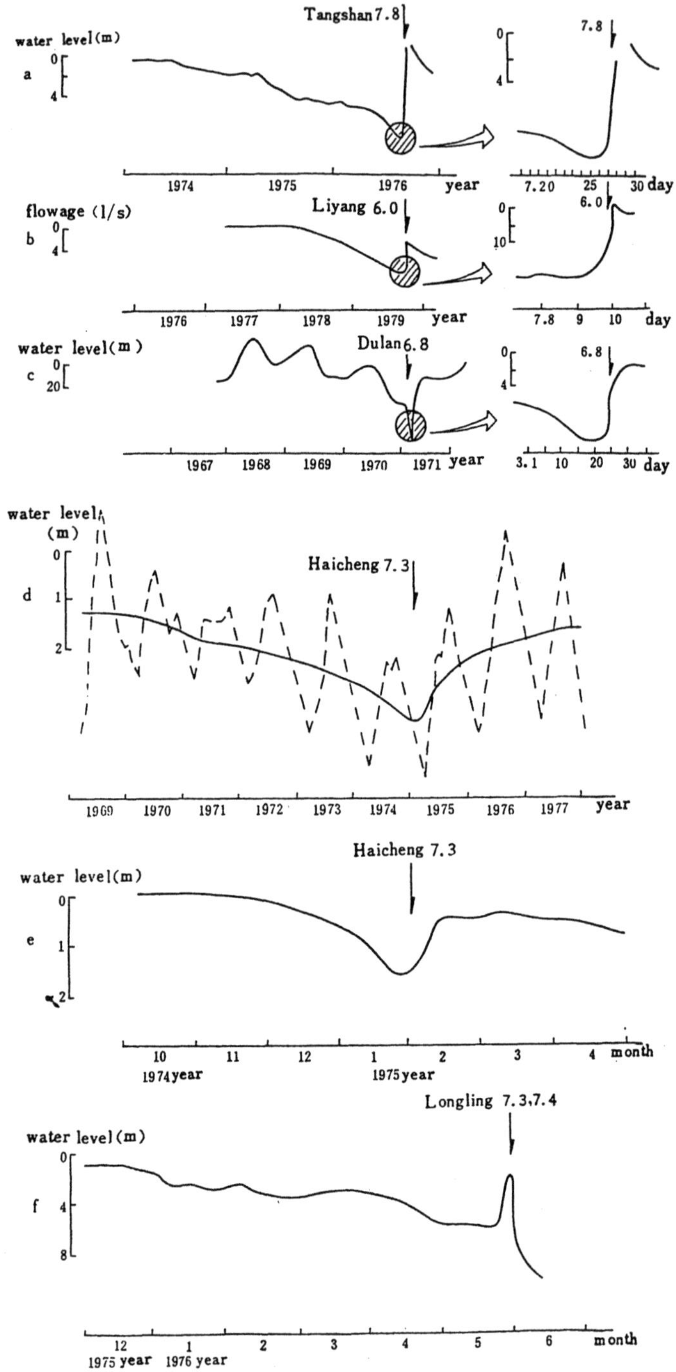

Fig. 7.1. Variations of ground–water level in the epicentral areas and their vicinities before and after the earthquakes; *a* Tangshan (Yue No. 42 well); *b* Liyang (Boat Hill Mine in Zhenjiang City); *c* Dulan (Tuosuo Lake); *d* Haicheng (Liaoyang well); *e* Haicheng (Borehole by No. 102 Geological Team of Yingkou); *f* Longling (a well in No. 1 Middle School of Baoshan)

at 1–2 m. In the spring of 1966 (before the quake), the water–level of this aquifer fell to about 3–5 m. After the mainshock the water–level rose in general, and many wells started to upwell or gush out automatically (Fig.7.2a).The Ningjin County government noticed the anomalous fall of ground water before the M7.2 earthquake of 22 March 1966 and were prepared for this large quake relying on the previous observations of falling water–level before the M6.8 earthquake of March 8. Before two strong aftershocks (M = 5.7 and 6.3) on 3 December 1967 and 16 January 1968, the whole process of ground–water level decrease was recorded by autographic instruments in numerous wells..

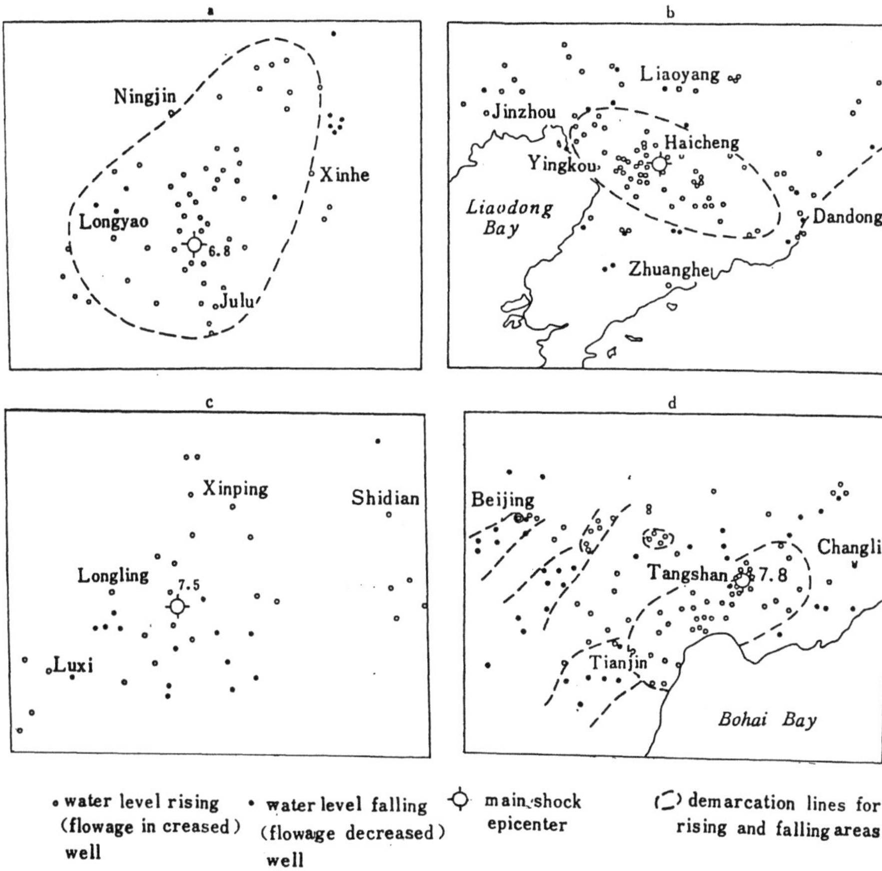

Fig. 7.2. Variations of ground–water levels before and after the earthquakes; *a* Xingtai; *b* Haicheng; *c* Longling; *d* Tangshan

3) Bohai earthquake.

As this earthquake was located under the sea, there were no observations in the epicentral area. However, many years of data on urban ground water in

Tangshan and the Liaodong Peninsula, and upwelling changes at Tangshan mine, indicate that the ground–water level in 1969 was at a low point of many years. This was the time at which the earthquake occurred.

4) Haicheng earthquake.

There were no observations collected for earthquake purposes around the epicenter. However, the long–period trend of the well used by the Anshan Iron and Steel Company close to Liaoyang City, and the trend of the well at the Running Water Company close to Yingkou City suggest that the Haicheng earthquake occurred near the lowest long–period water–level variation since 1969. The water–level began to rise quickly after the mainshock, and returned to the 1969 level within a year or so (Fig. 7.1d). As the Liaoyang well was only observed once a day, the details of the pre– and post–earthquake ground–water variations are hard to distinguish. A 300–m borehole in bedrock at Zhoujia Commune in Yingkou County ($\Delta =$ 10 km) was observed once every 10 days by No. 102 Geological Team of the Ministry of Metallurgy. The water–level began to fall swiftly 2 months or more before the mainshock, and rose several days prior to the mainshock (basically the same as the observations in the Tangshan epicentral area). The mainshock occurred shortly after the water–level reached its lowest point and turned to rise (Fig.7.1e). The water–level started to increase rapidly after the mainshock (Fig. 7.2b).

5) Longling earthquake.

Within 70–100 km of the epicenter, the water–levels of wells fell and the water flow of springs reduced 2 months before the mainshock. But the water–levels suddenly increased, and some ditches and ponds gushed, 2 or 3 days prior to the mainshock (Fig.7.2c). For the well at No.1 Middle School in Baoshan, about 80 km from the epicenter, the water–level data are complete and continuous since 1972. From March 1976, the water–level fell remarkably, then rose abruptly by about 2.5 cm 2 days before the mainshock (Fig.7.1f). The water flow of No.3 spring at No.2 Middle School in Tengchong, about 50 km from the epicenter, decreased gradually from March 1976, reached its lowest level on May 17, then began to increase drastically from May 18 – 20, and kept rising until May 29 when the mainshock occurred.

6) Songpan earthquake.

Since the summer of 1975, the water–level in the epicentral area tended to fall. After the spring of 1977, the anomalies became much more obvious. The water–level of most wells fell, springs dried up and it was difficult to get drinking water, and bamboo died of dryness in great quantities. The water–level of a few wells increased, and a few springs became full. As the epicenter was located in a remote mountainous area, there were no established fixed observation points. Outside the epicentral region, the ground–water level tended to fall beforehand, but to rise immediately before the mainshock. For instance, a drastic fall of well water–level in Dayi County, Sichuan Province, was noticed 5 months prior to the mainshock.

In order to discuss further the objective and general value of features of water–level anomalies before and after the earthquakes with M over 7, we will also use the following data on fairly systematic continuous recordings of earthquakes with M 6–7.

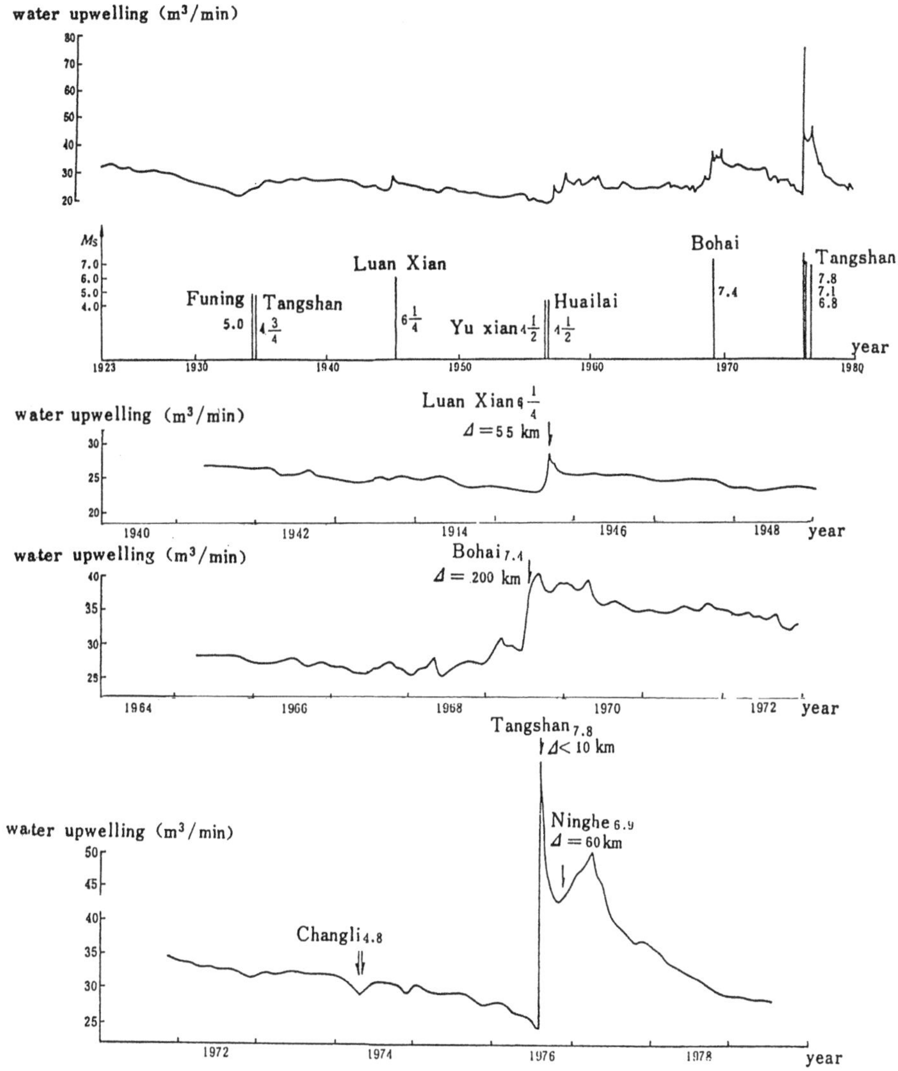

Fig.7.3. Relation between the earthquake and the trend of waterupwelling at Tangshan mine

Before and after the M6.8 Dulan earthquake of 24 March 1971 in Qinghai Province, the water–level at Tuosuo Lake hydrological station (Δ= 20 km) tended to fall relative to the stable background of the year (1970) before the mainshock. The level was apparently lower than at the same time in previous years. From 1 March to 15 March 1971, the water–level decrease accelerated, falling by 37 cm. From March 15 to 23, the water–level stayed at this low value. At

1000 hours on March 24 (12 h before the mainshock), the water–level suddenly rose by 21 cm. During the mainshock, the level rose fiercely, causing the lake water to spill in large quantities over the frozen surface. The water–level recovered slowly to its long–term average level after the earthquake (Fig. 7.1c).

Before the M6.0 earthquake in Liyang, Jiangsu Province, on 9 July 1979, the artesian well at Boat Hill Mine in Zhenjiang City, about 65 km from the epicenter, began to behave abnormally. The well had been observed since May 1977, and its upwelling flow rate had been stable at 100–130 ml / s for a long period of time. Starting in April 1978, the upwelling decreased gradually, and reached its lowest value (77 ml / s) in May 1979. The upwelling increased rapidly 10 h or so before the mainshock, with a flow rate of 82 ml / s increasing to 98 ml / s, and to over 100 ml / s after the mainshock (Fig. 7.1b).

Similar anomalies registered repeatedly in a given area, or at a single point for multiple earthquakes, undoubtedly have higher reliability than isolated anomalies. Figure 7.3 shows the relation between the upwelling variations at Tangshan Mine and three damaging earthquakes. The upwelling rate is normally very stable. During the 57 years of observations, there have been five periods of earthquake "highs" in Tangshan and its adjacent areas. The water flow rate obviously fell before the 1945 M6.2 Luanxian earthquake, the 1969 M7.4 Bohai earthquake, and the 1976 M7.8 Tangshan earthquake and the moderate earthquakes close to it. All the above–mentioned earthquakes took place around the low point of the upwelling rate. The flow rates in some mines suddenly augmented prior to the Tangshan, Luanxian and other earthquakes, and increased drastically following them.

The numerous cases described show clearly that there is a general phenomenon of water–level fall in the epicentral area and its vicinity before a large earthquake, and that the earthquake very often occurs near the time of lowest water–level stand in the local area. In addition, as the mainshock approaches, the ground–water level rises speedily, then more drastically after the mainshock. The ground–water behaviour can be summarized by the following process: falling (before earthquake) ——rising (before imminent mainshock) ——rising drastically (after).

7.1.3 Process of Variations in the Chemical Composition of Ground Water

1) Tangshan earthquake.

Before and after the Tangshan earthquake, a great number of observations of ground–water radon content were made in water wells with different epicentral distances, depths, geological structures and hydrogeological conditions. Some anomalous variations of chemical and gas compositions of ground–water have also been recorded. This earthquake has the most complete data since observations of hydrochemistry as an earthquake precursor started in China.

Around 1972, radon concentrations tended to rise anomalously in many wells in the northern part of North China. The Anguzhuang well in Luanxian (about 50 km from the M7.8 mainshock, and about 15 km from the M7.1

mainshock) is an artesian hot well of 490–m depth in fissured bedrock. The radon concentration of this well in 1972 was about $1036\times 10^3 Bq/m^3$; it increased gradually in the ensuing years, and amounted to $1221\times 10^3 Bq/m^3$ or more in early 1976. This phenomenon was really striking. It was not a unique example, and approximately synchronous variations were also observed in many professional wells located in Tianjin, Beijing and the northern part of Hebei Province, such as Tanggu No.4 well at Tianjin Textile Plant, Wenan, Guanzhuang and so on (Fig. 7.4). At Guanzhuang well, Beijing, about 135 km from the epicenter, systematic data are available since 1969. After 1972, the radon content began to increase gradually, and tended to further increase around the autumn of 1975. The anomaly amounted to 10% or more prior to the mainshock. Meanwhile, at Guanzhuang well, identical water–gas content (total content) and water conductivity variations were also observed. The total gas content increased by about 3 ml / l (14% of the mean value) from May 1975 to July 1976, and the water conductivity increased drastically since 1976 (Fig. 7.5).

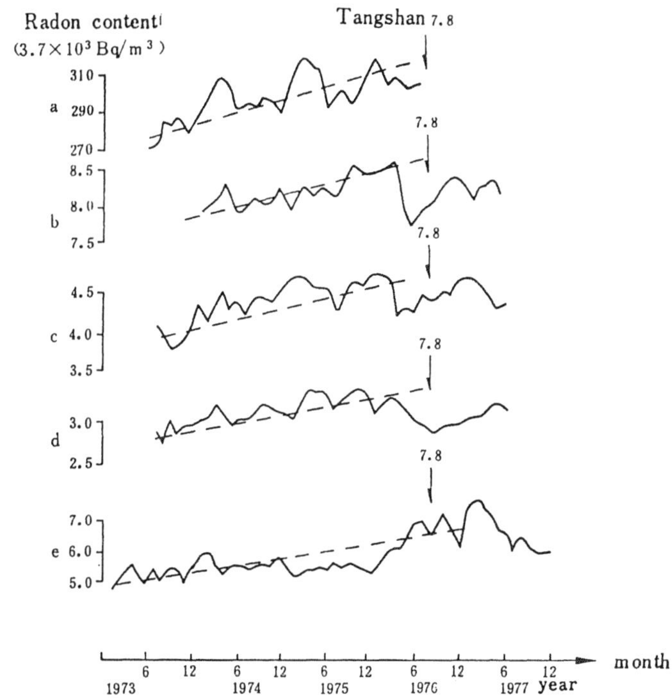

Fig. 7.4. Anomalies of water radon content before the Tangshan earthquake; *a* Anguzhuang well; *b* Tanggu well; *c* No.4 well at Tianjin Textile Plant; *d* Wenan well; *e* Guanzhuang well

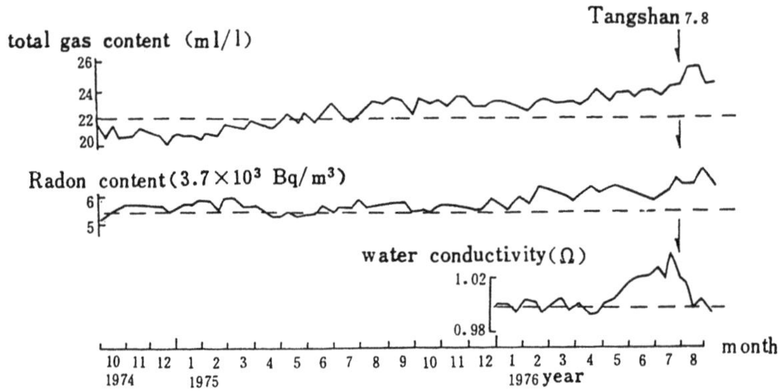

Fig. 7.5. Variations of total gas content, radon concentration and water conductivity at Guanzhuang well before the M 7.8 Tangshan earthquake

To sum up, potential rising variations of ground–water radon concentration in the Beijing–Tianjin–Tangshan region were observed for 1 year or more (even 3 to 4) before the Tangshan earthquake. The mainshock occurred near the peak of radon concentration variations (as at Guanzhuang well), or shortly after the peak was reached (as at Anguzhuang, Tanggu and No.4 wells). During a certain period of time after the mainshock, the radon concentration in many wells was not stable, but fluctuated drastically. Such anomalies were mainly concentrated within 200 km of the epicenter, with some individual wells about 500 km away also showing effects.

2) Bohai earthquake.

The Bohai earthquake occurred near the ground–water hydrochemical (radon) observation network in the northern part of North China, and anomalous variations in radon content were recorded in some wells in Beijing, Tianjin and Hebei Provience. For instance, at Tianjin Duck Farm about 210 km from the epicenter, the radon concentrations increased remarkably since November 1968, and had increased by 11 eman or more by June 1969. The Bohai earthquake happened right after the concentration decreased slightly. The nearby Tanggu, Baitangkou, Ninghe, Baodi and Langfang wells showed synchronous anomalies. Abrupt large changes were also observed immediately before the mainshock, and the concentration decreased and recovered to its normal value after the mainshock (Fig. 7.6b).

3) Haicheng earthquake.

There were a few radon observation points before the Haicheng earthquake, of which the Panjin well (Δ= 70 km) had operated since early 1974. The radon concentration of this well increased gradually, clearly exceeding the average annual baseline value. The content then decreased from November 1974, with the earthquake occurring as the content recovered close to its baseline value

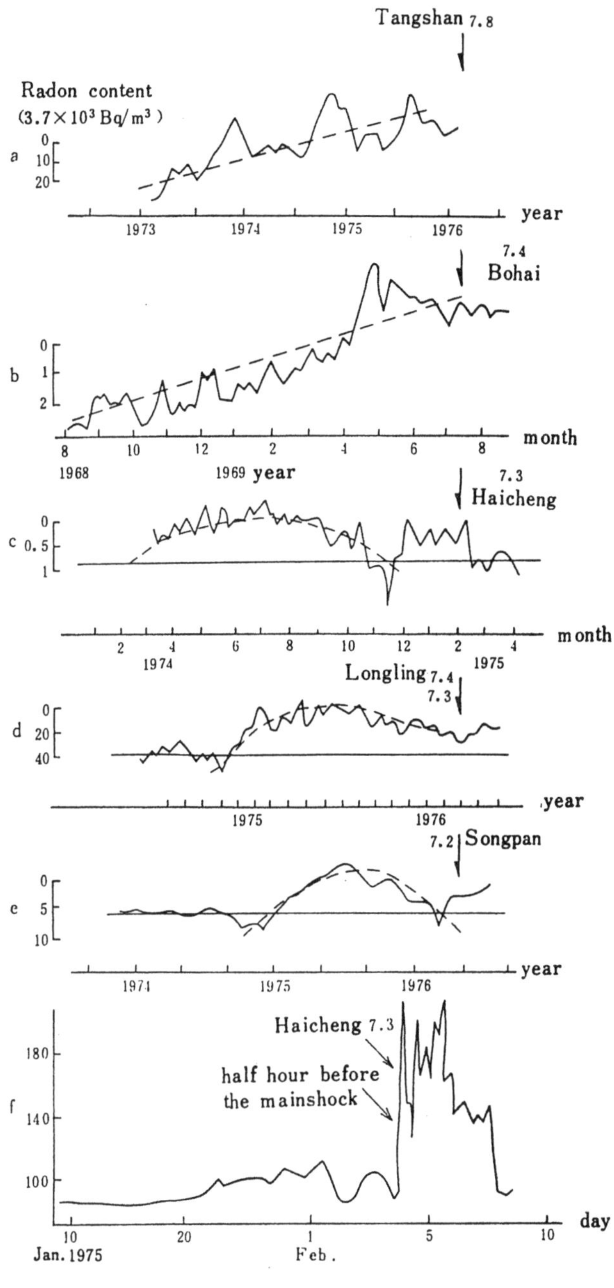

Fig. 7.6. Well–water radon content anomalies before some great earthquakes; *a* Tangshan(Anguzhuang well); *b* Bohai (Duck Farm well); *c* Haicheng (Panjin well); *d* Longling(Xiaguan well); *e* Songpan (Songpan spring); *f* Haicheng (Tanghe well, abrupt changes before impending mainshock)

(Fig.7.6c). Similar processes were also recorded at Jinxian well and Tangchi well in Dandong, away from the epicentral area. Abrupt radon con -centrationvariations were observed at the Tanghe well in Liaoyang before the impending mainshock; for 30 min before the shock the radon content varied from 71 to 130 eman, and reached 218 eman after the mainshock (Fig. 7.6f).

4) Songpan earthquake.

The observation point closest to the Songpan epicenter was the Songpan spring (Δ= 50 km). Although this spring is not well exposed, and is much disturbed, it still gave some anomalous variations before the mainshock. The radon content of the spring started to increase in 1974, the increase accelerated in 1975, and reached its peak value in October 1975. From April – May 1976, the radon content began to decrease, and returned close to its normal baseline value; 3 months later the mainshock occurred (Fig. 7.6e).

Within 200–300 km of the epicenter, there were some well–established observation points, such as Wushan (Δ= 260 km), Tongwei hot spring (Δ= 300 km), Lanshou (Δ= 360 km), and Mapao and Guan springs (Δ= 270 km) in Tianshui. The variations at these points are shown in Table 7.1. The Songpan mainshock occurred at a time when the radon content had increased toward the peak value, or had slightly decreased. The radon content at different points decreased to different extents.

To summarize, increases in hydrochemical concentration in ground–water may be a general anomalous phenomenon before a large earthquake. The entire variation can be summarized as follows: increase – decrease (prior to mainshock) – fluctuate (after mainshock).

Table 7.1. Yearly average variations of radon content in some spring wells before the Songpan earthquake

Eman / Year / Spring Point	1970	1972	1973	1974	1975	1976	1977	The first half of 1978
Wushan	51.15	50.10	—	65.20	78.70	79.21	78.00	77.54
Songpan	38.91	38.80	37.50	40.40	42.65	41.91	38.39	39.27
Tongwei	—	22.60	22.40	24.46	25.54	26.69	24.89	23.60
Lanzhou	—	7.90	7.76	7.92	8.20	8.31	8.23	7.91
Guanquan	—	—	—	6.62	7.60	7.68	7.48	7.51
Mapaoquan	—	—	5.97	6.77	7.40	7.46	7.62	7.50

7.2 Process of Impending Abnormal Variations of Ground Water (Gas) Prior to a Large Earthquake

In the above section, we have discussed ground–water (gas) anomalies of several years before large earthquakes. The imminent stage before an approaching earthquake generally has large amplitude variations, with complicated and diverse shapes. During the observations of the last decade or so in China, the data on imminent anomalies before large earthquakes have been much more extensive and carefully collected than those on the long term variations, so confidence in the data is high. In this section, the imminent anomalies to the nine great earthquakes and some moderate earthquakes will be further studied and analyzed.

7.2.1 Morphology of Anomalous Variations

1) Five main shapes for imminent ground–water level anomalies.

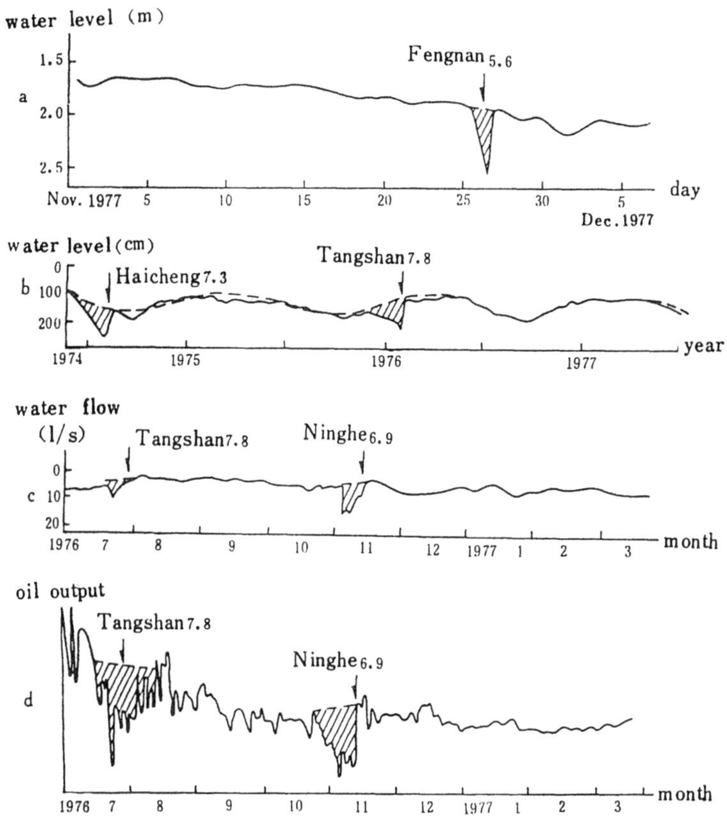

Fig. 7.7. Imminent anomalies of gradual decrease of ground–water level; *a* the strong M5.6 Tangshan aftershock (Yue No. 42 well); *b* the Haicheng and Tangshan earthquakes (Panshan well) ; *c* the Tangshan earthquake (Zunhua spring); *d* the Tangshan earthquake (oil well in Bohai Oil Field)

a. Gradual variations imply that there exists a distinct process of continuous variations for the generation, development and termination of the imminent anomalies. This process usually lasts tens of days but can be as short as several hours. Such variations are the most common, particularly the imminent variations dominated by gradual decreases. The earthquakes often occurred as the decreases were approaching their minimum, or as the recovery began. Figure 7.7a shows a water–level decrease of 29 cm recorded at Yue No.42 professional observation well before the 27 November 1977 M5.6 strong aftershock of the Tangshan quake. This well is about 706 m deep, and is located on the outskirts of Tangshan City, about 20 km from the mainshock epicenter. Figure 7.7b presents the response of Panshan well in Liaoning Province, to the Haicheng earthquake (about 50 km away), and the Tangshan earthquake (about 300 km). The water–level dropped intensely by 89 cm before the Haicheng earthquake, and again decreased by 40 cm or more before the Tangshan quake. For both great earthquakes, the anomaly shapes were rather similar, exhibiting the pre–seismic drop, and the post–seismic recovery of the water–level. It is very clear in Figs. 7.7c and d that sudden decreases of oil production at a well in Bohai Oil Field, and of water flow from Zunhua spring (both close to Tangshan City) occurred before the M7.8 Tangshan earthquake and the M6.9 Ninghe earthquake, respectively.

In addition to the gradual decrease anomalies, gradual rise phenomena were sometimes observed at certain tectonic positions. For example, before the Tangshan quake, gradual rise anomalies with amplitudes of 20 cm or more were recorded in June 1976 at the observation well at Luannan Weather Station ($\Delta = 48$km). This is located close to the Changli–Ninghe fracture at the south boundary of the Tangshan fault depression rhombic block (Fig. 7.8).

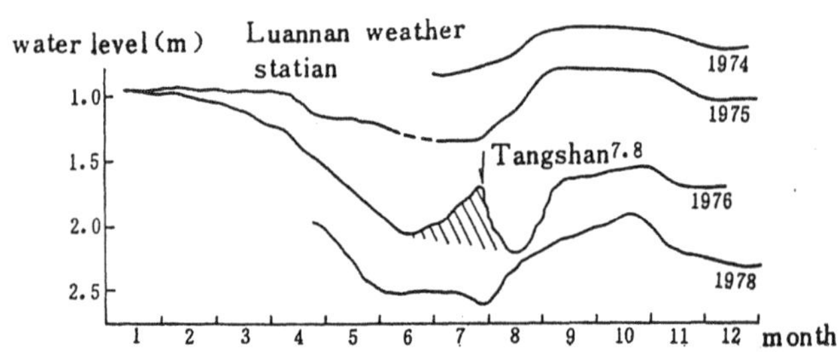

Fig. 7.8. Imminent anomalies of gradual rise of ground–water level

b. Step–like anomalous variations.

The shape of some anomalies seems to be step–like (Fig. 7.9). For the well at Luannan Weather Station, Hebei Province, the normal daily variation of water level is about 1 cm, but the water–level decreased abruptly in a step–like manner 10 h before the Tangshan mainshock, and varied in such a mode by 30 cm and 9 cm respectively during the M7.8 Tangshan and the M7.1 Luanxian earthquakes.

Such a shape and amplitude is unusual in the long–term observations. The Tangshan observation well (about 140 m deep) also showed clear multiple step–like variations in water–level before the nearby M4.5 earthquake (Fig. 7.9).

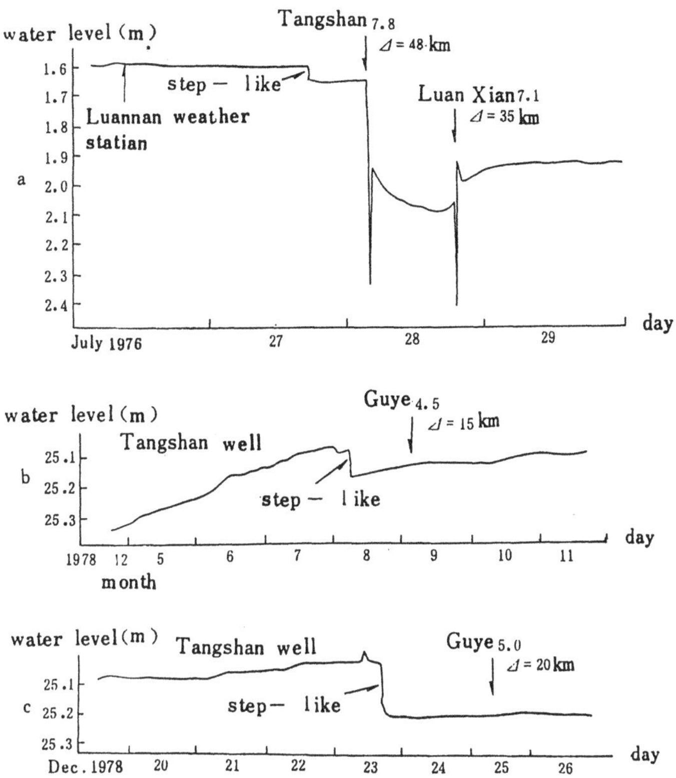

Fig. 7.9. Imminent step–like anomalies in ground–water level

c. Oscillatory anomalous variations.

Before the earthquakes occurred, there were drastic fluctuations and repeated oscillations of the water–level, water pressure, and water flow in a few wells. These had varying period and amplitude, and they died out gradually after the earthquakes. Oscillatory anomalies were observed in a 110–m deep Jiangyou borehole ($\Delta =$ 140 km) before the August 1976 Songpan earthquake, Sichuan Province (Fig. 7.10). The water–level of this borehole was usually stable, but it oscillated with a 8–cm amplitude half an hour before the August 22 M6.7 earthquake (Fig. 7.10a), and with 10–cm amplitude 3 h before the August 23 M7.2 shock (Fig. 7.10b).

d. Distortions of earth tide.

It is well known that some wells respond precisely and sensitively to the earth tide fluctuations. However, it is less noticed that the shape of the tidal curve recorded by such wells near the epicenters of large earthquakes becomes somewhat distorted before the earthquakes. Numerous distortions have been recorded in China during the ground–water observations of the past decade. There are two common kinds of distortion, one being phase shift, and the other tidal amplitude change; sometimes the two are combined.

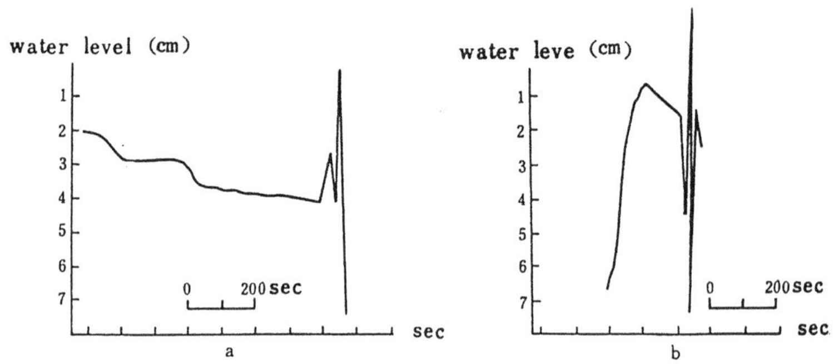

Fig. 7.10. Oscillatory anomalies of water level before the impending 1976 Songpan earthquake; *a* before the August 22 M6.7 earthquake; *b* before the August 23 M7.2 earthquake

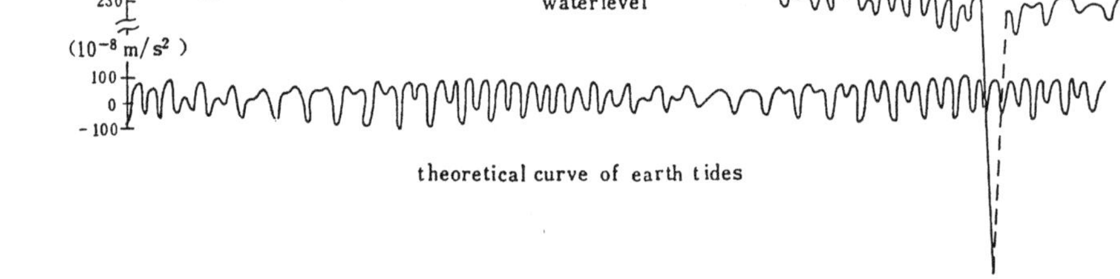

Fig. 7.11. Distortions of earth tide signal in ground–water level prior to the earthquakes

e. Morphological anomalies in annual variations.

There is a professional observation well of 94.6 m depth located at Huozhuang village, Tianjin, for observing the salt water concealed in the sedimentary layer of paleolagoon. The water level had remained stable since the well was put into operation, without any annual variations. However, since April 1976, prior to the Tangshan quake, the well ($\Delta = 90$km) displayed the annual variation morphology shown in Fig. 7.12. Similar situations were also discovered at Jinggu village, Tangshan, and Longjiadian, Changli. The water–flow conditions of such water–bearing strata were destroyed suddenly before the earthquakes, altering the opening extent, so that water–level changes and hydrochemical and water temperature variations were produced. For example, the many years′ steady state radon content in Shuijingzi well in Kangding, Sichuan Province, was abruptly changed before the Longling and Songpan earthquakes to the drastic yearly variations shown in Fig. 7.12. Annual variations of water temperature in Zhedutang Pond in Kangding, Sichuan Province, were also caused before the M7.9 Luhuo earthquake.

2) Two main shapes of imminent ground–water radon anomalies.

a. Step–like anomalous changes.

As the large earthquakes approached, the long– and medium–period (potential) radon anomalies developed into the imminent stage. There were evident turns in the recorded curves tens of days to months before the mainshocks; most of these were sudden flattening or decreasing. For instance, 3 months or so before the Tangshan earthquake, namely in April and May 1976, the Tiantan (Δ= 55 km), Tanggu (Δ= 80 km), Tianjin No.2 (Δ= 90 km) and Tianjin Texile Plant No.4 (Δ= 100km) professional wells showed sudden decreases in radon content, bringing rising anomalies of approximately 1 year back to the normal baseline (Fig.7.13a). During this period, 11 observation wells within 200 km of Tangshan showed turns of different extent in radon content, with the average amplitude reaching 14% or so.

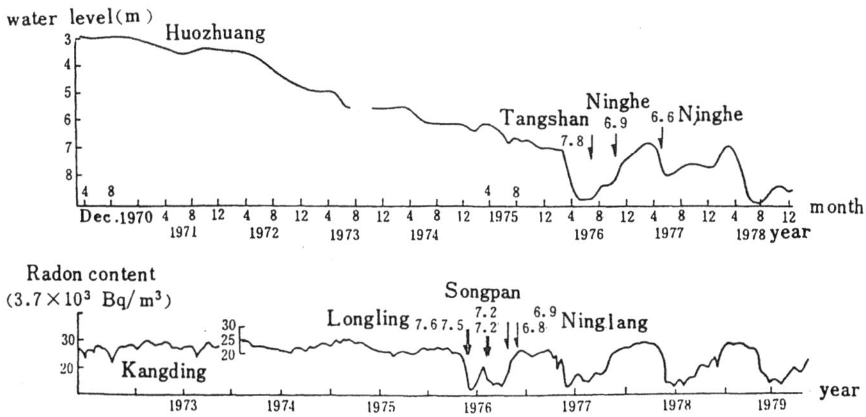

Fig. 7.12. Morphological anomalies of annual ground–water variations just prior to the large earthquakes

By comparing the data observed before many great earthquakes, it is found that the imminent phenomenon of step–like turns occurs fairly generally. Examples are the variations of the Jinxian and Dandong wells (Fig. 7.13f) before the Haicheng earthquake, the Tianjin Duck Farm and Ninghe wells (Fig. 7.13g) before the Bohai earthquake, and the variations of the Songpan spring at the epicenter before the Songpan earthquake, and of the Xiaguan hot spring before the Longling earthquake.

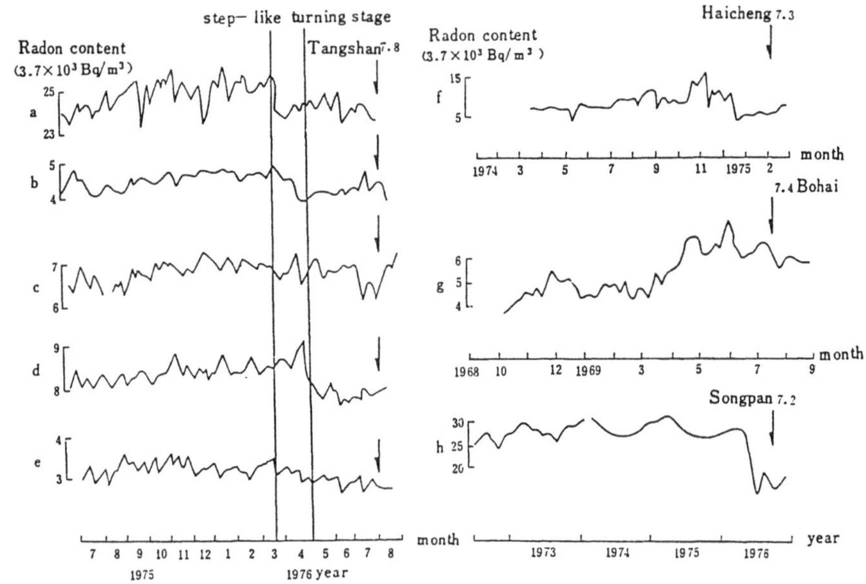

Fig. 7.13. Step–like turning radon–concentration anomalies before the Tangshan earthquake: *a* Tiantan well; *b* Tianjin Textile Plant No.4 well; *c* Tianjin No.2 well; *d* Tanggu well; *e* Wenan well; the Haicheng earthquake; *f* Dandong well; the Bohai earthquake: *g* Ninghe well; the Songpan earthquake; *h* Kangding well

b. Pulse–like anomalous variations.

The other common imminent anomalies are pulse–like abrupt changes, which last from several days to only several hours.

The amplitudes of pulse–like abrupt changes seem to be larger than those of the potential anomalies or the step–like turning anomalies, and they are observed over much larger areas, with their appearance time much closer to the earthquake occurrence time. Taking the Tangshan earthquake as an example, rather rare large–amplitude pulse–like radon variations were observed far away at Shengyang, Liaoning Province, Qufu, Shandong Provinceand even at Qingjiang, Jiangsu Province ($\Delta =$ 640 km) (Figs. 7.14a, b and c), as well as at the wells in Tianjin and other areas near the epicenter.

There still exist many problems with the methods for detecting radon content variations in ground water, and significant errors can result if there is carelessness in any of the many steps, such as water sampling, degassing, instrumental operations, and so on. To verify whether such pulse–like abrupt changes were random, artificial or associated with other natural factors, we have conducted the following research.

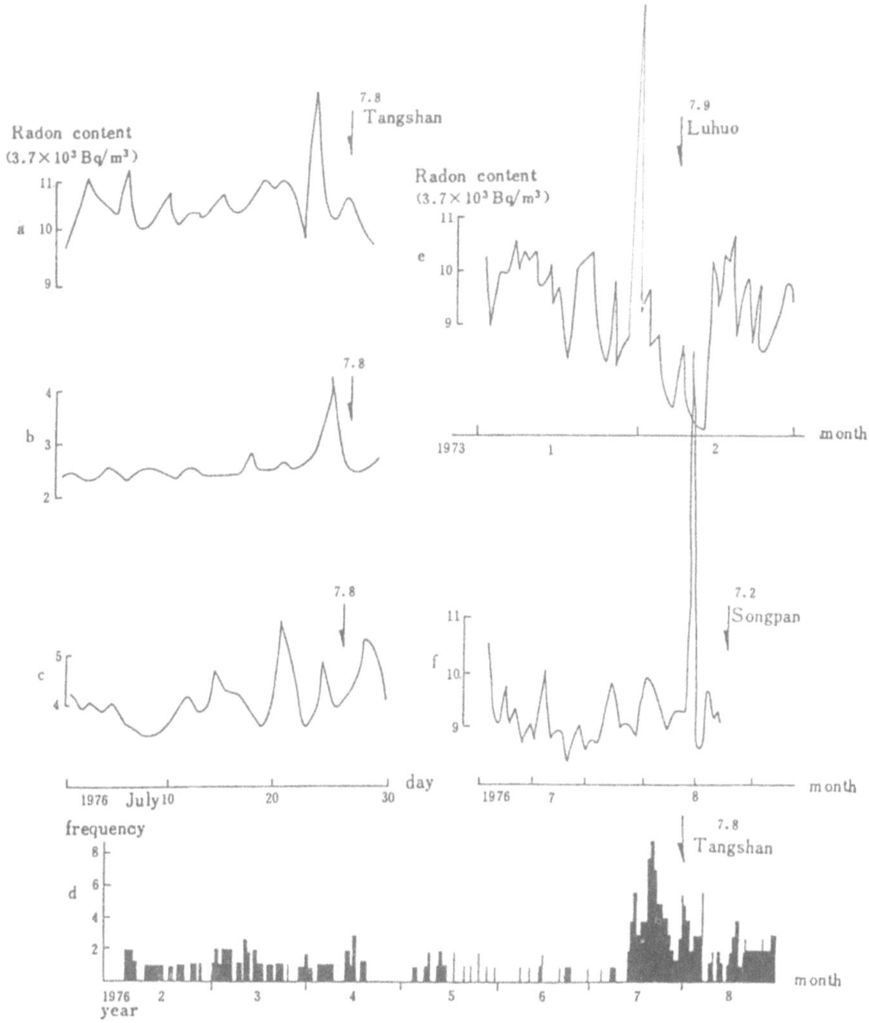

Fig. 7.14. Pulse–like abrupt radon–concentration anomalies before some great earthquakes; *a* Shengyang well before the Tangshan quake; *b* Qufu well before the Tangshan quake; *c* Qingjiang well before the Tangshan quake; *d* statistical frequency of abrupt changes of radon content in observation wells in North China before the Tangshan quake; *e* Guzan spring before the Luhuo earthquake; *f* Guzan spring before the Songpan earthquake

α. Larger–area and long–duration statistical studies. Based on the unified criteria of the anomalies, the data have been processed from several tens of wells observed for many years in North China. The results indicate that the regional abrupt frequency before the Bohai, Haicheng and Tangshan earthquakes was several times higher than that of the usual background. It can be seen that such variations are non–random (Fig. 7.14d).

β. The Guzan well in Sichuan Province shows a fairly good correlation with earthquake occurrences (Figs. 7.14e and f). Systematic analysis of the data obtained from this well since 1972 suggest that major earthquakes tend to occur 15 days after the abrupt variations appear. The correlation coefficient amounted to about 0.7.

γ. Continuous observations made by automatic radon detectors also show pulse–like abrupt changes before impending earthquakes. At the Longling earthquake site, the FD–128 instrument was used. The maximum pulse–like variations, with 130% amplitude and 3–h duration, occurred 39 h before an M >6 earthquake that happened during the radon observations.

3) Shapes of imminent ground–water temperature anomalies.

Sudden increases of ground–water temperature have been observed in the epicentral area before some impending large earthquakes. Before the Longling quake of 29 May 1976, the water temperature at the Balazhang hot spring suddenly increased from 81℃ to 91℃ on April 18, and remained between 90℃ and 92℃. The spring is exposed in cracks in granite and granite–porphyry close to the epicenter. Such variations are similar to the variations of water–level and radon content before a number of large earthquakes. After the Longling earthquake, the temperature fell promptly, and after the strong aftershock of M6.0 on June 9 the water–temperature returned more or less to its normal 81℃ level (Fig. 7.15a).

However, the most common imminent anomalous phenomena are step–like decreases of ground–water temperature. Figure 7.15b shows the temperature changes of the deep well at Xiongxian before the M7.8 Tangshan earthquake. The temperature of this well, beginning on 21 July 1976, decreased suddenly to below 68℃ from its original temperature of 71℃. After the earthquake, the temperature recovered again to 70℃ or so by August 2. Figure 7.15c shows the sudden changes of water temperature in the well located at the People's Park, Tangshan City before the 2 May 1977 M6.5 strong aftershock.

Such water temperature decreases were observed many times in the above wells before the Tangshan mainshock and its numerous strong aftershocks.

4) Gas eruptions before imminent earthquakes.

Fifteen days before the 15 November 1976 M 6.9 Ninghe earthquake, the concentration of soluble hydrogen in the well ($\Delta = 130$ km) at Beijing Guanghua Dyeing and Weaving Mill increased drastically from 0.0137% to 0.18%. The rise continued to 0.24% after the earthquake. The hydrogen content quickly decreased after December 1976, and remained at a value between 0.0027% and 0.0005% after April 1977.

Before the M7.8 Tangshan earthquake, the dry well at Wangquan village, Beijing ($\Delta = 165$ km) suddenly erupted gas on 26 July 1976, with a deafening sound that could be heard 200 m away. The eruption suddenly ceased about 5 h before the mainshock. After that, every gas eruption was accompanied by strong aftershocks. The eruptions of 8 August 1976 amounted to 2.5 m in height, with a

94 dB sound level and 38 m / s airflow rate. The concentration of carbon dioxide in the gas was 13.4%, the oxygen content was decreased to 4% – 5%, and the argon to nitrogen ratio reached 1.57%. Three days afer the two strong M6 aftershocks on 8 and 9 August 1976, this ratio fell towards the normal value in air (1.18%). Up until half a month before the M6.9 Ninghe earthquake of 15 November 1976, the carbon dioxide content in this well increased gradually from 1% to 3%, then decreased again after the quake.

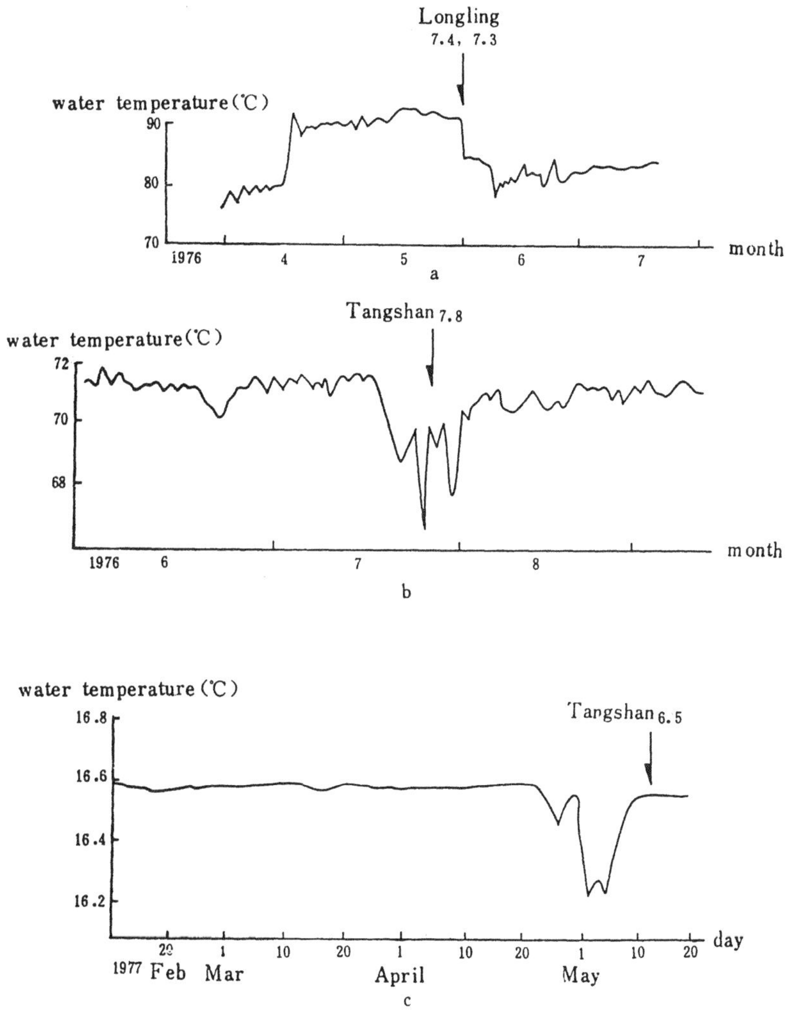

Fig. 7.15. Ground water temperature changes before some earthquakes; *a* Balazhang hot spring (Δ= 10 km)before the Longling earthquake; *b* Xiongxian well (Δ= 190 km) before the Tangshan mainshock; *c* The People's Park well, Tangshan City (Δ= 65 km) before a Tangshan aftershock

A well of 56–m depth at Yangguanlin Commune, Fengrun County, Tangshan region, began to emit gas from small holes in the cement cover of the well about 20 days before the Tangshan mainshock. The gas emission was at most 2–3 days before the mainshock, when the sound could be heard more than 20 m away, and the emission died down gradually several days after the mainshock. Since then, such gas emissions have been followed by aftershock activity. Normally, the gas eruption reached its peak 2–9 days before the strong aftershocks. The occurrence times of several of the strong aftershocks were predicted fairly precisely based on these anomalies.

7.2.2 Inhomogeneous Areal Distribution of Anomalies

1) Ground–water level.

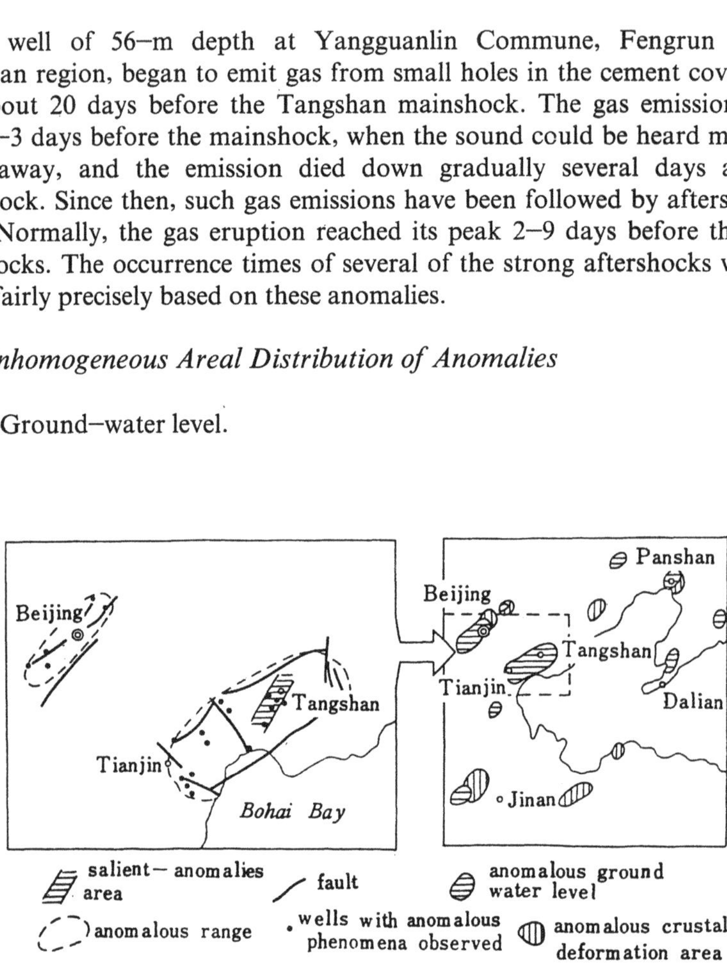

Fig. 7.16. Spatial distribution of water–level anomaly before the Tangshan earthquake

The ground–water level observations before and after some great earthquakes demonstrate that those wells that show anomalies are distributed inhomogeneously in space. The inhomogeneity is manifested in two ways: firstly, not all wells show anomalous changes; secondly, the anomaly amplitudes do not attenuate uniformly with epicentral distance. Before the Tangshan earthquake, there were individual areas of water–level anomalies scattered in Beijing and south Liaoning Province, as well as the anomalies concentrated in Tangshan and adjacent areas (Fig.7.16). These anomalous areas were not connected spatially, but their distributions were often constrained by tectonic elements. For example, the Tangshan–Ninghe area of water–level changes was approximately the size of the Tangshan rhombic block, and no obvious rise or fall of water–level was observed in many wells outside the block. However, outstanding water–level anomalies were observed in the wells near the Babaoshan fault and the Tancheng–Yingkou fault zone.

The inhomogeneity of water level observations was also distinct before several moderate earthquakes, such as the M5.2 Fengnan earthquake of 25 May 1970. The anomalous wells were distributed near faults, while wells in non–tectonic zones exhibited no obvious anomalies. Further, the characteriztics of the water level anomalies were different in each tectonic zone, while their shapes were different in different segments along the same tectonic zone. In general, the anomalous areas within the epicentral region were characterized by many anomalous observations, with large variations, long duration and fairly consistent shape.

Such inhomogeneous spatial patterns, constrained by tectonic conditions, are also noticed in many other earthquake cases, both at home and abroad. Therefore, the cases of very simple and uniform variation of water–level reported previously may give a distorted impression because of inadequate data. Only if the data are sufficiently abundant can the complex features be clearly seen.

2) Hydrochemistry of ground water.

Stations for observing the chemical composition of ground water (including radon) are far fewer than for water–level. Therefore, the spatial distributions of hydrochemical anomalies are not well understood at present. Nevertheless, numerous indicators suggest that their distributions are also inhomogeneous, and are constrained by the tectonic and hydrogeological conditions.

Data from the Tangshan quake show that there were 28 observation points within 200 km of the epicenter, of which 11 (roughly 36%) displayed long– and medium–term potential anomalous variations. There were very few points showing anomalies outside the epicentral area. Wells showing anomalies were usually spatially mixed with those not showing anomalies, both inside and outside the epicentral area. Wells that showed potential anomalies and that displayed imminent anomalies appeared alternately.It generally appeared that the anomalous wells were concentrated near the epicentral area, with fewer outside, mostly along active tectonic zones.

The imminent and sudden anomalies were more scattered, scarce and irregular. Simple statistical results indicate that anomalous variations in wells near the epicenter are no clearer than in those far away. For example, the closely monitored deep well at the Running Water Factory in Tangshan showed no anomalies although it was located in the meizoseismal region. At the same time, the wells in Qufu, Shandong Province ($\Delta=450$ km), and Qingjiang, Jiangsu Province ($\Delta=$ 640 km), far from the epicenter, showed distinct imminent anomalies. Also, two wells (Running Water Factory and Power Supply Plant) observing the same aquifer within the Tangshan meizoseimal area showed different results, thus illustrating the major role played by local conditions at each well.

It seems that the factors influencing the hydrochemical (gas) anomalies were much more complicated than those affecting water–level. As well as the factors of earthquake location and local well conditions, the sites of the matter from which the chemicals or gas originated might also influence the hydrochemical anomalies. The observed spatial distributions of chemical anomalies before the large earthquakes showed a complicated non–uniform pattern.

7.2.3 Relation Between the Magnitude of Anomalies and Earthquakes

In order to analyze systematically the relation between the magnitude (amplitude and duration) of imminent ground–water level anomalies and earthquake parameters (occurrence time, location and magnitude), we have collected observed data from 60 earthquakes in China, of which 32 were damaging ones with M > 6. Continuous records were obtained from many of the wells using autographic recorders (Table 7.2).

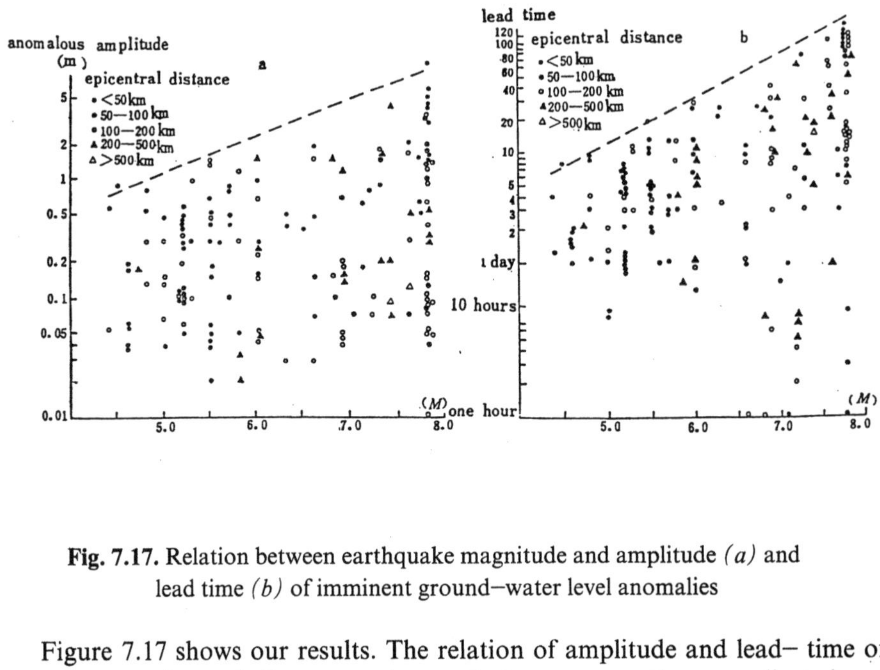

Fig. 7.17. Relation between earthquake magnitude and amplitude *(a)* and lead time *(b)* of imminent ground–water level anomalies

Figure 7.17 shows our results. The relation of amplitude and lead– time of anomalies to epicentral distance and magnitude appears rather complicated and obscure. As regards ground–water level, higher magnitude quakes tend to have longer lead–time anomalies (10–20 days for M5, 2–3 months for M7). However, such statistical regularities are not of great importance for practical earthquake prediction, as such large quantities of data from a concentrated set of wells could not be collected before one earthquake. Comparisons of anomalies between various arbitrarily selected wells showed almost no regularity, perhaps because the tectonic environments of the well networks were complex. We conclude that such anomalous phenomena may be different in different areas, for different wells and for different earthquakes. Anomalies observed at the same epicentral distance may differ greatly from one another before various earthquakes of roughly the same magnitude (Fig. 7.17). It appears that the variations of pre–seismic anomalies are far more complicated than it has been previously thought.

The relation between hydrochemical anomaly magnitudes and earthquake parameters is also very complicated and uncertain. For instance, before the Tangshan, Haicheng, Songpan and Longling events, it was found that no imminent anomalies were observed in some wells that had previously shown potential (long–term) variations. Also, many wells that showed imminent anomalies had recorded no potential variations. This might imply that the durations of the anomalies were very inhomogeneous. For a particular earthquake, the anomalies measured in several wells with similar epicentral distances, were greatly different. Therefore, both the detection of anomalies and their size are dependent both on the common element of earthquake source region and the individual element of local well condition. The former element tends to make the anomalies conform to a statistical law, while the latter induces very complicated and inhomogeneous behaviour.

7.3 Discussion

7.3.1 Unity of Ground–water Anomaly Processes

Before and after the large earthquakes, there were diverse shapes and amplitudes of ground–water level, hydrochemical, water temperature and gas concentration anomalies. However, the advent times of the different types of anomalies are fairly well synchronized with the stages of the variation process, and the anomaly sites were relatively concentrated.The anomalies may result from variations in porosity and fissures in water–bearing rock acted on by tectonic stresses.

Taking Tangshan as an example, the character of the ground–water level anomalies (decrease–rise–quake–drastic rise) corresponds fairly well with the hydrochemical (radon) anomalies (rise – fall – quake –fluctuation). Also, the variation processes are fairly synchronous, both starting in 1972–1973. Both sets of anomalies changed direction during the same period, mid–1975 to April or May 1976. Ten days or so before the mainshock, the imminent anomalies occurred in both cases. These were (Figs. 7.1, 7.2, 7.3, 7.14 and 7.15) sudden changes and rising of water–level, abrupt radon variations, water temperature fall (Tangshan and Xiongxian), water–flow reduction (Tangshan Mine and Zunhua Spring) and gas eruptions (Fengrun, Changli and Beijing).

The distribution of wells in which anomalies occurred shows that the most intense variations were in the epicentral area. Otherwise, anomalies were located predominantly in the Cangdong fracture zone and the Babaoshan fault zone to the west, and in the nearby Tanlu fracture zone to the east (Fig. 7.16).

From those wells at which several parameters were monitored, it is often found that the times of occurrence of different anomalies are approximately coincident (for example, radon variations, gas content and water conductivity at Guanzhuang, shown in Fig. 7.5).

Table 7.2. List of earthquakes with associatedwater–level observations

Earthquakes	Closest distance to point (km)	Methods	Earthquakes	Closest distance to point (km)	Methods
1. The M 7.8 Tangshan earthquake of Hebei Province, on July 28 1976	●	△	15. The M 6.8 Yanyuan earthquake of Si–chuan Province, on Dec. 13 1976		
2. The M7.7 Tonghai earthquake of Yun–nan Province, on Jan. 5 1970	●		16. The M 6.8 Dulan earthquake of Qinghai Province, on Mar. 24 1971	●	△
3. The M7.3 and 7.4 Longling earth–quakes of Yunnan Province, on May 29 1976	○		17. The M 6.7 Songpan earthquake of Sichu–an Province, on August 22 1976		△
4. The M7.4 Bohai earthquake on July 18 1969		△	18. The M 6.7 Puer earthquake of Yunnan Province, on April 28 1971		
5. The M7.4 Taiwan earthquake on July 23 1978		△	19. The M 6.6 Ninghe earthquake of Tian–jin, on May 1977	●	△
6. The M7.3 Haicheng earthquake of Liaoning Province on Feb. 4 1975	●		20. The M 6.4 Yushu earthquake of Qinghai Province, on Mar. 29 1979		
7. The M7.2 Songpan earthquake of Sichuan Province, on August 6 1976		△	21. The M 6.3 Horinger earthquake of Inner Mongolia on April 6 1976	●	
8. The M 7.2 Songpan earthquake of Sichuan Province, on August 23 1976		△	22. The M 6.1 and 6.3 Lulong earthquakes of Hebei Province, on August 31 1976	●	△
9. The M7.1 Zhaotong earthquake of Yunnan Province, on May 11 1974	○		23. The M 6.0 Taiwan earthquake on May 5 1935	○	△
10. The M 7.1 Tangshan earthquake of Hebei Province, on July 28 1976	●	△	24. The M 6.0 Jinghe earthquake of Xin–jiang Region, on June 3 1973		
11. The M 7.0 Taiwan earthquake on April 21 1935	○	△	25 The M 6.0 Lulong earthquake of Hebei Province, on Mar. 7 1977	●	△
12. The M 6.9 Yanyuah earthquake of Sichuan Province, on Nov. 7 1976			26. The M 6.0 Haicheng earthquake of Liaoning Province, on May 18 1978	●	△
13. The M 6.9 Ninghe earthquake of Tianjin, on Nov. 15 1976	●	△	27. The M 6.0 Kuqa earthquake of Xin–jiang Region, on Mar. 29 1979		
14. The M 6.9 Dawu earthquake of Sichuan Province, on Feb. 24 1981	○		28. The M 6.0 Liyang earthquake of Jiangsu Province, July 9 1979	●	△

29. The M 6.0 Wuyuan earthquake of Inner Mongolia, on August 25,1979	●		46. The M 5.2 Fengnan earthquake of Hebei Province, on May 25,1970	●	△
30. The M 5.8 Shulu earthquake of Hebei Province, on Dec. 3,1967	●	△	47. The M 5.2 Tangshan earthquake of Hebei Province, on Feb. 7,1980	●	△
31. The M 5.8 Haicheng earthquake of Liaoning Province, on August 28,1975	●	△	48. The M 5.0 Eshan earthquake of Yunnan Province, on Jan. 14,1970	●	
32. The M 5.8 Korla earthquake of Xing-jiang Region, on April 22 1978			49. The M 5.0 Jiangchuan earthquake of Yunnan Province, on Jan. 27,1970	●	
33. the M 5.8 Kangding earthquake of sichuan Province, on Sept. 30,1972	○		50. The M 5.0 Guzhen earthquake of Anhui Province, on Mar. 2,1979		△
34. The M 5.7 Eshan earthquake of Yun-nan Province, on Feb. 5,1970	●		51. The M 4.8 Jiangchuan earthquake of Yunnan Province, on July 17,1972	●	
35. The M 5.7 Xiji earthquake of Ning-xia Region, on Dec. 3,1970	●		52. The M 4.8 Chengwu earthquake of Shangdong Province, on July 9,1977	●	
36. The M 5.7 Songpan earthquake of Sichuan Province, on Jan. 16 1974			53. The M 4.7 Mianduhe earthquake of Heilongjiang Province, on August 31,1977		△
37. The M 5.6 Tangshan earthquake of Hebei Province, on Nov. 27,1977	●	△	54. The M 4.7 Tangshan earthquake of Hebei Province, on Nov. 23,1978	●	△
38. The M 5.5 Shulu earthquake of Hebei Province, on Jan. 16,1968	●	△	55. The M 4.7 Tangshan earthquake of Hebei Province, on Dec. 6,1978	●	△
39. The M 5.5 Longyao earthquake of Hebei Province, on May 15,1968	●	△	56. The M 4.7 Tangshan earthquake of Hebei Province, on Dec. 1,1978	●	△
40. The M 5.5 Baoshan earthquake of Yun-nan Province, on Feb. 5,1971	●		57. The M 4.6 Ningjin earthquake of Hebei Province, on Jan. 7,1968	●	△
41. The M 5.5 Shibing earthquake of Yun-nan Province, on Jan. 23,1972	●		58. The M 4.6 Tangshan earthquake of Hebei Province, on Dec. 25,1978	●	△
42. The M 5.5 Liyang earthquake of Jiang-su Province, on April 22,1974	●		59. The M 4.4 Horinger earthquake of Inner Mongolia, on June 24,1975	○	
43. The M 5.5 Chuxiong earthquake of Yunnan Province, on Jan. 12,1975	●	△	60. The M 4.4 Dacheng earthquake of Hebei Province, on April 22,1976	●	△
44. The M 5.3 dacheng earthquake of Hebei Province, on Dec. 31,1973	●	△			
45. The M 5.3 Tangshan earthquake of Hebei Province, on Mar. 12,1980	●	△			

7.3.2 Ground–water Anomalies and Fissure Evolution

At present, there is little understanding of ground–water variations before and after large earthquakes. We suggest that a reasonable explanation is the development of microfissures during the seismogenic process, since variations of pressure and of water channels within the strata are the most fundamental cause of changes in water–level, hydrochemistry, temperature and gas content. The pressure is what alters the ground–water trend, and the channels allow such variations to be displayed in observation wells. Changes in porosity and fissures within the rock body may provide the most direct way to satisfy both. In–situ observations and laboratory studies of rock mechanics have proved that pore pressure tends to fall (water–level fall) while radon content tends to increase with growth of fissures. However, during the process of stressing to failure, the changes in fissures are comparatively complex in morphology, time and space. The temporal fluctuations of the ground–water changes seem to be governed by variations in lithological and physical properties of the rocks, and by the evolution of fissures. The diversity of anomalies may be a reflection of the diversity of fissure evolution. For example, the five shapes of imminent ground–water level anomaly may reflect the five types of fissure evolution in the source regions and its vicinity just prior to the earthquake (Figs. 7.7 – 7.12).

Table 7.3. Morphology of imminent anomalies and their analysis

morph ology	Possible origion	Possible locations
I. gradual	fissures in-creased slowly in water-bearing bodies in a large scale	appeared vastly in epicenters and their vicinities, or at some parts of active tectonic zones outside
II., Step-like	Fissures developed into fault creeping or local rupturing	in some weakened parts of epicenters or their adjacent areas
III. oscella-tory	rapid variations of porosity pressure due to breathing effects of develo-ping fissures	where being fastly adjusted during fis-sure growth processes in epicenters and their vicinities
IV. Distorted shape of earth tides	medium status altered owing to fissure developments	Ditto I
V. changes of an-nual morpho-logy	vertical fissures developed, linked with surfaces, making aquifers connected	where the fissures fairly developed and waterproof layers easily destroyed in epicenters and their vicinities

In hydrochemical (radon) observations, gradual, step–like and sudden oscillatory changes have also been noted (Figs. 7.12 and 7.13).

Upwelling of material from depth only occurs under certain tectonic conditions. Fissures develop extensively before and after the large earthquakes, but the gas eruptions and water temperature increases were seldom accompanied by deep mass transport. Rather, shallow fissure activities are often observed before the earthquakes. The phenomena usually observed result from the mixing of shallow water with gas emitted from the shallow rock layers due to the evolution of fissures within those layers. As springs and deep confined wells were often selected for observations, the water–level would decrease anomalously when the shallow layers were not linked with aquifers or connected to the surface (Figs. 7.7 and 7.15).

7.3.3 Imminent Rebound of Ground Water

We here emphasized the imminent rebound phenomenon for ground–water level and radon variations (Fig. 7.18).

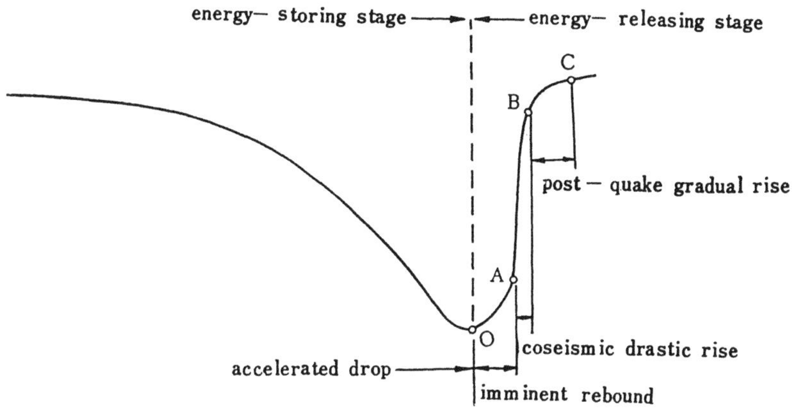

Fig. 7.18. Imminent rebound phenomenon for ground–water level

Figure 7.18 can be divided into three stages. In the first stage (OA) before the earthquake, the water–level turned to rise at a gradually increasing rate, following a drastic fall. In the second stage (AB), as the fissures became linked the water–level rose fiercely. In the third stage, as the dislocation was over, the water– level kept rising at a slow rate. These three stages describe the short–term rupturing process of a water–bearing rock body. Rock mechanics experiments on rupturing confirm that the instability begins with fairly slow rupturing, then accelerates to the critical velocity until total collapse. Therefore, the imminent rebound of the ground–water level might be due to a slow creep of fractures prior to the earthquake.

CHAPTER 8

Unusual Atmospheric Phenomena Before Large Earthquakes

Numerous heavenly and earthly phenomena and biological anomalies have been described and recorded in historical books in China. However,most of these records have been human observations, so that it is hard to eliminate personal and regional differences. The reports have generally met with puzzlement or skepticism in the scientific community.

Before and after the nine great earthquakes, a variety of anomalous phenomena occurred on the earth, in the atmosphere and in the biological world, thus providing a rather rich data set and the chance for us to study the connections between these phenomena and earthquake occurrences. Because the Haicheng and Tangshan earthquakes occurred in densely populated cities, many people saw such phenomena before and after these earthquakes. Furthermore, some of the phenomena before and after these earthquakes were actually registered by certain seismic, astronomical, meteorological, oceanographic and radio–broadcasting instruments. These data may be of great scientific value in studying seismogenic processes.

In this chapter, those phenomena that occurred similarly before all of the nine earthquakes will be treated in detail with some additional description of unique individual phenomena.

8.1 Variations of Atmospheric Pressure

It is well known and generally accepted that variation of atmospheric pressure is an element in triggering earthquakes. We have found that there are statistical indications for short–term and imminent variations of atmospheric pressure before the nine mainshocks within the epicentral areas and their vicinities.

8.1.1 Short–term Variations

About 1 month before the mainshocks, the monthly average atmospheric distributions in the areas near the epicenters have two kinds of characteristics:

1) The maximum variation of monthly mean pressure occurred within the epicentral area and its adjacent areas; moreover, the center of departure of the monthly mean pressure basically coincided with the epicentral area (Fig. 8.1). This was the case before the Xingtai, Haicheng, Longling, Tangshan, Songpan and Luhuo earthquakes. The center of departure of atmospheric pressure was also close to the epicenter before the Tonghai and Yongshan earthquakes, with an exception for the Bohai earthquake.

2) The value of center of departure of monthly mean pressure was generally

the maximum or sub–maximum atmospheric pressure variation ever recorded in the local area. For the Tangshan earthquake, the monthly mean pressure of July 1976 in Tianjin was the greatest for the past 40 years; there were similar effects for some of the other earthquakes.

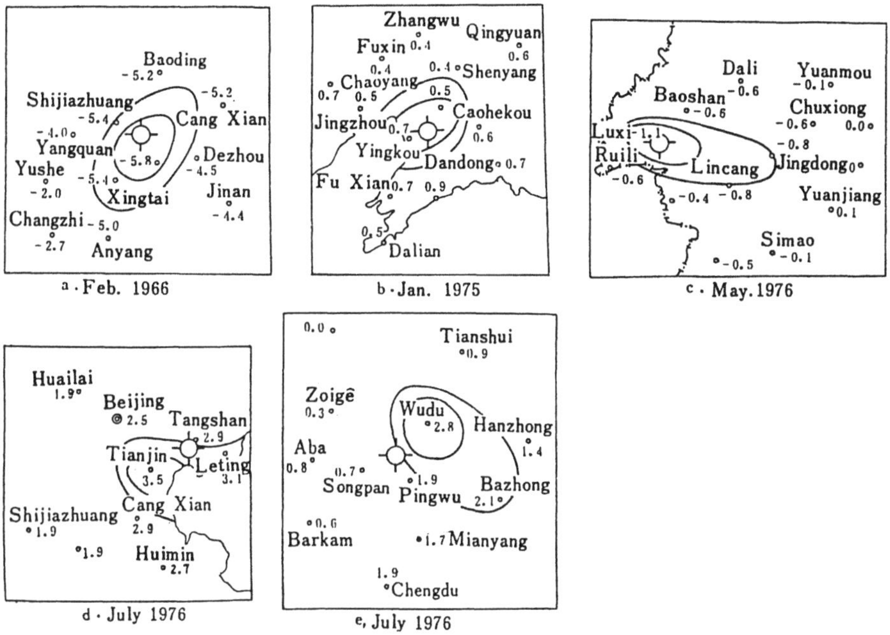

Fig. 8.1. Distributions of departures of variations from monthly mean pressure before some great earthquakes; *a* the Xingtai earth–quake (8 March 1966); *b* the Haicheng earthquake (4 February 1975); *c* the Longling earthquake (29 May 1976); *d* the Tangshan earthquake (28 July 1976); *e* the Songpan earthquake (16 August 1976)

8.1.2 Imminent Variations

There ought to be intense weather activity before a large earthquake. In addition to the possibly unique cold or hot, strong wind, and rain or fog caused thereby, the daily or hourly atmospheric pressure variations in epicentral areas were characterized by two evident features. In Fig. 8.2, the curve d presents the hourly variations of atmospheric pressure, the curve h gives the daily variations at 0200 hours Beijing time, and the other curves all show daily average changes of atmospheric pressure:

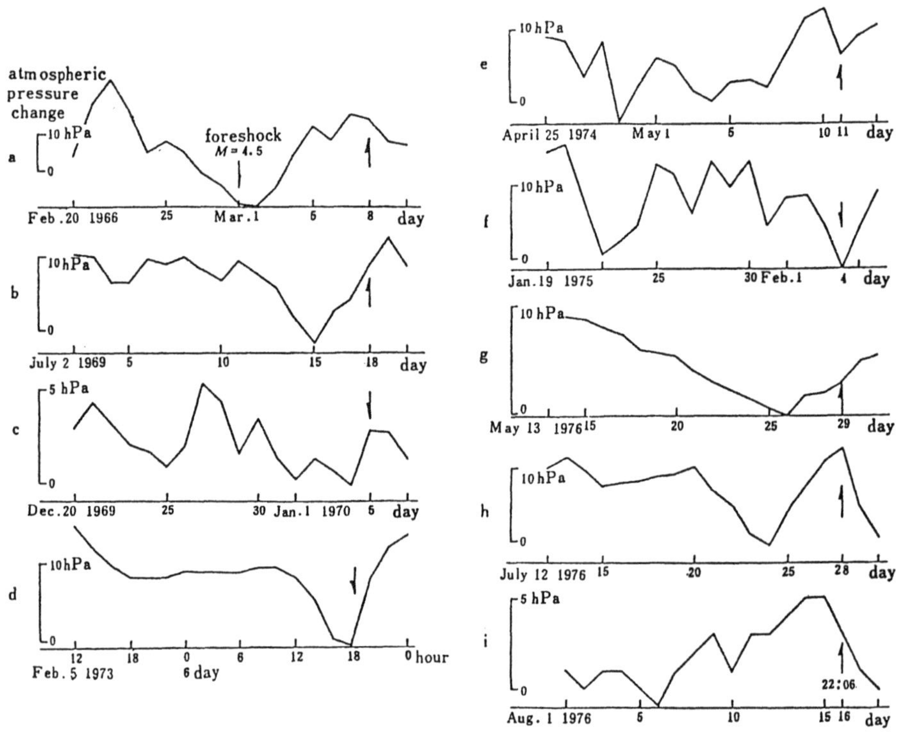

Fig. 8.2. Short–term variations of atmospheric pressure before the great earthquakes; *a* Xingtai earthquake; *b* Bohai earthquake; *c* Tonghai earthquake; *d* Luhuo earthquake; *e* Yongshan earthquake; *f* Haicheng earthquake; *g* Longling earthquake; *h* Tangshan earthquake; *i* Songpan earthquake

1) The earthquakes occurred at a time when the atmospheric pressure kept changing, reached a maximum, or a while after. For example, before the Xingtai earthquake, the atmospheric pressure kept decreasing for 8 days, and reached its lowest point on March 2. At that time, the seismic activity with M over 4 began. Before the Tonghai earthquake, the pressure in the epicentral area reached its lowest point on January 4, and the mainshock took place at 0100 hours on January 5. Before the Luhuo earthquake, the pressure in the area decreased steeply, reaching a minimum at 1800 hours on February 6, shortly before the mainshock. The Haicheng mainshock occurred when the atmospheric pressure had decreased to its lowest point. Before the Longling earthquake, there was a continuous fall in atmospheric pressure for 14 days running, a very rare occurrence in this area. The mainshock occurred as the pressure turned to rise. One hour prior to the Tangshan mainshock, at 0200 hours on July 28, the atmospheric pressure in this area rose to a maximum. These are typical earthquake cases. It seems that atmospheric pressure changes with amplitude over 10 millibar and a duration long-

er than 4 days may be important imminent precursory indicators under certain conditions.

2) For the nine great earthquakes, the atmospheric pressure fell to a minimum prior to those mainshocks that occurred in winter (December to March) and rose to a maximum for those that occurred in summer (May to August).

8.2 Variations of Rainfall

During a seismogenic process, particularly during reservoir–induced and impoundment–induced earthquake preparation, the important role of water and other fluids has aroused the concern of many seismological workers. After systematically analyzing the effects of atmospheric rainfall on the occurrence of the nine earthquakes, we have found the following two facts:

1) Large earthquake activities in North China and Southwest China are correlated with heavy rainfall before the earthquakes. From Fig. 8.3, it can be seen that the large earthquakes in North China occurred within 3 years after periods of heavy rains in the region.

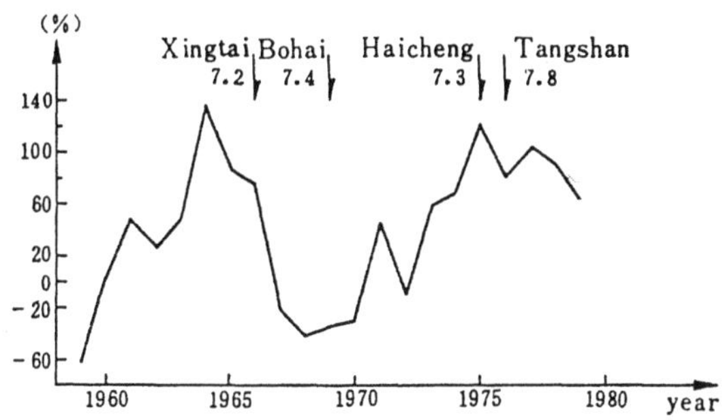

Fig. 8.3. Relation between the sum (3–year sliding value) of departure percent of precipitation at five stations (Tianjin, Yingkou, Dalian, Yantai and Hengshui) in North China and the occurrence time of large earthquakes

Figure 8.4 gives the situation in which the five large earthquakes in Southwest China migrated along with movements of climate areas. In Fig. 8.4, the five large earthquakes are divided into three regions, the North Sichuan, West Sichuan and Yunnan regions. Two or three typical weather stations are selected in each region. After the criterion for rainfall anomalies is defined, based on the yearly rainfall variations, the rainy areas can be delineated, as shown by the solid line in the figure. It is seen that the five large earthquakes in the three regions, after the advent of rainy weather in the Sichuan–Yunnan region since 1965, not only migrated spatially with the movement of the rainy areas, but also occurred within 3 years after rainy weather, similar to the situation in North China.

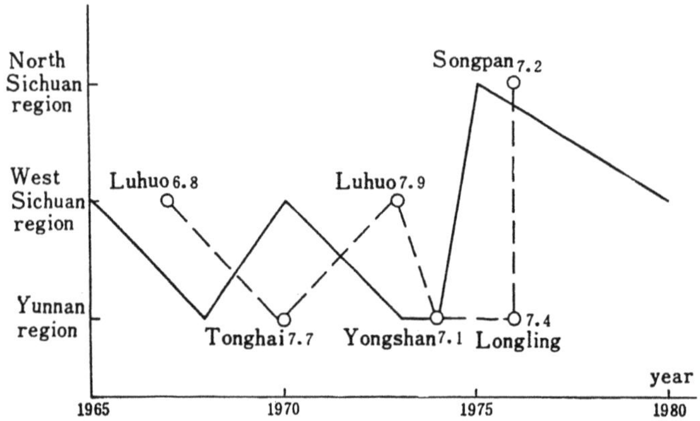

Fig. 8.4. Correlations (1965–1980) between the movement of rainy areas in Sichuan–Yunnan region north of 24 ° N and west of 105 ° E (*solid line*) and the migration of large earthquakes (*dashed line*)

2) Within about one month before the mainshocks, the areas with heaviest rainfall would be the forthcoming mainshock epicenters. This is the case for the Luhuo, Tangshan, Longling, Haicheng, Yongshan, Xingtai and Songpan earthquakes, as shown in Fig. 8.5. It seems that heavy rain may trigger earth–quakes under certain conditions.

8.3 Variations of Physical Meteorology

In the last decade or so, pre–seismic and coseismic macroscopic investigations have shown that there may exist peculiar physical phenomena that can be directly sensed by human beings or observed by special instruments of astronomy, meteorology, oceanography, electronic communications and so on. These phenomena include various effects of sound, light, electricity, magnetism, heat, radiation, atmospheric nature and so on. These phenomena are not directly caused by astronomical and meteorological factors, although they exist or are displayed in the atmosphere; also these phenomena are out of the range of principal research of their own disciplines (as these random and peculiar phenomena are often regarded as disturbances of no importance in their own fields). We suggest that research into these phenomena should be placed under studies of

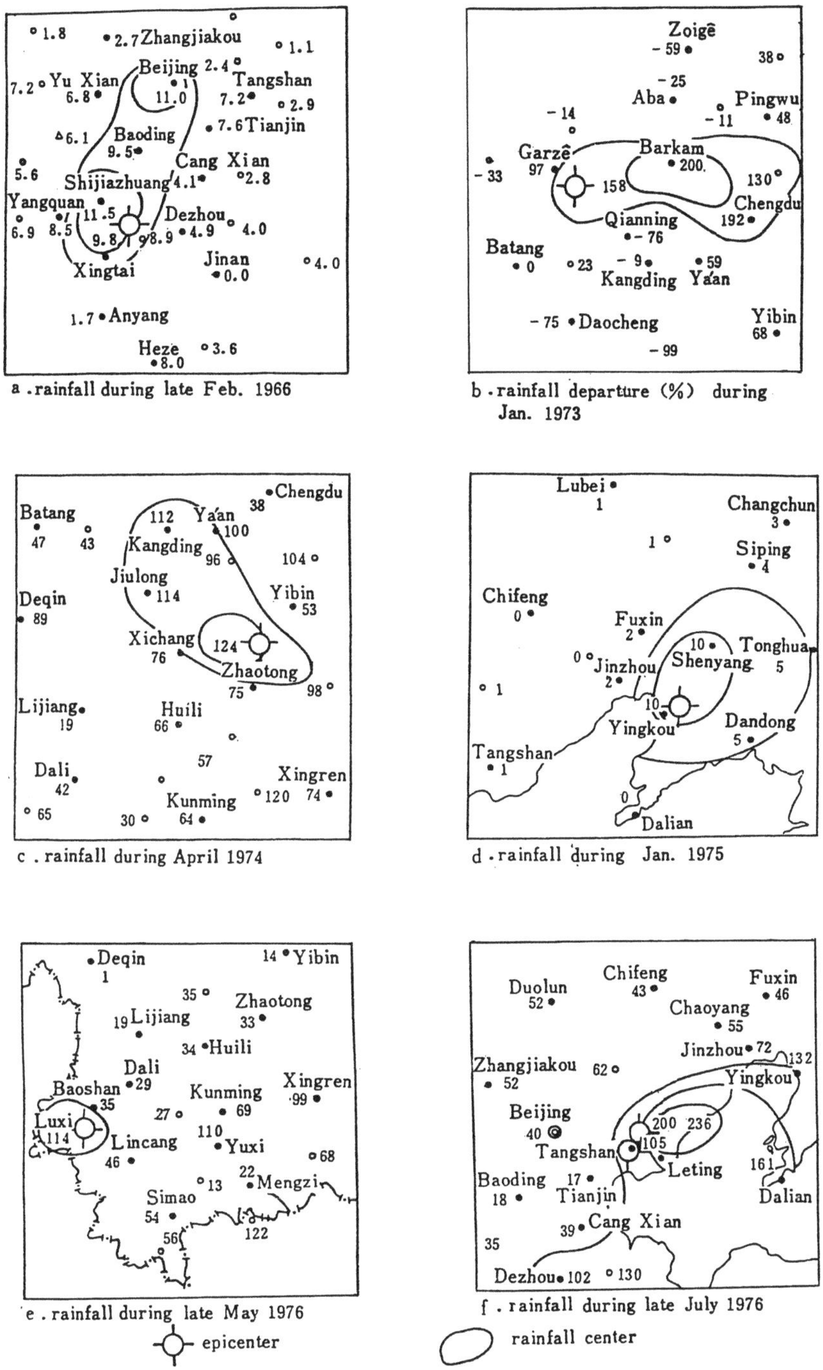

a. rainfall during late Feb. 1966

b. rainfall departure (%) during Jan. 1973

c. rainfall during April 1974

d. rainfall during Jan. 1975

e. rainfall during late May 1976

f. rainfall during late July 1976

Fig. 8.5. Rainfall situations before some great earthquakes; *a* Xingtai; *b* Luhuo; *c* Yongshan; *d* Haicheng; *e* Longling; *f* Tangshan "physio–meteorology". Such physio–meteorogical phenomena might be the outcome of inter–influence of internal and external physical processes of the crust before or during earthquakes. It should be acknowledged that the majority of such phenomena, and the predictions of earthquakes from such phenomena are considered as "scientifically groundless". However, in our opinion, phenomena that cannont be explained by modern science should not necessarily be taken as non–scientific. In order to develop earthquake prediction science, it is still necessary to consider these pre–seismic physio–meteorological phenomena.

8.3.1 Geosound

The geosound heard by people before the nine great earthquakes is multiple and diversified. It has been described as vehicle rumbling, wind roaring, drum beats, crackling of tree rupture, whistling of torn cloth, thundering and so on. Its occurrence often produces panic.

Before the Songpan earthquake of 16 August 1976, a geosound monitor was designed and installed in the Longquanyi Middle School in Chengdu City. Before August 16, a single–syllable sound, as pronounced in Chinese like "clattering", "clanking", "clucking" was heard first, then came drum beats, and even the singing of a frog. Finally a long duration of thundering, lasting about 10 s, was heard at 2024 hours on August 16, and 1 h later the Songpan mainshock occurred.

The geosound can be heard mainly within the future area of relatively high seismic intensity. The geosound was mostly heard about 10 min before the Tangshan mainshock (Table 8.1).

Table 8.1. Investigations of Geosound

Time before the mainshock	6 – 1(hour)	1 hour – 10 minutes	10 minutes	total
Cases investigatedd	7	18	89	114
Proportion(%)	6	16	78	100

The pre–seismic geosound is different in different regions at different times and for different earthquakes, but the general feature is that the sound gradually strengthens, with distinguishable direction and continuity; it becomes loudest just prior to the forthcoming mainshock. The sound causes panic because of its difference from familiar sounds.

8.3.2 Earth Light

This is called earthquake light abroad, and it is fairly mysterious and difficult to analyze scientifically among earthquake–related phenomena. Studies have been conducted for about a decade in other countries. The phenomena were well described by T. Terada (1931) and Derr (1973). Terada said: "With regard to all these testimonies of witnesses, it must be always kept in mind that people are naturally alive to all kinds of phenomena observed at the time of a severe earthquake, and are apt to regard them as something connected with the catastrophic occurrence, while they forget that the same phenomena are frequently observed on many occasions not at all connected with earthquakes". After studying photographs of earthquake light at Matshushiro, Japan, Derr said: "The existence of luminous phenomena, or earthquake lights, is well established. The luminosity occurs in the air close to the ground, generally over certain areas in the epicentral region, principally during, but also before and after the earthquake".

Table 8.2. Earthquake Light Phenomena

earthquake \ type	advent time	light shape	light colour	light area	nature and miscellaneous
Xingtai	7 hours before the mainshock	sheet-like	white	northeast of Xingtai, low sky	non-evening sky light
Bohai	daytime, just prior to the mainshock	smoke-like about 100 meters in height	white	sea area northwest of Penglai	unclear
Tonghai	seconds before the mainshock, during earthquake	flashing, fire ball	blue, white, red	scattering	unclear
Yongshan	4 hours before the mainshock	sheet-like	fire-like		unclear
Haicheng	right before the mainshock	bright flame	purplish red	near Haicheng county	unclear
	2–3 seconds before the mainshock	belt-like	strong white light in the night sky	near Haicheng county, visual angle of 20–30°, 0.5km high	rich in magnesite, multiple ground cracks
	during the mainshock	light erupted from cracks	blue white, fire	Haicheng crack area	burnable
	before, during and after the mainshock	fire ball	like signal rocket	scattered	flammable gas
Longling	one hour–8 minutes before the mainshock	light screen	red, orange red	meizoseismal area	unclear
	10 minutes before the mainshock and during the quake	fire ball as large as basin, fire erupted from cracks	fire-like, from white to red, red	Shankou and mountain area meizoseismal area	unclear burning bed rock
Tangshan	during the quake	fire area at ground	red	local areas	crops burned
	one day before the mainshock	Bright in sea, fire-dragon like belt in sea		Qinhuangdao sea area	"S_3" sea light rarely seen in a decade
Songpan	21:00 on July 22	fire ball as big as basin, about 10m in height		Yuqian commune Jinzhu county	like organic phosphorus burning

Out of the nine great earthquakes in China, seven occurred at night, so the earthquake light has been described in many different ways. Unfortunately, no photograph was taken of any earthquake light. Based on data available at present, the particular light phenomena for each earthquake have been listed in Table 8.2 by selecting those reports deemed reliable. For further details, readers can consult the summary report for each earthquake.

The time distribution of the occurrence of various kinds of earthquake light was investigated statistically by the Working Team of Tangshan Earthquake (Table 8.3).

Table 8.3. Earth Light Investigations

Time before mainshock	6 – 1 (hour)	1 hour – 10 minutes	10 minutes	total
Cases investigated	42	54	124	230
Proportion (%)	19	23	58	100

From the above summarization, it is believable that earth light exists before earthquakes, with the light most concentrated and strongest during the earthquakes themselves, even though the cause and nature of the light is mostly ambiguous. It should be particularly noted here that the earth light is usually noiseless and heatless, although the light was reported by a few researchers to be noisy, and even to burn people. This may be an important clue for studying the nature of earth light. If the mechanism of earthquake light can be deduced, a very effective method for imminent earthquake prediction may be found by developing the corresponding monitoring methods and instruments.

8.3.3 Electromagnetic Phenomena

Like earth light, electromagnetic anomalies are also fairly mysterious. Before the nine earthquakes, many disturbances occurred to propagating waves, in particular radios were disturbed, with noise increased, frequency selection and volume character affected. In addition, the atmospheric electromagnetic field itself also exhibited anomalous phenomena.

The compass in the airplane which was in the air just above Songpan for the coming disaster relief mission was strongly disturbed tens of seconds immediately before the Songpan mainshock of M7.2 on 23 August 1976.

On the afternoon of 27 July 1976 (just before the Tangshan mainshock), a full crystal data amplifier in No. 701 Institute, Beijing, fluctuated greatly, and could not be used to process experimental data although a Faraday electrostatic shield was used.

The quality of the microwave transmission line between Beijing and Tianjin decayed seriously 2–3 days before the Tangshan earthquake.

One hour before the strong Tangshan aftershock (M6.9 Ninghe earthquake) at 2153 hours on 15 November 1976, the atmospheric potential rose significantly, as measured by an atmospheric potential monitor at Qingxian station, about 100 km from the epicenter (Fig. 8.6a). Before another strong aftershock (M6.2 Ninghe shock) on 12 May 1977, the atmospheric electric potential decreased slowly every day, and on the day of the earthquake, the normal daily variations of potential at Tianjin station completely disappeared (Fig. 8.6b).

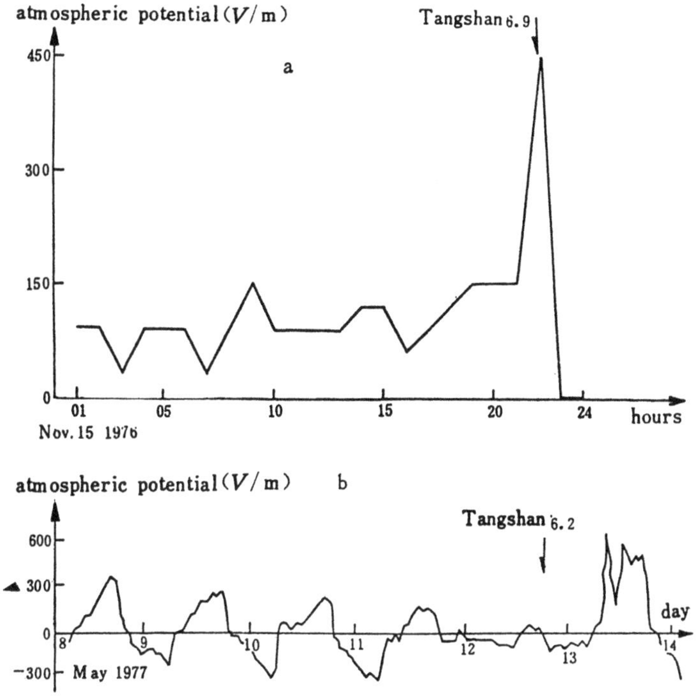

Fig. 8.6. Variations of atmospheric electric potential before strong Tangshan aftershocks; *a* Qingxian Station; *b* Tianjin Station

In July 1976, the Beijing Institute of Meteorology was using radar to monitor rain in the northwest part of Beijing. Before the Tangshan mainshock, a fan–shaped and finger–like distorted echo with intensity 65 dB, an area about 78x80 square km, about 6 km in height over the plane area, southeast of the observation point, was observed by the Institute. The shape of such an echo was obviously different from that due to geography, thunderstorm, or flying birds. This echo lasted for 2 months or so, during which time its shape did not vary much, apart from to–and–fro variations of its azimuth and area, and a slight decrease in its intensity and height. Some professionals from the National Central Meteorological Bureau were invited to investigate it on 10 September 1976 together with the photo taken for the shape of distorted echo. The analysis results showed that this echo in the SE direction was a phenomenon of extra refraction produced by a certain origin as the radar wave travelled. Why it appeared before the Tangshan earthqake, and why it came into being in the Tangshan earthquake region were not known.

The above phenomena suggest that before large earthquakes occur, there might exist a not–yet–understood physical process which either causes the nature and state of the atmosphere above the surface to change, thus influencing wave propagation, or alters the natural status of the original electromagnetic field at the surface.

Before the Tangshan earthquake, disturbances in the ionosphere were observed at the Beijing Ionosphere Observation Station. Enlightened by this fact, we investigated variations in the ionosphere shortly before other earthquakes, and found that the ionosphere above the epicentral areas was indeed sometimes disturbed for 1 day before the mainshocks. From the recordings, it can be seen that there were a great number of tiny inhomogeneous structures in the E and F layers, and the backwave was dispersed before the Tangshan and Songpan earthquakes. The disturbances appeared as multiple reflections and continuous expansion before the Yongshan, Songpan, Luhuo and Tonghai earthquakes. In addition, the daily advent time (Beijing time), t, for the maximum frequency value of the F_2 layer was clearly shifted backwards 1 day before the Xingtai earthquake. This was also the case for the Bohai and Longling earthquakes.

The above ionospheric variations might affect the propagation of radio waves, as well as produce changes in the surface electromagnetic field. However, it is not clear at present why simultaneous variations in ionosphere occurred before the earthquakes or how they could be connected with crustal movements or physical processes at the sources. During a large earthquake, the seismic surface waves could propagate upward to influence the status of the ionosphere, and such effects could extend thousands of kilometers. This was previously observed during the great 1964 Alaska earthquake of M8.4.

The electromagnetic anomalies were mostly observed within 1 day right before the mainshocks.

8.3.4 Thermal Phenomena

Pre–seismic thermal phenomena are very attractive subject matters for research. Since the Xingtai earthquake, almost every earthquake investigation has reported the sultriness felt by people, as well as other thermal anomalies. By comparing and analyzing the main thermal anomalies before the nine great earthquakes and the local weather variation data, we have found some problems which merit further study (Table 8.4).

From Table 8.4, the following three points can be appreciated:

1) The thermal anomalies reported before individual earthquakes may be related to the output of underground heat within the epicentral regions. For example, before the Longling earthquake, the temperature of Baliezhang spring increased by 10℃ or more, thus indicating the intense underground movement of heated material. As for the Xingtai earthquake, although the Handan area is further south in latitude compared with Xingtai, with the frozen soil layer much shallower than in Xingtai every year, still in March 1966 the frozen soil in Handan was 7 cm deep, while before the earthquake there was no frozen soil in Xingtai, which demonstrates that underground temperature increased within the

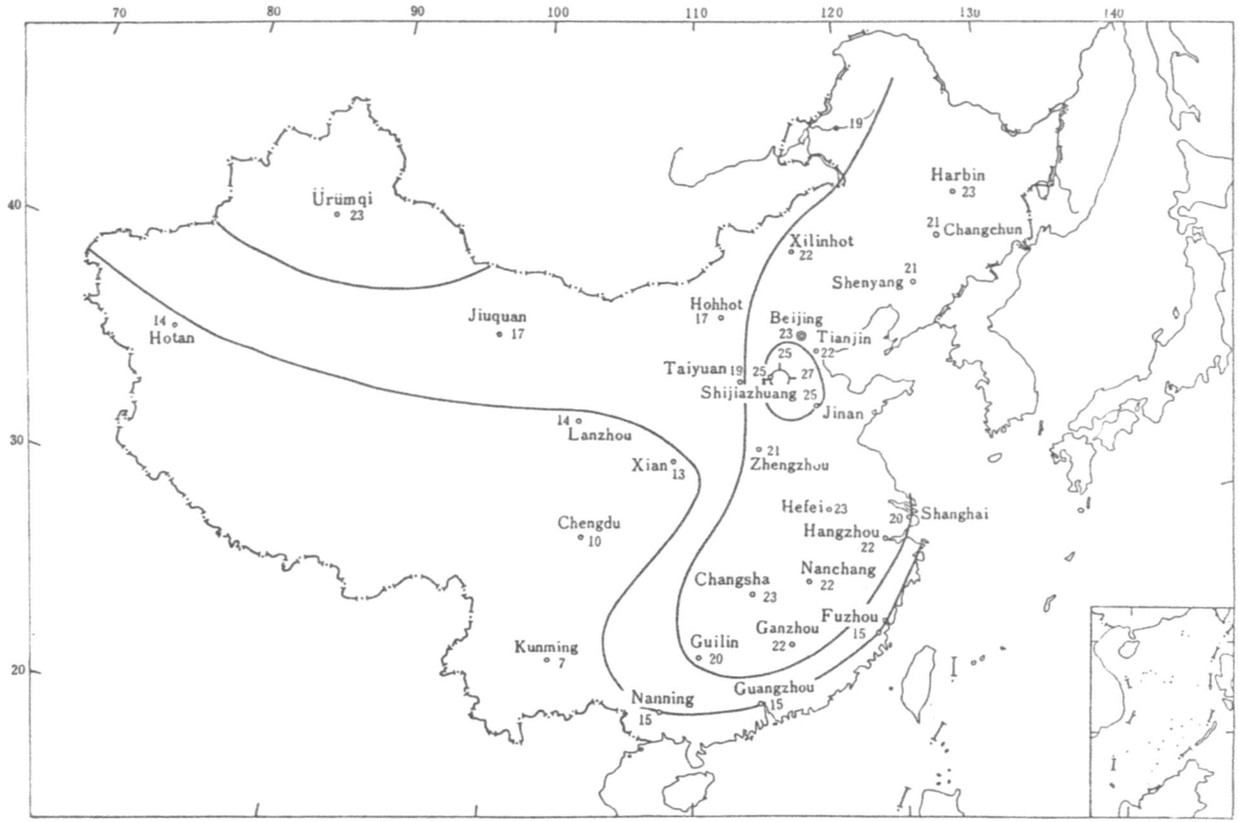

Fig.8.7. Distribution of air temperature increases (unit.℃) before the Xingtai earthquake(22 February 1966–2March 1966)

Table 8.4. Geothermal Phenomena

earthquake	reports on thermal phenomena	meteorological investigation	causal analysis
Xingtai	Spring came early, ground surface became wet	In Xingtai area, in Feb. 1966, the frozen layer was very shallow, and no frozen soil in March, forming the least days of frozen siol. From Feb. 22 - Mar. 2 1966, the temperature around the epicenter kept increasing, amounting to 27°C(Fig.8.7)	possible output of groundheat
Bohai	fog covering the small island in the sea, which appeared deformed, no reports on thermal anomalies	atmospheric pressure increased during the earthquake, temperature rose after the heavy rain	meteorological effects
Tonghai	sultry at night before the quake, thermal turbulent flow over mountains	high barometric pressure, temperature increase, thick cloud; ground(20cm in depth) temperature increased markedly (Fig. 8.8)	weather effects, but the output of underground heat could not be ruled out
Luhuo	temperature reached peak several days before the mainshock	intense low barometric pressure system moved to the epicenter	weather effects
Yongshan	especially sultry and hot days before the mainshock	atmospheric pressure low, humid, rainy, rarely seen in years, couple of days before the quake	weather effects
Haicheng	particularly hot in Jan., frozen surface melted, ice melted; grass grown on the shady mountain; temperature increased steeply and night temperature fluctuated	getting warm nation-wide in January and the monthly maximum temperature appeared first in Shengyang area since 1960. On Feb.3 1975, the strong warm low pressure moved to the epicenter, surface temperature as high as 14.8°C	weather effects (urban hot island effect), but the output of underground heat could not be ruled out
Longling	May 28-29 1976, people felt sultry and hot, ill and tired	barometric pressure minimized and then rose, less changes in temperature, and rainy. Water temperature of springs in meizoseimal area increased by 10°C or more before the mainshock	weather effects, possible output of underground heat
Tangshan	warm winter for 1975-1976, few reports on thermal anomalies before the mainshock	nationwide warm winter plants blossomed and bore fruits twice in general. Atmospheric pressure high, cool and rainy before the mainshock. The ground(0.8m in depth)temperature was lower than usual, but increased suddenly three days before the mainshock, with its center at the Tangshan area	weather effects, the cause of the ground(0.8m)temperature increase was not clear
Songpan	few reports on thermal anomalies before the mainshock	high pressure, low temperature, and low humidity	weather effects

epicentral area.

2) The thermal anomalies before some earthquakes might be associated with weather effects, for example the Luhuo and Yongshan earthquakes. However, the output of underground heat could not be ruled out for the thermal anomalies before the Tonghai and Haicheng earthquakes.

3) There were few reports on thermal anomalies in case the barometric pressure was high, temperature low and humidity low for weather conditions, being identical with the natural status. This was the case for the Tangshan and Songpan earthquakes, and for the Bohai earthquake.

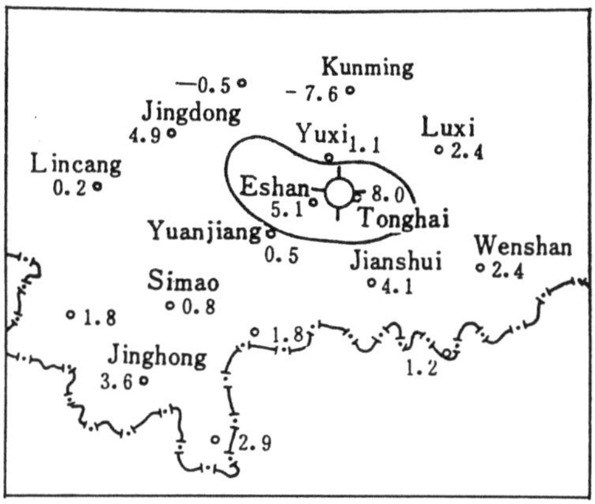

Fig. 8.8. Underground (10 cm in depth) temperature anomalies and distributions (unit in ℃) before the Tonghai earthquake (1969)

8.4 Abnormal Behaviour of Animals

Pre–earthquake abnormal animal behaviour has been documented both at home and abroad, in the past and at present, and has been used as a precursor for preventing seismic disasters. As the nine great earthquakes occurred in densely populated urban and rural areas, the behaviour of various kinds of animals was observed continuously. The investigations have considerably enriched the descriptions of such "strange" phenomena, and the scientific value has been improved by the participation of some biologists in field survey and experimental research.

It can be said that abnomal animal behaviour was the most general, vivid, realistic and swift macroscopic phenomenon observed before the nine earthquakes. In accordance with work done by the Biophysical Institute of the Academy of Sciences of China, and by Jiang Jinchang and the summary reports of each earthquake, we have summarized the pre–seismic animal anomalies as follows:

1) Before land earthquakes, the abnormal responses were most usually shown by cattle, dogs, rats, fish, chickens, birds (doves and parrots), cats, pigs, snakes, and so on, while before marine earthquakes, varieties of sea animals, seagulls and so on behaved anomalously, in addition to anomalies of land animals along the coastline.

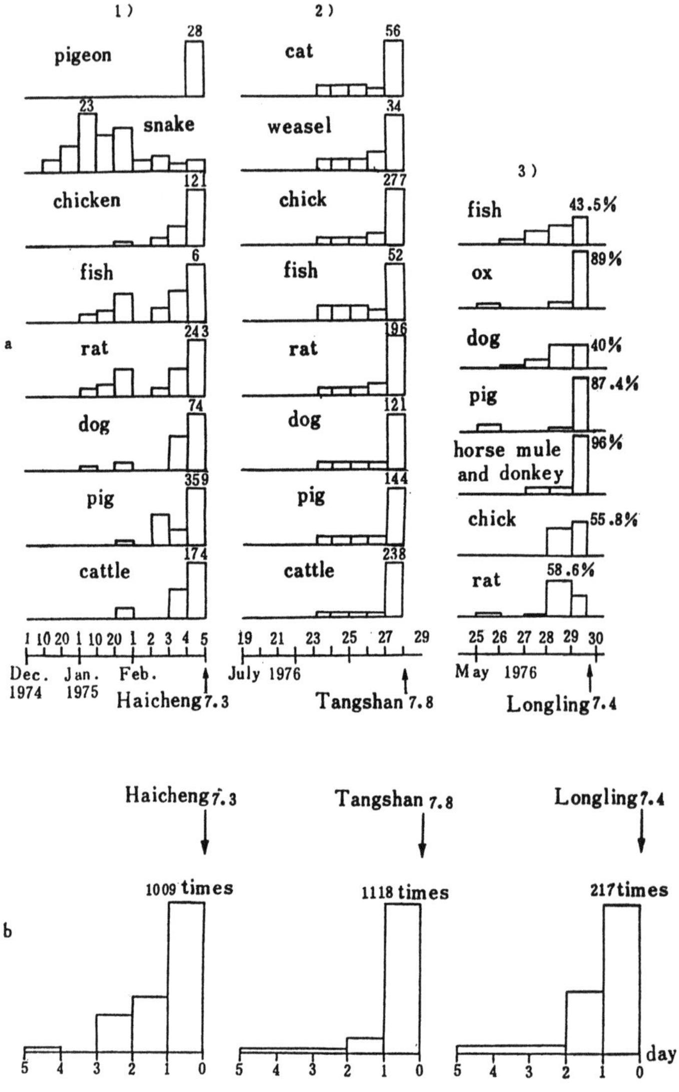

Fig. 8.9. *a.b.*Time distributions *(a)* and sum distributions (times / day) of various animal anomalies before some great earthquakes *(b)*; *1* Haicheng quake; *2* Tangshan quake; *3* Longling quake

2) Abnormal animal behaviour, as a typical feature, spread panic, for example penned cattle jumping out of their stable in terror, dogs barking and weeping madly, rats running away in flocks in daytime, hibernating snakes coming out of their holes, and goldfish jumping in frenzy; and also panicky constraint, for ex-

ample certain animals flocking together foolishly without moving when encountering people, allowing themselves to be easily caught. Such uneasiness and panic of animals would appear suddenly and die away rapidly, lasting only minutes or hours, and scarcely seen for days. Here it should be particularly pointed out that pre–seismic abnormal animal behaviour was different in different cases. The same animal did not have the same anomalous response to different earthquakes. There were various special phenomena before earthquakes in different seismic regions. For example, before the Tangshan earthquake, weasels ran away in flocks; before the Songpan earthquake, frogs gathered together and migrated; before the Haicheng earthquake, hibernating snakes came out; before the Yongshan earthquake, the bees flew away from the bee–hives; before the Bohai earthquake, the birds and animals in Tianjin Zoo became panicky and frightened and the little jellyfish floated on the water. Naturally, these phenomena could be related to the differences in wildlife conditions in the various regions.

3) The main animal anomalies seemed to be concentrated within a day before the mainshocks (Fig. 8.9), mostly within 11 h before them.

4) The areas in which the animal anomalies appeared were about 100 or 200km away from the forthcoming epicenters, or as far as several hundred kilometers away. Besides, the spatial distributions of animal anomalies were inhomogeneous:

a. The anomalies were distributed in stripes along the strikes of the earthquake– generating fault zones. For the Haicheng earthquake, over two –thirds of animal anomalies were focussed along the NE Panjin–Xiuyan–Dandong zone, nearly consistent with the strike of the earthquake–producing tectonic zone.

b. The animal anomalies were mostly focussed in areas of high seismic intensity, as for example the Longling and Tangshan earthquakes. About 70% of animal anomalies before the Tangshan earthquake were concentrated in areas with seismic intensity over 7 ° .

c. The animal anomalies were distributed along the main active faults, and migrated toward the epicenters. Before the Songpan earthquake, there were three periods of " high tide" of animal anomalies and other macroscopic anomalies since mid–June, with the locations centered along the Longmenshan fracture zone. These moved gradually toward the epicenter, and on August 14 (2 days before the mainshock) a great number of animal anomalies appeared in the area very close to the epicenter.

5) The inhomogeneity and temporal concentration for the distribution of animal anomalies before the earthquakes were roughly in agreement with the sudden intense variations in the natural electric field, electromagnetic wave disturbance, ground water, escape of ground gas, geosound, earthquake light and existent conditions, indicating that there might exist a physical or chemical process which stimulated the nervous organs and bodies of animals to behave unusually. Certainly, individual animal anomalies or group anomalies of the same animal were by no means unique phenomena before the earthquakes. Animals may also behave abnormally under weather conditions; this is the crux of this complicated problem. However, the practical significance of animal anoma-

lies for predicting earthquakes furthering can be improved by the analysis of group animal anomalies, based on regional statistics.

The four main kinds of atmospheric anomalies have been described above, neglecting several other anomalous phenomena that have also been observed. Discussion of these anomalies, their possible association with earthquake occurrence, and the investigations of the causal mechanism will be important matters for further research in the future.

CHAPTER 9

Large–scale Regional Anomalous Phenomena

Before and after the nine earthquakes, some anomalies were also observed within much larger regions, with some anomaly points having epicentral distances of over 1000 km, in addition to the anomalies discussed earlier with ranges of 200–300 km. As such long–distance anomalies before large earthquakes are sporadic, little research has been done on the relation between them and the earthquakes, and they have often been neglected or doubted. However, the data obtained from the nine great earthquakes have revealed that the large–scale regional anomalies may contain information on seismogenic processes. Therefore, in our opinion, study of the time and space distributions before large earthquakes should not be confined to the epicentral areas and their vicinities, but to much larger–sized regions. The following will mainly discuss the data obtained from the North China seismic region.

9.1 Large–scale Regional Connections of Seismicity

9.1.1 Large–scale Regional Seismicity Migration

1) Spatial contraction of large regional seismicity before major earthquakes.

Some pertinent data on pre–seismic minor seismicity have been introduced for the nine earthquakes in Chap. 4. Before a large earthquake, there were anomalies in frequency of occurrence, energy release and b values of minor earthquakes within a 200–300 km range. There might also exist a relationship between the minor seismicity within a larger regional range and the seismicity around the forthcoming mainshock epicenter, under the action of a large regional tectonic stress field. With the development of the seismogenic process and tiny variations in the large regional stress field, the small earthquake activities within the large regional range would display an anomalous distribution pattern. For example, the earthquakes with $M_L > 4$ within the northern part of North China (37 ° –42 ° N, 116 ° –123 ° E) were concentrated toward the epicenters before the Haicheng and Tangshan earthquakes. From the distribution of earthquakes with $M_S > 4$ during 1966 through 1978, the earthquakes were mainly concentrated in the northwest part of Beijing and the Hejian area from 1966 to 18 July 1969 when the Bohai earthquake occurred; after the Bohai earthquake and up to 1972, they were focussed in the Bohai, Tangshan and Liaodong Peninsula regions; from 1973 to 1974, they were further centered around the Tangshan and Haicheng epicentral areas (see Chap.4) and minor earthquakes became frequent, particularly in the south Liaoning region, where about 580 earthquakes were recorded in 1974. Meanwhile, seismic activity outside of the Tangshan and Haicheng areas was relatively calm. Therefore, it is suggested that the earthquake activities within the large regional range migrated toward the hypocentral area from outside before these two great earthquakes.

Statistical studies were made by Jian Xiue on the latitude dependence of earthquakes with M over 4 within the North China region (32 ° –43 ° N, 108 ° –124 ° E) from 1964 to 1976 (Fig. 9.1). From this figure, it is seen that earthquakes were distributed in different latitudes during different periods, and approached the latitudes of the mainshock epicenters gradually before the Xingtai, Haicheng and Tangshan earthquakes. This not only illustrates the connections between the various seismic activities within the large–scale range, but also demonstrates the effects of the latitudinal seismic zone on the earthquakes (see Chap. 3).

The above–mentioned data imply that the minor seismicity within the large–scale regional range migrates toward the forthcoming mainshock epicenter and its nearby latitudes before large earthquakes in North China. This is consistent with the results obtained in Chap.4 on the migrations of small earthquakes within small areas toward the hypocenter.

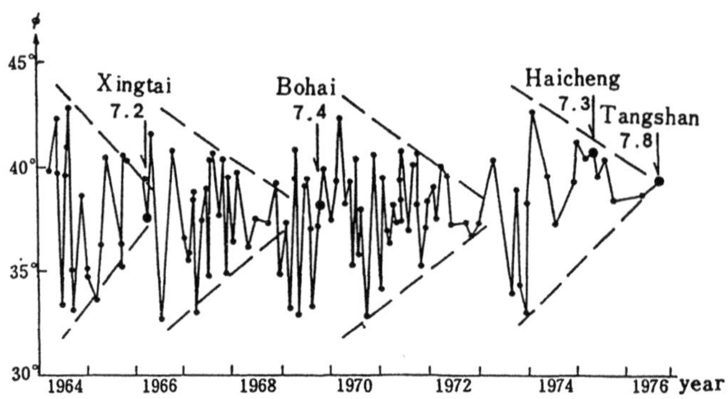

Fig. 9.1. Latitude–dependent distributions of earthquakes of M_S>4 in the North China region

2) Spatial dispersion of seismic activity after large earthquakes.

It is noticeable that small earthquakes diffused outward from the mainshock epicenters after large earthquakes. The epicentral distribution of earthquakes with M_L>4 from August 1976 to December 1978 after the Tangshan earthquake is given in Fig. 9.2. The earthquakes spread from the mainshock epicenter in a NW direction from Ninghe to Shanhe and in a NE direction from Qianxi, Hebei Province to Chaoyang and Beipiao, Liaoning Province. The Baodi County, Tianjin M5.5 earthquake of 24 August 1976, and the M5.1 Chaoyang, Liaoning Province earthquake of September 1976 are examples. In addition, an M 4.5 earthquake occurred in Xinghua, Jiangsu Province. These earthquakes took place in tectonic localities associated with the geological structures of the Tangshan earthquake region, and closely followed the Tangshan mainshock. Therefore it may be said that these earthquakes were produced due to the adjustments of the stress field after the Tangshan earthquake, and they can be termed "tractional" earthquakes.

9.1.2 "Window Effects" of Large Regional Seismicity

Wang Zegao (1979) studied fluctuations in aftershock activity of the Xingtai earthquake in the recent decade or so. He suggested that the frequency and energy release of the Xingtai aftershocks were closely associated with the strong seismicity in the whole North China region. The aftershock rate generally attenuated exponentially, but the aftershock frequency was conspicuously higher than the normal background value. It seemed that the Xingtai aftershock activities may mirror the changes in stress state of the North China region, hence playing a "window" role in monitoring the large regional seismicity.

9.1.3 Correspondence of Seismicity Between Large–scale Regions

Sun Jialin analyzed the recent seismicity in the Helingeer and Bayinmuren areas, Inner Mongolian Autonomous Region, and Tanghsan and other areas, all with latitude close to 40 ° N. He suggested that remarkable correspondence existed between the seismic activities in these areas although the distance between them reaches 1200 km (Fig. 9.3).

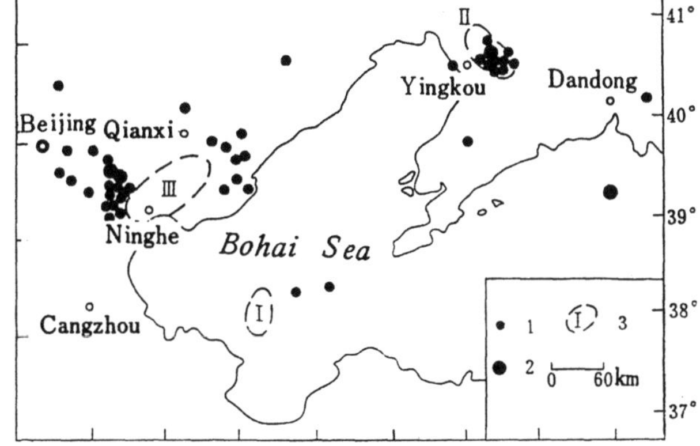

Fig. 9.2. Epicentral distributions of earthquakes with $M_L > 4$ after the Tangshan mainshock (Aug. 1976 – Dec. 1978); *1* earthquakes of $4 < M_L < 5.2$; *2* earthquakes of $5 < M_L < 6.5$; *3* aftershock areas (I. Bohai, II. Haicheng, III. Tangshan)

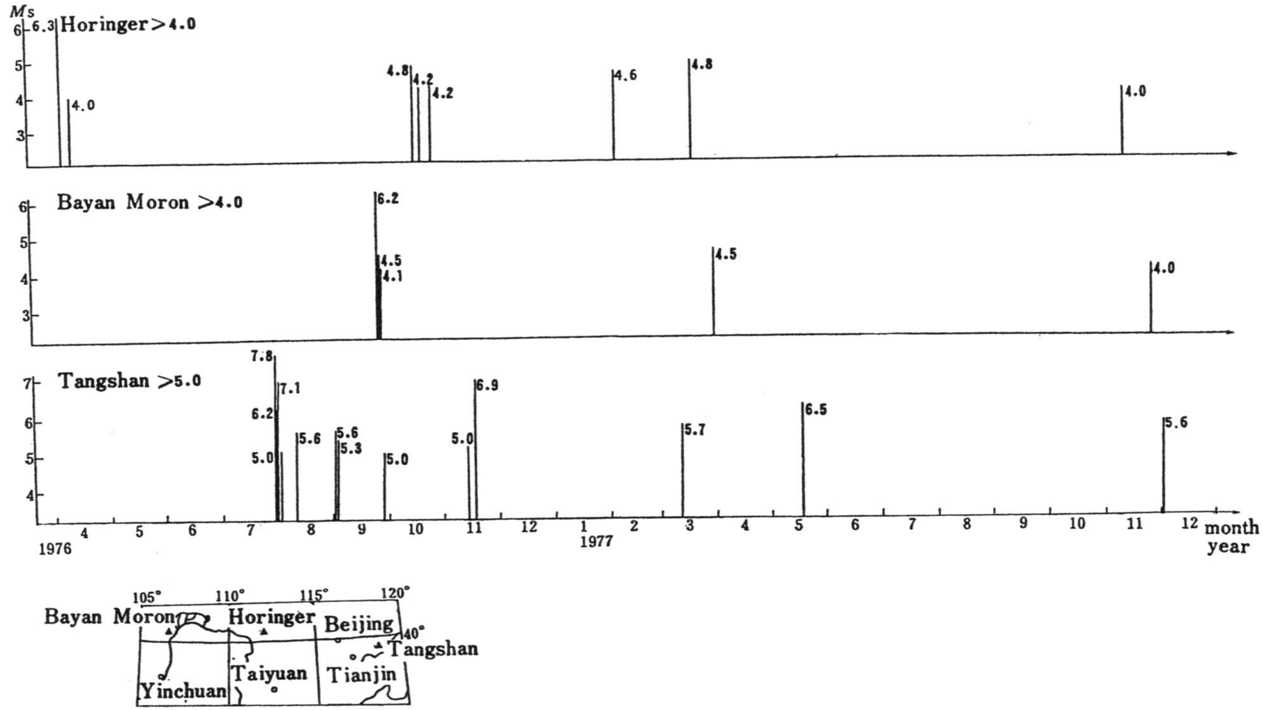

Fig. 9.3. Correspondence between earthquakes in Helingeer and Bayinmuren and Tangshan areas

The seismicity level is fairly low in the Northeast China region, but there also existed a temporal correspondence between the $M_S > 3$ seismicity along several NE tectonic zones, such as the Fuxin–Zhenan tectonic zone, the Baicheng–Xiqi tectonic zone, and the Yinan–Shunan tectonic zone [as represented by No. (1), (2) and (3) in Fig. 9.4]. In addition, a certain correspondence exists between the earthquakes in these areas and the Tangshan aftershocks (Fig. 9.4).

Such correspondences between the earthquakes in a large–scale region indicate correlations in the stress variatios in various areas under the influence of driving forces. Correspondences between the historical earthquakes will be further discussed in Chap. 13.

9.1.4 Consecutive Occurrences of Strong Earthquakes in a Large Region

Li Ziqiang et al. (1980) and Li Qinzhu et al. (1980) studied the clustering of strong earthquakes in the North China region. We shall divide into several groups of so–called "consecutive" earthquakes those of them which were close to each other in space and time, as follows:

the first group: Xingtai (March 1966, M = 7.2)— Hejian (March 1967, M = 6.7)—Bohai (July 1969, M = 7.4);

the second group: Haicheng (February 1975, M = 7.3)— Helingeer (April 1976, M = 6.3)—Tangshan (July 1976, M = 7.8);

the third group: Tonghai (January 1970, M = 7.7)—Dayi (February 1970, M = 6.3);

the fourth group: Longling (May 1976, M = 7.4)—Songpan (August 1976, M = 7.2)—Yanyuan (November 1976, M6.9).

There exist certain similarities in rupturing mode, kinematic features, precursory phenomena and other aspects for each of these groups. For example, the three earthquakes in the first group occurred in this order from west to east at the intersection points along the latitudinal tectonic zone and the NE tectonic zone in the central part of North China, and the source ruptures were right–lateral dislocations of NE orientation, showing the unity of mode for the causative stresses. In the second group the earthquakes all occurred within the Yinshan latitudinal tectonic zone, and the ranges of the seismic anomalies and precursory anomalies related to them overlapped each other. As said before, a fairly close correspondence seems to exist between the Helingeer earthquake and the Tangshan aftershocks. Earthquakes in the third group took place one after another on the east side of the North–South seismic zone, with an interval of only 50 days between each. The fourth group of earthquakes occurred at the Northeast seismic zone from Longling to Songpan, before which the regional seismic activities were apparently focussed at this seismic zone.

Therefore, it can be seen that each individual earthquake of the consecutive groups might not be an independent event. Rather, they could be commonly developed and intrinsically related events, originating in the large–scale unified stress field.

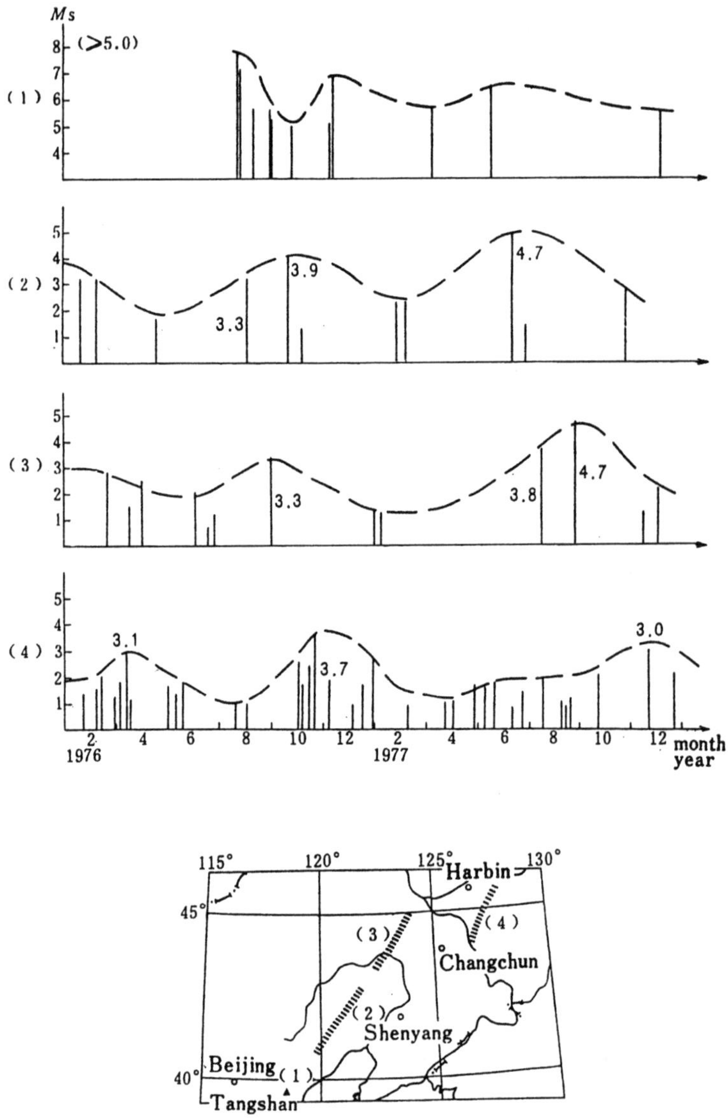

Fig. 9.4. Correspondence between earthquakes in the Northeast China region; *1* Tangshan seismic area; *2* Fuxin–Zhenan tectonic zone; *3* Baicheng–Xiqi tectonic zone; *4* Yinan–Shunan tectonic zone

9.2 Large–scale Regional Microscopic and Macroscopic Anomalous Phenomena

9.2.1 "Sensitive Points" Phenomena in Precursory Observations

As described in Chap. 4 through 8, relatively concentrated anomaly areas appear in the epicentral regions before the mainshocks. The anomaly sites were distributed unevenly in these areas, and were not dependent only upon the epicentral distances. For example, before the Haicheng earthquake, obvious anomalies were observed at Jinxian deformation station about 220 km away from the epicenter; anomalies were observed at Dahuichang deformation station in the Beijing region ($\Delta = 180$km) before the Tangshan earthquake; anomalies in radon content and tilt were observed at Kangding station before the Luhuo earthquake, and radon anomalies were observed in the Tianjin region before the Bohai

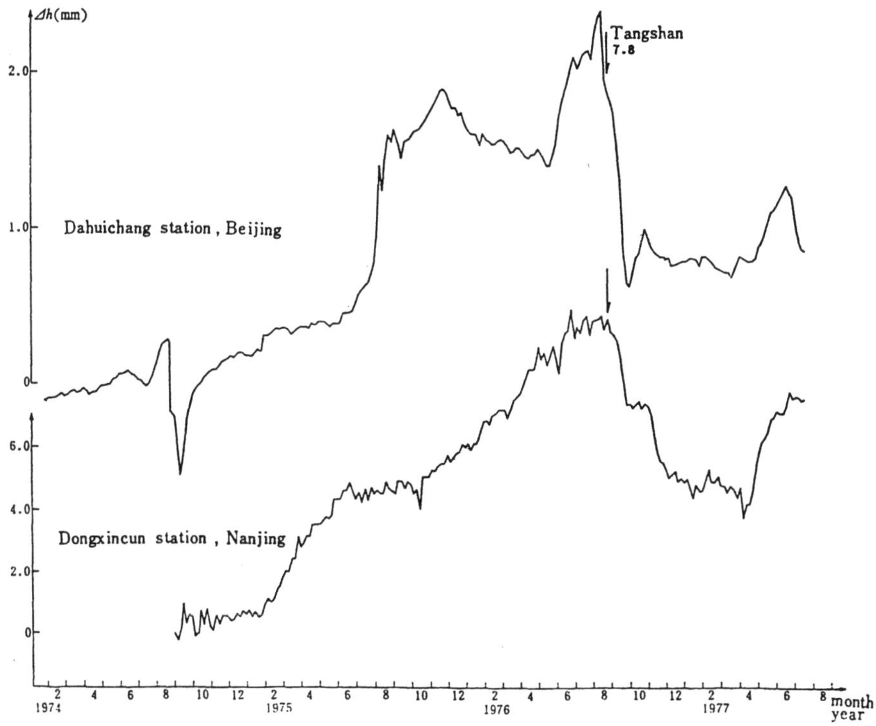

Fig. 9.5. Comparison between the short–levelling measurements made at Dongxincun station at Nanjing and Dahuichang station in Beijing

earthquake. In addition to the anomaly points about 200–300km from the epicenters, there were also some sporadically distributed anomaly points within much larger regions. For instance, anomalies similar to those observed at Dahuichang station were also observed at Nanjing short–levelling station, Jiangsu Province, about 900 km from the epicenter, before the Tangshan mainshock (Fig. 9.5).

Also before the Tangshan earthquake, radon anomalies were observed at Qingjiang station, Jiangsu Province, about 640 km from the epicenter, with an amplitude no smaller than that in the epicentral area. Before the Haicheng earthquake, radon anomalies appeared at Huhehot station, Inner Mongolia, about 840 km from the epicenter, and a slow drop of resistivity was observed from 1972 till 1974 at Baodi station, Beijing region, about 510 km from the Haicheng epicenter. (This possibly indicates the influence of the seismogenic preparation of the Tangshan earthquake.) The resistivity was rapidly recovered after the Haicheng earthquake (Fig. 9.6), mirroring the influence of the Haicheng earthquake (for explanation see Chap. 12). It then continued to fall until the Tangshan earthquake. Besides, sudden variations of radon content were observed at Anguzhuang station, Tangshan City, after the Haicheng earthquake occurred. Such distant anomalous phenomena were also recorded at Huailai tilt station, Hebei Province ($\Delta = 1750$ km) before the 1973 M7.9 Luhuo earthquake and tilt anomalies were also observed at Hongshan station, Longyao County, Hebei Province ($\Delta =$ 1800 km) before the 1970 M7.7 Tonghai earthquake. In some foreign reports, the relatios between precursory variations of vectors of some tiltmeters in Japan and earthquakes in the Xizang (Tibet) and Taiwan regions of China were also discussed (Nishimura 1953).

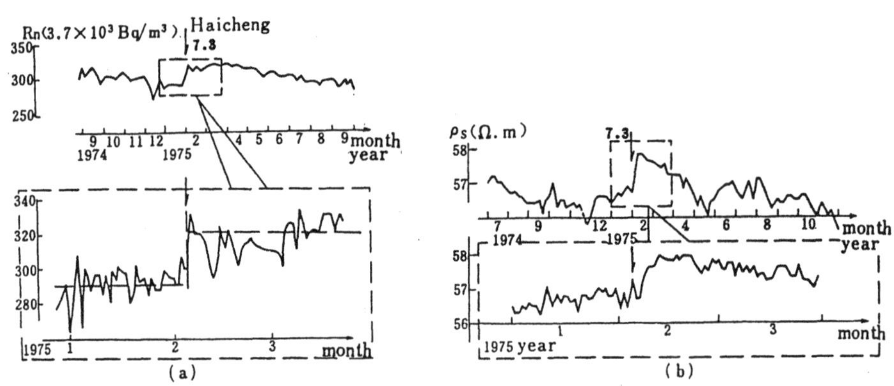

Fig. 9.6. Resistivity and radon content observations after the Haicheng earthquake; *a* variation of radon content at Anguzhuang point, Tangshan City; *b* variation of resistivity at Baodi station

We have termed the conspicuous anomaly points that occur far from the epicenter during the seismogenic preparation and occurrence processes of an earthquake, the sensitive points of the earthquake. This term implies that these points in a given region are more sensitive to stress changes than others surrounding them during the seismogenic preparation and occurrence processes, or that these points are more prone to high stress accumulations than others in the same region. Generally speaking, no other anomalies might be observed in the region, but there might exist certain undetected connections within some particular tectonic localities or the deep crust. The stress variations pertinent to the seimogenic process might then be observed at a distant point, making this point a sensitive one responding to the seismogenic process. The same point can be sensitive to one earthquake, but not to another, because whether this point could be a sensitive one is related to the position of the point with respect to the hypocenter and to the stress state in the focal region, as well as to the geological structures and deep structural conditions at the point (see Chap. 12).

From the data of the nine great earthquakes, the sensitive points or "windows" seem to be located at certain peculiar tectonic segments, or at the sites of past large earthquake epicenters.

The above–mentioned distant anomalous phenomena may naturally be doubted, or may be interpreted as the result of nonseismic factors. However, considering that such anomalies repeated before and after many earthquakes, and taking into account the close temporal relationship between the anomalies and the earthquake occurrences, such phenomena are still worthy of description and analysis.

9.2.2 Distribution of Large–regional Macroscopic Anomalies

1) Ground fissures in large–regional range.

Macroscopic anomalies are mainly concentrated around the epicenter before and after large earthquakes. However, based on observations of the past 10 years or so, macroscopic anomalies also occur in the large–scale range. Taking the North China region as an example, about 20,000 aseismic ground fissures were observed in this region from 1974 to 1975, distributed in Henan, Anhui, Shandong, Hubei and Hebei Provinces, mainly along the Liaokao, Tanlu and Haian–Yangzhou active faults. Obvious hints at tectonic movement might be also given by fissures in some of these areas. Then the Haicheng earthquake occurred at the north segment of the Tanlu fracture zone, and the Tangshan earthquake occurred at the extension of the Liaokao fracture zone. These macroscopic anomalies might be indicators of large–regional crustal tectonic movements before these earthquakes. The cracking phenomena died away gradually as the seismic activity decayed after the Tangshan earthquake.

2) Macroscopic anomalies in large–scale range.

Figure 9.7 shows the frequency of macroscopic anomalies in seven provinces and municipalities in East China from 1974 to 1977. The figure shows various macroscopic anomalies observed on many separate occasions in Liaoning and Shandong Provinces before the Haicheng earthquake, and in Jilin and Liaoning Provinces before the Tangshan earthquake. In addition, various macroscopic anomalies appeared in a large–regional range after the Haicheng and Tangshan earthquakes. Particularly after the Tangshan earthquake, the anomalies extended southward into Hebei, Shandong and Jiangsu Provinces, and northward into Liaoning and Jilin Provinces.

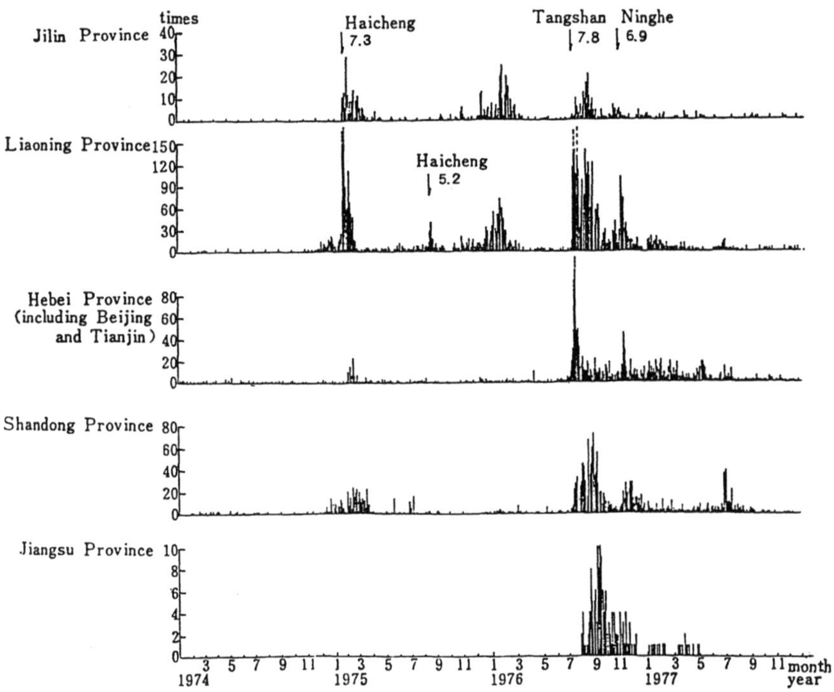

Fig. 9.7. Frequency diagram of macroscopic anomalies in seven provinces and municipalities in East China from 1974 to 1977

The areas of anomaly concentration before and after the Haicheng and Tangshan earthquakes are consistent with certain tectonic zones of large–scale and intense activity in East China, such as the NE Tanlu and Cangdong tectonic zones and the NW Chaoyang–Dandong tectonic zone. It seems that these tectonic zones have been not only the sites of intensive crustal movements in East China in recent years, but are also the segments which are most sensitive to pre– and post–seismic anomalies and at which the anomalies are most concentrated. Other areas of anomaly concentration are: Heze area in Shandong Province, Yangzhou area in Jiangsu Province and Shulan area in Jilin Province. Within the relatively stable blocks bordering active tectonic zones, the seismicity level is low, and the anomalies associated with large earthquakes are seldom seen.

A concentrated zone of macroscopic anomalies roughly in conformity with the earthquake–producing tectonic zones, with some other sensitive areas and points in addition, constitutes a complicated pattern of unevenly–distributed anomalies in a large–scale region before and after a large earthquake. This illustrates that the Tangshan and Haicheng earthquakes cannot be considered as the result of movements of the given fault segments alone, but are possibly the products of overall crustal movements in a large–scale region involving numerous principal tectonic zones in East China.

9.2.3 Spatial Migration of Large–regional Macroscopic Anomaly Areas

1) Preseismic macroscopic anomalies migrated toward epicenters from outside.

In the spring of 1976 before the Tangshan earthquake, along the narrow zone of NNE extension through Tangshan, there were great numbers of

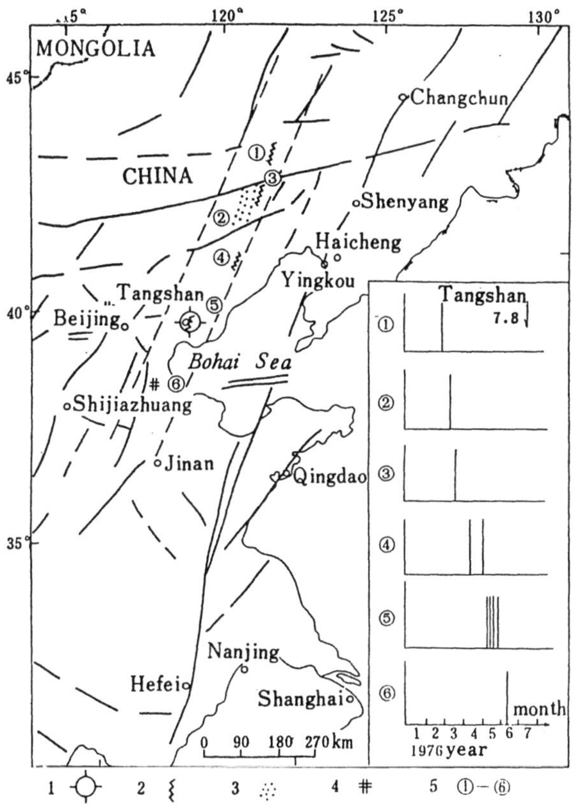

Fig. 9.8. Time–space variations of some major macro scopic events before the Tangshan earthquake; *1.* mainshock (M7.8) epicenter; *2* ground fissures, mine bursts and well collapses; *3* felt ground shaking; *4* upwelling of waste wells; *5* (1)–(6) locations and time order of macroscopic events

macroscopic anomalies, mainly of ground water (gas). There also appeared other macroscopic anomalies such as felt ground shaking, ground fissures, and mine bursts and collapses. Figure 9.8 gives the following major macroscopic anomalous events:

a. Fairly sizable NE ground fissures appeared first on 8 March 1976 at Kulunqi, south of Tongliao City, Jilin Province, about 500 km from Tangshan. b. Many earth shakings occurred in the Chaoyang area, Liaoning Province. The strongest one, on 11 March 1976, was felt in 22 people's communes, in three counties, over an area of about 1680 sq km. c. On 14 March 1976, a strong coal burst happened in Beipiao Coal Mine, Liaoning Province, with a great deal of gas emitted, causing serious casualties. d. From March–April 1976, local collapses occurred at Binggou Coal Mine, Jianchang County, south Liaoning Province. Furthermore, the time of large collapses in this mine was fairly coincident with the time of minor seismic activity in Chaoyang area before the Tangshan mainshock, and the number of local collapse events in this mine was related to the fluctuations in the Tangshan aftershocks (Fig. 9.9). e. In May 1976, the seismic observation well was deformed at No. 10 Middle School of Tangshan City near the Tangshan earthquake fault (No. V fault of Kailuan Mine). The cement well wall was broken, crumbled off fractionally, and a great deal of soil and sand poured in, 3 days before the Tangshan mainshock. A seismic fault appeared near the well after the earthquake, with a horizontal dislocation amounting to 1.5 m.

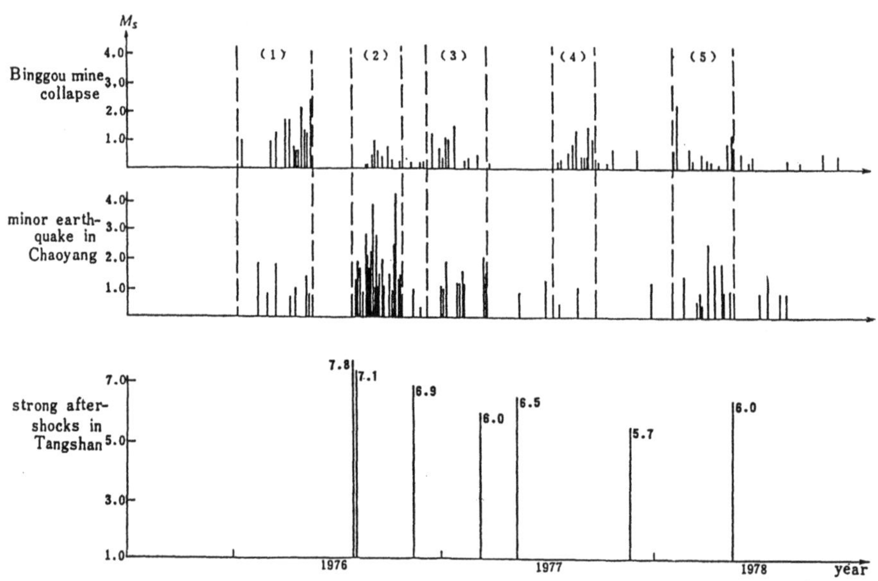

Fig. 9.9. Comparison between the collapses of the Binggou Coal Mine, Jianchang County and earthquake activities in the Chaoyang and Tangshan areas; *1–5*the active times of minor earthquakes in the Chaoyang area and of collapses of the Binggou Coal Mine

Meanwhile, in May 1976, ground fissures appeared in Wangtan area, Luoting County, Tangshan region, causing the irrigation channel to be displaced, and a horizontal displacement of 0.8 m appeared after the earthquake. f. Finally, in mid–June 1976, waste–oil wells erupted in Qingxian, Hebei Province. From Fig. 9.8 it is seen that all these events occurred several months before the mainshock, mainly concentrated along the narrow zone of northward extension of the Tangshan earthquake fault. Analysis of the order of occurrence of these events shows they first appeared in outside areas, then approached the epicenter gradually, and were concentrated around the epicenter roughly in April–May 1976. This time is consistent with that of the short–term anomalies recorded in the Bei jing–Tianjin region around Tangshan. Finally, a large number of macroscopic anomalies appeared successively in Tangshan and its surrounding areas.

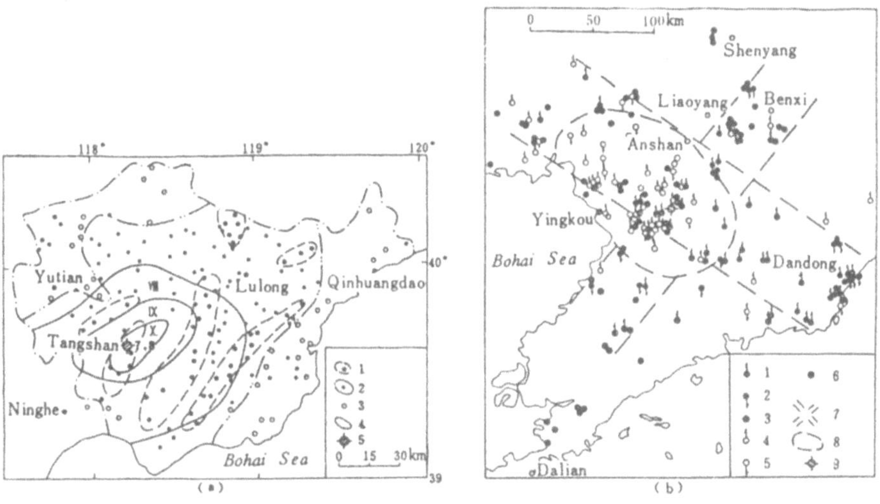

Fig. 9.10. *a.b.*Migration of the short–term and impending anomalies of ground–water level toward the epicenters from the outside before the large earthquakes; *a* The Tangshan earthquake: *1* the anomalous range 16 h before the main shock; *2* the anomalous range 1–2 days before the mainshock; *3* the anomalous range 3–5 days before the mainshock; *4* isoseismal line; *5* the mainshock epicenter; *b* The Haicheng earthquake before Jan. 1975: *1* rise; *2* fall; *3* muddled, from the end of Jan. 1975 up to the time of the mainshock; *4* rise; *5* fall; *6* muddled; *7* the anomalous range; *8* boundary between rise area and the fall area; *9* the mainshock epicenter

The distribution of macroscopic anomalous points in ground–water level also migrated toward the epicenters from outside before the Tangshan, Haicheng and Songpan earthquakes. Figure 9.10 shows the case before the Tangshan and Haicheng earthquakes. From this figure it is seen that the anomalies were principally focussed in the Chaoyang, Benxi, Dandong and Luda areas before the end

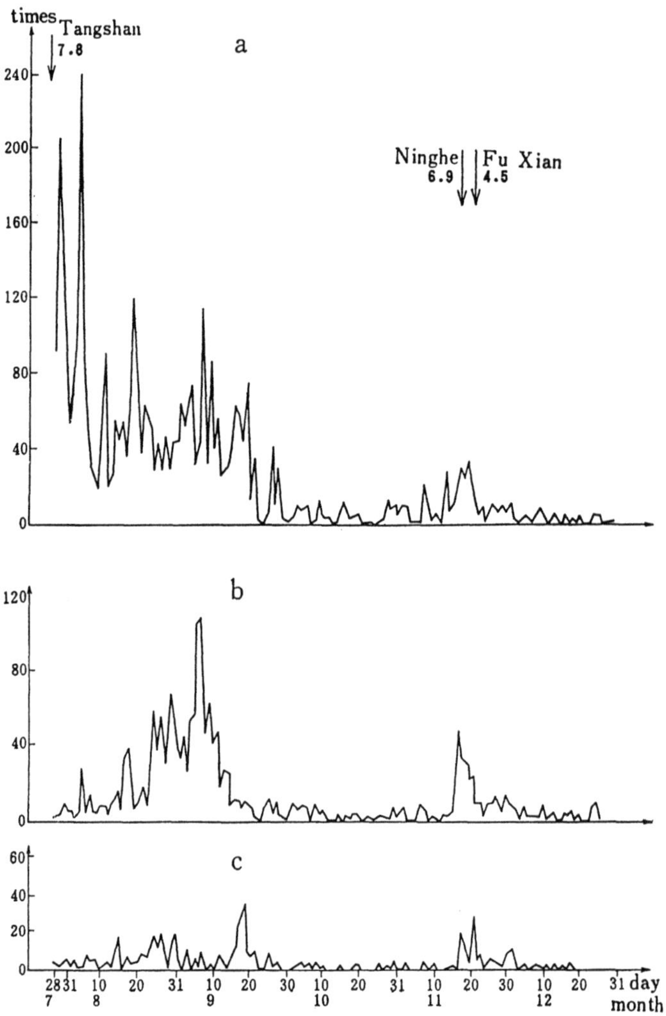

Fig. 9.11. Spreading of macroscopic anomalies in Liaoning Province after the Tangshan earthquake; *a* the west part of Liaoning Province; *b* central part; *c* eastern part

of January 1975. The ground–water level anomalies were mainly concentrated in the Yingkou, and Haicheng areas from the end of January to 4 February 1975 (1 day before the mainshock). Before the Tangshan earthquake, the anomalies first appeared in the Yutian, Qinglong, Jianchang and Changli areas (Fig. 9.10a), then migrated gradually from the outside towards the epicenter several days before the mainshock, and were mainly focussed in the seismogenic fault of the Tangshan quake and in some nearby NE faults 16 h before the mainshock. The migration of the macroscopic phenomena before the Songpan earthquake will be dealt with in Chap. 10. The data on migrations of macroscopic anomalies of ground–water level prior to the Haicheng, Tangshan and Songpan earthquakes have shown that the migration velocity of imminent macroscopic anomalies before these earthquakes were 2–3 km / day, 2–3 km / day and 3–6 km / day, respectively.

2) Outward spreading of post–seismic macroscopic anomalies.

The fact that the regional macroscopic anomalies following the Tangshan and Haicheng earthquakes were diversified and widely distributed is discussed in this section. Furthermore, macroscopic anomalies are also spread outward from the epicenters. Figure 9.11 compares the time–dependence of macroscopic anomalies in west, central and east Liaoning Province, closest to the Tangshan region, following the Tangshan earthquake.

Fig. 9.12. Shift of areas of artesian water wells before and after the Tangshan earthquake; *1–6* several minutes before the mainshock to hours before the mainshock; *7–8* and *22* July 28; *9–10* July 29 to 30; *11–17* July 31 to August 2; *18–20* August 13 to 16; *21* Sept.1; *13* Sept.7

From this figure the initiation time of the anomalies and the time when large quantities of the anomalies appeared can be seen. The anomalies appear earlier close to the mainshock epicenter, with later appearances at greater epicentral distances. Furthermore, the macroscopic anomalies also spread out gradually after some strong aftershocks in the Tangshan earthquake area, though the ranges and quantities affected by them were less than those of the mainshock. After the Tangshan earthquake, the tractional earthquakes outside the Tangshan area very often occurred at the time when the macroscopic anomalies reached a peak, or slightly later. Within a certain time period after the Tangshan earthquake, the macroscopic anomalies had expanded into many provinces and cities in the eastern part of China as the Tangshan aftershocks fluctuated. Such post–seismic anomalies were one of the main reasons why so many people in many provinces stayed outdoors after the Tangshan earthquake.

Considering that these macroscopic anomalies may be caused by many possible factors, an easily identified phenomenon is chosen for comparison in order to confirm the objectivty of the macroscopic studies. Figure 9.12 gives the order of outward migration of automatic self–flow of ground–water from the epicentral area after the Tangshan earthquake.

Table 9.1. The time–space migration of macroscopic anomalies before and after the Tangshan earthquake

time	direction	earthquake	Observed data	apparent speed (km/day)	remarks
before earthquake	Towards epicenters from outside	Tangshan earthquake	1. time difference between the appearence of anomalies in the epicentre and that of anomalies in outside areas before the impending main-shock.	5 - 20	
			2. time differences among "high-tide" periods of anomalies in Beipiao, Chaoyang and Jianchang areas from Feb. to April 1976.	2 - 3	
		Haicheng earthquake	1. time differences between high-peak of anomalies in Dandong area and that of anomalies in Haicheng area.	3	
			2. time differencestween high-peak of anomalies in Panjin area and that in Haicheng area	2 - 3	
		Songpan earthquake	time differences between high-peak of anomalies in Qionglai area and that in Beichuan area	1 - 3	
after earthquake	outward from epicentral areas	Tangshan earthquake	1. time differences between high-peak of anomalies in Tangshan and that in Chaoyang	40	along tectonic zones
			2. time differences among high-peak of anomalies in west, central and east Liaoning Province	5 - 8	crossing tectonic zones
			3. time differences among high-peak of anomalies in Chaoyang, Tongliao and Baizhen areas.	40-60	along tectonic zones
			4. time differences among high-peak of anomalies in Tongliao, Siping and Yanji areas.	5 -10	along tectonic zones
			5. time differences among high-peak of anomalies in Tangshan, Dezhou and Heze areas.	40-60	along tectonic zones
			6. time differences between high-peak of anomalies in Xinyi area and that in Yangzhou area	20	along tectonic zones
			7. order of self-flow of ground-water along Tangshan—Shandong region	60-70 6 -10	along tectonic zones Crossing tectonic zones

It can be seen from Fig. 9.12 that the self–flow of ground–water appeared around 28 July 1976 in the Tangshan and Tianjin areas, while the flow occurred successively from July 29 to 30 in the Cangzhou District, from July 31 to August 1 in the Linqing and Dezhou Districts, Shandong Province, and on August 10 in the Jinan and Zibo areas, Shandong Province.

Based on the above facts, some features of the time–dependent migration of the large–scale regional macroscopic anomalies before and after the large earthquakes can be described as follows:

a. Migration direction. The anomalies concentrated toward the epicenters from outside before the mainshocks, but spread out from the epicentral areas afterwards.

b. Migration speed. The pre–seismic anomalies approached the epicenters at a velocity of several kilometers per day (Table 9.1). The migration speed was characterized by an anisotropic nature. For example, the anomalies moved along tectonic zones at a velocity of tens of kilometers per day after the mainshocks, much faster than they crossed the tectonic zones (Table 9.1).

c. Migration manner. The anomalies mainly moved in a discontinuous jumpy manner.

In summary, within a large–scale range before and after a large earthquake, either seismic activity or precursors and macroscopic anomalies appeared as a whole in an inhomogeneous spatial distribution. The anomalies were mainly concentrated within active tectonic zones or near intersection points of tectonic zones. Hence, the regions of anomaly concentration and the isolated sensitive points constitute a complicated and inhomogenous distribution pattern. The anomalous regions migrate toward the epicenters from outside before the mainshocks, and spread outward from the epicenters after the mainshocks. These phenomena may have mirrored large–scale regional crustal deformation, and accordingly it is possible to delineate the range of tectonic movement and the shapes of tectonic blocks associated with seismogenic preparation of a large earthquake, based on these anomalies. The origin and mechanism of the anomalies appearing in a large–scale range will be further dealt with in Chap. 12.

CHAPTER 10

Comprehensive Analyses of Time–Space Variations of Anomalous Phenomena Before and After Large Earthquakes

The previous chapters have respectively dealt with the variational processes of seismicity and various precursory phenomena in the epicentral areas and their vicinities before and after the nine large earthquakes. In this chapter, a comprehensive generalization will be made of the time process and space distribution of these anomalies, and analyzed accordingly from a mechanical viewpoint. The purpose of comprehensive comparison studies of the anomalies is to understand more completely the essentiality of earthquake preparation process.

As there are many differences in the data obtained before and after the nine large earthquakes, the so–called similarities and differences between the anomalies have been summarized from some events with fairly complete data.

10.1Comprehensive Analyses of Temporal Processes of Anomalous Phenomna

10.1.1 Similarities of Temporal Processes of Anomalous Variations

1) Several years before the large earthquakes, the frequency of regional seismicity increased year by year. Meanwhile, the frequency–magnitude relation (b value) in an epicentral area and its surroundings shows a rise–fall process.

2) Years before the mainshocks, a variational process of decrease and increase for regional wave velocity was observed.

3) Years before the mainshocks, there were anomalous decreases of apparent resistivity in shallow and deep crust, and a slow recovery after the mainshocks took place.

4) Within several years before the mainshocks, the deformation rate increased gradually, and a year or 6 months before some mainshocks, the rate accelerated or slowed down anomalously.

5) Years before the mainshocks, the ground–water level fell, while the radon content increased slowly. In addition, variations in earth stress, geomagnetism and gravity were also observed for years running.

6) Several months prior to the mainshocks, there were obvious turns in most of the observation curves for various precursors (variation rate accelerated or slowed down), and there appeared also a few abrupt anomalies.

7) Prior to the impending mainshocks, there were abrupt changes (pulse– or

step–like) observed in many fields (such as ground water,radon content,water temperature,ground gases and well water ejecting). Also, there were a great number of macroscopic anomalies and atmospheric pressure variation and rainfall anomalies.

8) Large earthquakes were mainly accompanied by strike–slip dislocation at nearly vertical fault planes; coseismic great changes in ground–water level and fault displacements were recorded by some observational means in the epicentral areas.

9) After large earthquakes, frequent aftershock activities were observed, and some anomalies recovered to their normal state(like ground–water level and apparent resistivity); surface subsidence in epicentral areas was stronger than uplift.

The above–listed similarities of anomalies may be of significance in distinguishing earthquake anomalies in the future. Of course, as precursory observations and research are still at the studying stage, with insufficient data,the above summarization is very preliminary and not completely accurate.

10.1.2 Comprehensive Analyses of Anomalous Processes

We have comprehensively compared the time processes of various anomalous variations, and divided them into different anomalous stages.

The following gives comprehensive diagrams for variational processes of various anomalies on three time–scales before the mainshocks (years, months, days), and discussions on the behaviour of anomalies with different time–scales.

1) Variational process of anomalies during years.

Figures 10.1–10.6 show variational processes of various anomalies observed several years before the Bohai, Luhuo, Haicheng, Longling, Tangshan and Songpan earthquakes. From these comprehensive comparison curves it seemed that the variational process several years before the mainshocks might be divided into two stages:

a. Normal variation stage. During this stage, the curves were rather stable, almost without any changes, just oscillating at normal values, and the normal values (mean values) usually could be derived.

b. Anomalous variation stage. After the normal stage, the observation curves departed clearly from the normal values, and tended to rise or fall anomalously.

For instance, before the Haicheng earthquake (Fig.10.3), no obvious variations were observed in weak seismicity, levelling and tilting during 1972–1973, constituting the normal variation stage; the anomalous variation stage arrived in 1974. Two such stages also appeared in many observation curves for other events.

Several years before these large earthquakes, the anomalous variation stages for various items featured the following two characteristics:

α. The anomalous stages mostly happened within 1–4 years before the mainshocks.

β. The general process of anomalies showed either continuous increase or continuous decrease, or fluctuating changes against a potential rising(falling) background.

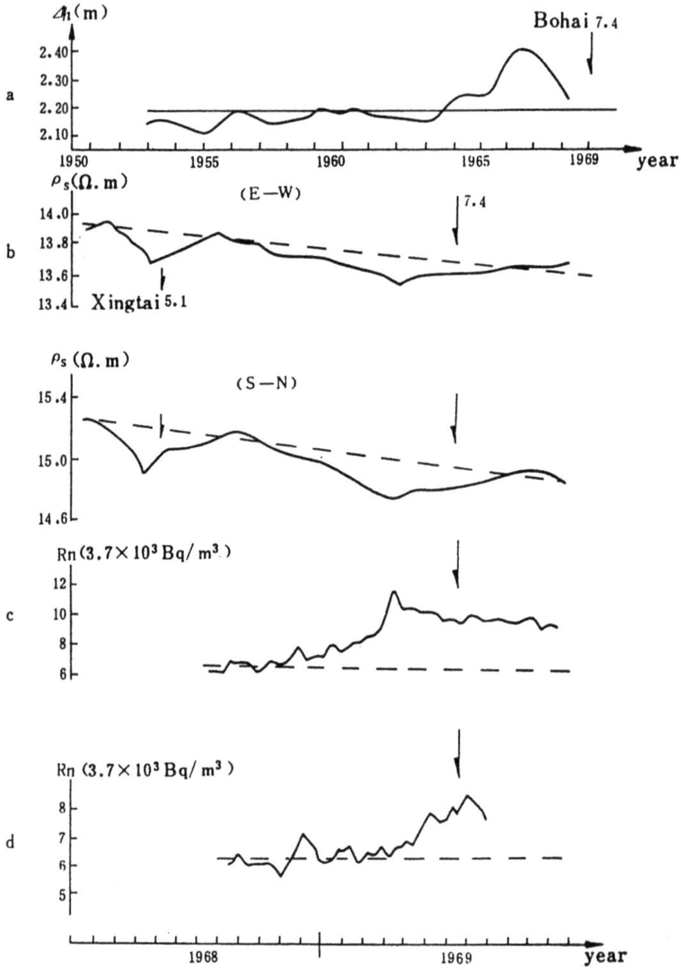

Fig. 10.1. Anomalous variations several years before the Bohai earthquake; *a* Yantai sea level (Δ= 220km); *b* Longyao resistivity (Δ= 460km); *c* radon content at Duck Farm, Tianjin (Δ= 210km); *d* radon content at Baodi (Δ= 270km)

2) Variational process for anomalies during several months.

Figure 10.7 gives a comprehensive diagram of the curves observed several months before the Tangshan earthquake. In addition, from the posterior parts of some observation curves as shown in Figs. 10.1–6 , the variational process of anomalies several months before several other shocks can be seen. The variational process of anomalies for several months before the mainshocks featured the following two points.

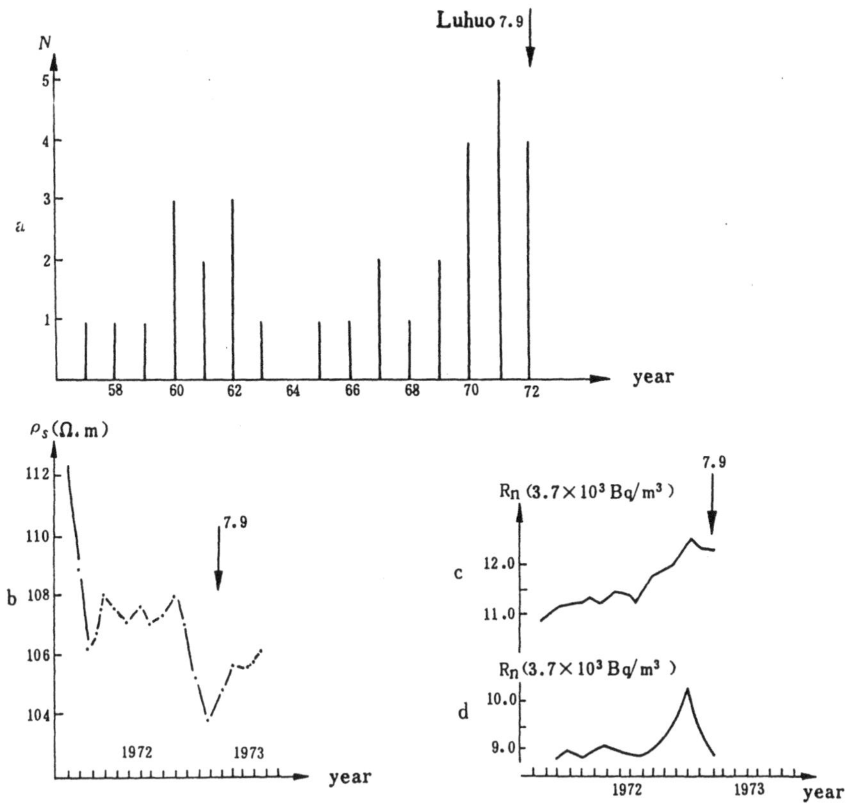

Fig. 10.2. Anomalous variations several years before the Luhuo earthquake; *a* frequency of earthquakes with M>5 in Sichuan Province; *b* resistivity changes at Ganzi station (Δ= 40km); *c* radon content variations at Xichang station (Δ= 400km); *d* radon content variations at Kangding (Δ= 200km)

a. Within several months before the mainshocks, the mean variation rates of anomalies would accelerate or slow down clearly compared with those of several years before the mainshocks, and the short–term anomalies had a very large variational amplitude. For example, before the Tangshan earthquake, multiple anomalies kept appearing after 1972–1973 (Fig. 10.5), and around the spring of 1976, some anomalies clearly slowed down in variation rate, and others further accelerated (Figs. 10.5, and 10.7).

b. Several months before the large earthquakes, the curves of anomalous variations were very complex in shape, with continuous basic variation, and sometimes showed sudden changes.

3) Variational process of anomalies during several days.

a. On the background of anomalous variations several months before the mainshocks, a number of abrupt variations (pulse–like or step–like) were observed several days or more than 10 days prior to the impending mainshocks (Figs. 10.8–10.11). Imminent rebound changes in water–level were also observed just before the Tangshan earthquake (see Chap. 7).

b. Several days prior to the mainshocks, a large number of macroscopic anomalies were observed, including abnormal animal behaviour and physico–climatic anomalies (see Chap. 8). For instance, 2 or 3 days before the Haicheng earthquake the macroscopic anomalies in ground water (gas),animal behaviour and earthquake light and sound were maximized in the epicentral area (Fig. 10.8), synchronized with the anomalies of foreshocks, tilts, telluric current and radon content.

The features of representative anomalous processes for three different time–scales are described above. However, for a certain item of observation, the recorded curve which perfectly showed the typical features of three stages was unusual. At numerous stations, quite remarkable tendency variations for years were recorded, but no impending sudden changes were observed; whereas the case at other stations was just the opposite. Usually, the stress–strain curves obtained from laboratory rupture experiments on rock samples may show whole rupture processes of rock sample, which reflect separate rupturing stages, and so it is expected to see such stages in field observations. As a matter of fact, a much closer look into various parts of a rock sample may reveal discrepancies among the stress–strain curves for these parts, thus showing that the complete phasic variation processes of anomalies observed in the field were related to the location and the stress–strain mechanisms of the stations, and the inhomogeneity of regional stress fields.

In addition to the three above–mentioned time–scales, the anomalous characteristics of the data on seismicity and geodesy several years or 10 years before the mainshocks were also discussed in Chaps. 4 and 6. Here, we may select four time–scales to elaborate the variational processes of anomalies together with their characteristics. For a large earthquake, it might be simply held that during the seismogenic process there exist four developmental stages: long–term (over several years), mid–term (several years), short–term (several months), and imminent (several days), each of which perhaps represents a different separate stress–strain state. As shown in Fig. 10.12, the curves of ground–water level and apparent resistivity changed before the Tangshan earthquake and they give typically the entire developmental process of usual anomalies and its stage divisions before the mainshock.

Based on the data available at present on variational processes of earthquake precursors, Ma Zongjin et al. (1987) compiled a table listing the divisions of long–, mid–, short–, and imminent– term stages of a whole seismogenic process, as well as the behaviour of various anomalies (Table 10.1). The table also shows coseismic and post–seismic situations and possible physical mechanisms for various stages.

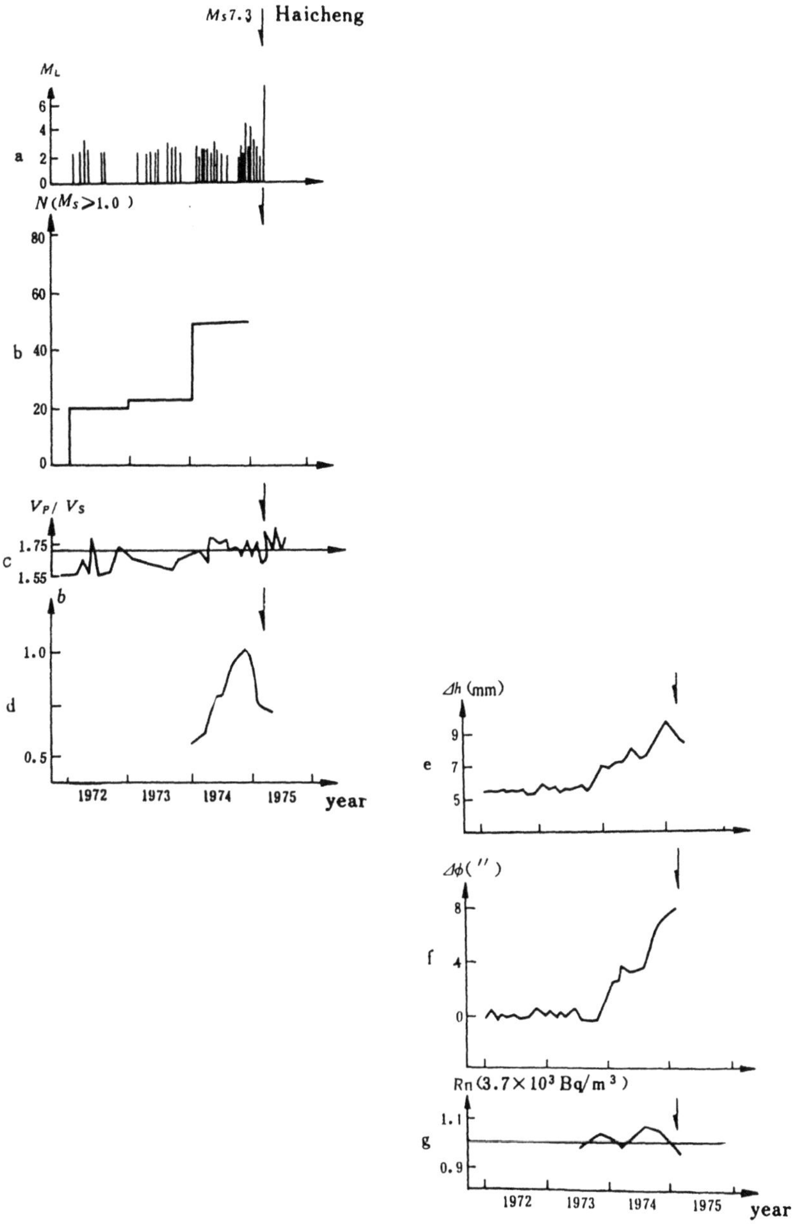

Fig. 10.3. Anomalous variations several years before the Haicheng earthquake; *a* seismic activity in Liaoning Province; *b* frequency of small earthquakes; *c* seismic wave velocity ratio; *d* b value; *e* levelling in Jinxian ($\Delta = 190$ km); *f* tilt at Yingkou ($\Delta = 180$ km); *g* radon content variations (weighted)

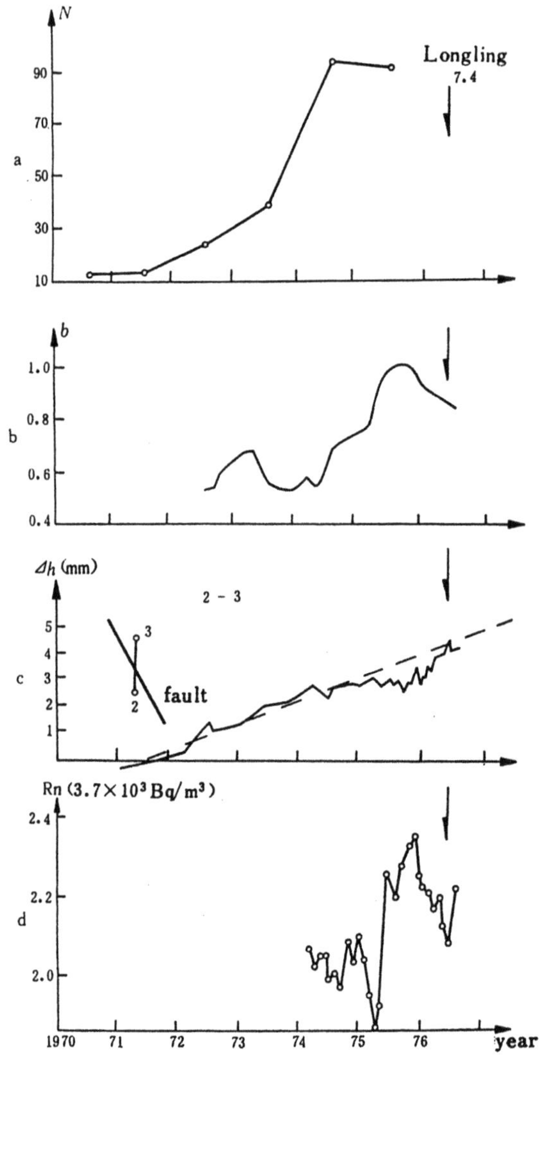

Fig.10.4. Anomalous variations several years before the Longling earthquake; *a* frequency of earthquakes in Longling area and its vicinity; *b* b value; *c* baseline at Xiaguan (Δ= 170 km); *d* radon content variations at Xiaguan (Δ= 170 km)

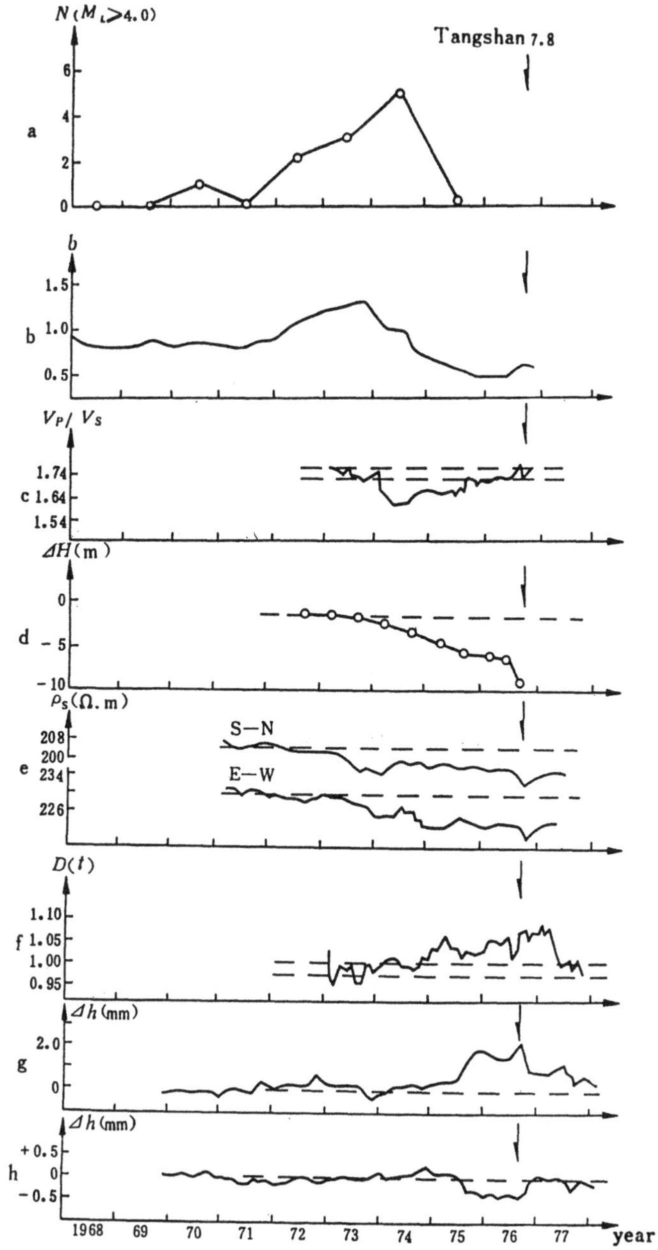

Fig.10.5. Anomalous variations several years before the Tangshan earthquakes; *a* frequency of earthquakes in Tangshan area and its vicinity; *b* b value; *c* seismic wave velocity ratio; *d* ground–water level at Tangshan; *e* resistivity at Changli (Δ= 70 km); *f* mean variations of radon content of well water in Beijing –Tianjin– Tangshan region; *g* levelling at Dahuichang station (Δ= 180 km); *h* levelling at Niu– kouyu station (Δ= 180 km)

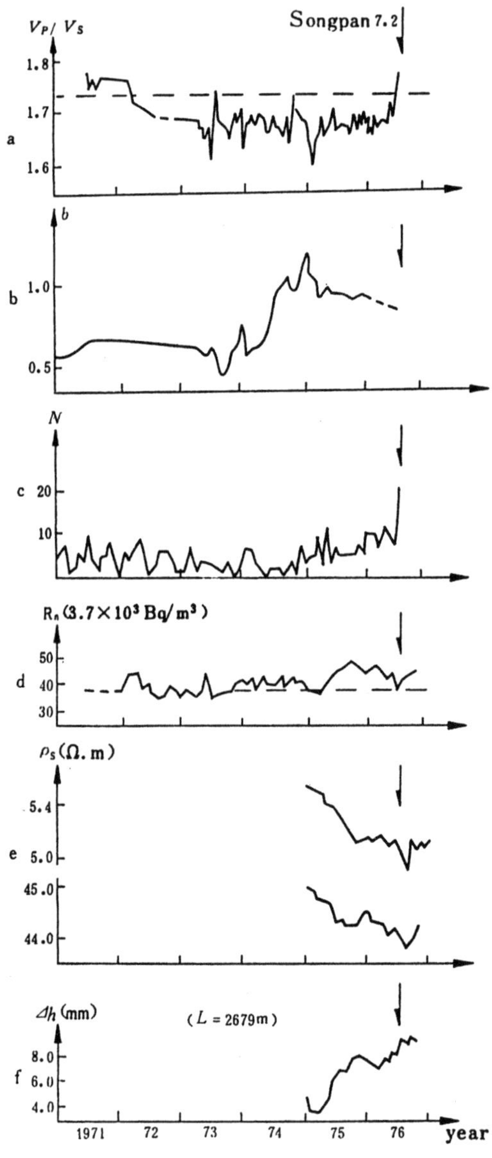

Fig. 10.6. Anomalous variations several years before the Songpan earthquake; *a* velocity ratio in Songpan area; *b* b value; *c* frequency of small earthquakes in Longmenshan area; *d* radon content at Songpan (Δ= 50 km); *e* resistivity at Wudu (Δ= 100 km); *f* levelling at Songpan (Δ= 50 km)

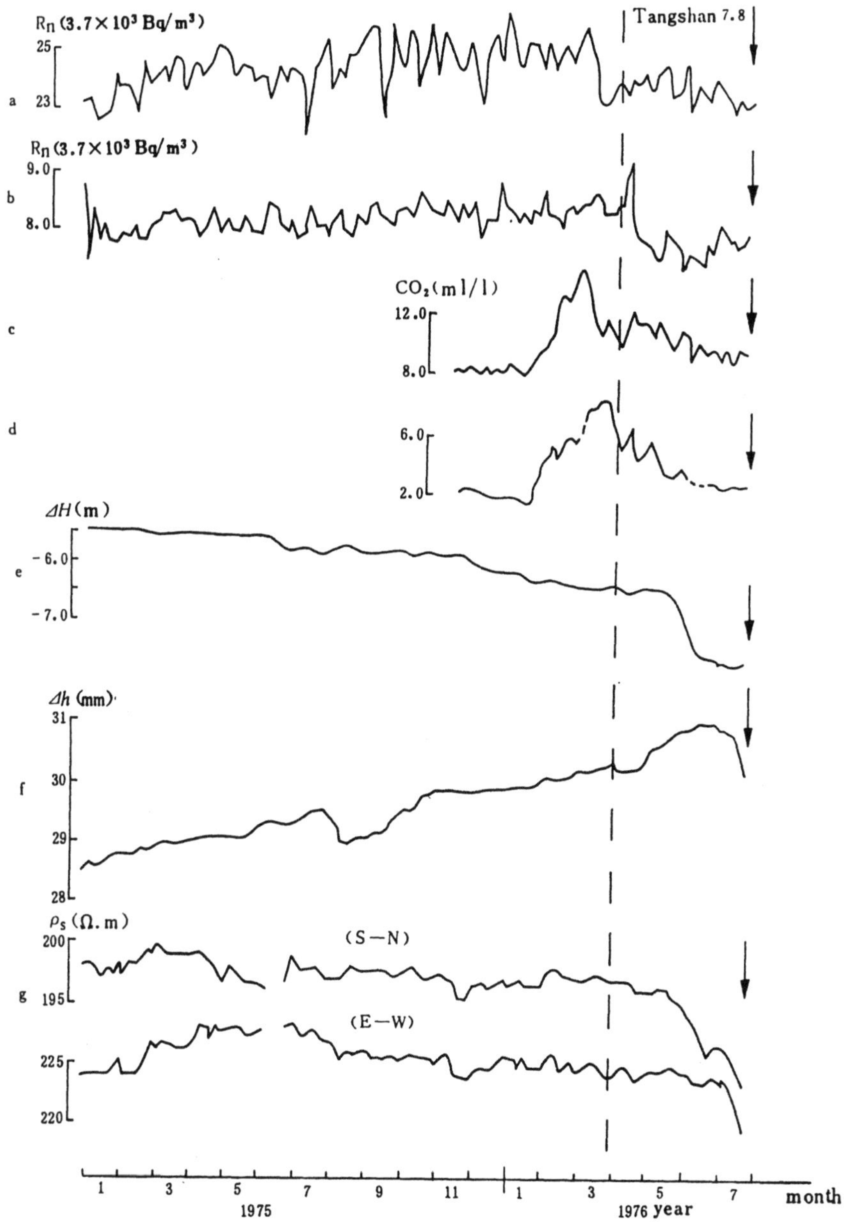

Fig. 10.7. Anomalous variations several months before the Tangshan earthquake; *a* radon content at Tiantuan; *b* water radon content at Tanggu; *c* carbon dioxide variations in No.2 well in Tianjin; *d* carbon dioxide changes in No.4 well in Tianjin; *e* water level variations in Biaokou well in Ninghe; *f* levelling at Dahuichang station; *g* resistivity at Changli

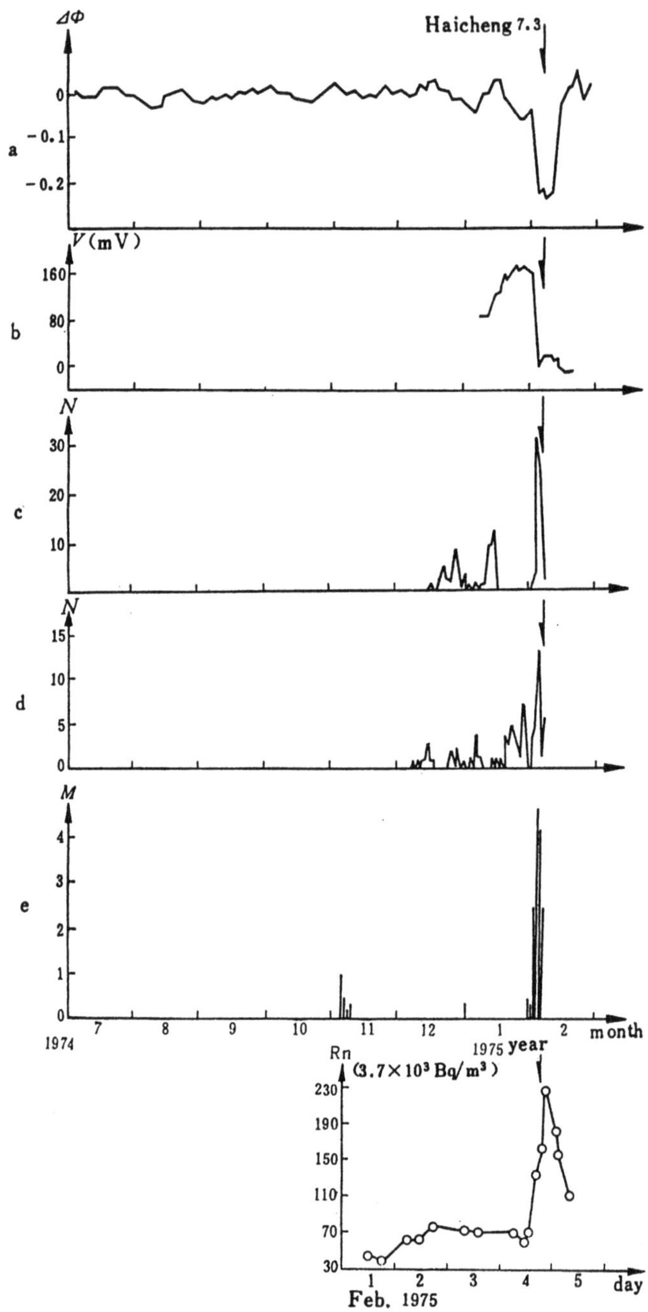

Fig. 10.8. Sudden anomalous variations observed several days before the Haicheng earthquake; *a* tilting at Shengyang (Δ= 150 km); *b* earth electric potential changes at No.102 Team, Ministry of Metallurgy (Δ= 15 km); *c* anomalous frequency of animal behaviour in Dandong area (Δ= 150 km); *d* frequency of ground water level anomalies in Dandong area; *e* foreshocks; *f* water radon content changes at Tanghe (Δ= 72 km)

Table 10.1. Stage division of a whole seismogenic process and the anomaly behaviour in the various stages

	stage / type of anomalies	long-term (tens of years–years before)	Med-term (years–months before)	short-term (months–days before)	imminent-term (within several days)	during earthquake	after earthquake
1	seismicity	studies of large earthquakes: stages, regionalization, types, time-space regularity and energy releases		foreshock series features, space features, and signal shocks		mainshock	aftershocks
	seismic wave		wave velocity variations in source regions, changes of focal mechanism solutions	precursory waves, earthquake sound (high, low frequency)		focal mechanism intensity effects of seismic waves	
2	deformation	anomalies in geodesy, fault displacements, earth tilts		anomalies in baseline, leveling routes, earth tilts	tilt storm	ground deformation	recovering and adjustments
	geology	features of seismotectonic zones and machanical analyses		geological background for precursors		coseismic geological changes	geological remains
	stress	Palaeostress field	anomalous curves of borehole stress measurement				vatiations of stress fields
3	geophysical prospecting	sounding of shallow and deep crustal structures	anomalies of mobile survey of gravity and geomagnetism	gravity, geomagnetism, telluric current, resistivity	anomalies→sudden variations luminescences		different meterials
4	groundwater		anomalous areas of ground water flowage	anomalous changes of water level	sudden changes of water level		different aftermath
	hydrochemistry			anomalies in contents of Rn, F, carbon dioxide, conductivity, and nitrous acid and others			different-extent aftermath
5	statistics	statistical analyses of historical earthquakes (seismic probabitity), correlation studies of seismic data and astronomical cycles					
6	meteorology		analyses of mean values and extremal values of rainfall and temperature analyses of atmospheric waves		meteorological anomalies: wind, rosy clouds, fog		
7	astronomy		solar activity, earth rotation and polar drift stress		lunar corresponдance	variations in polar drift and rate of earth rotation	
8	biology			botanical anomalies	unusual animal behaviour		
9	physical mechanism	elastic strain accummulation	non-elastic strain, fissure developments, fluid and thermal material movements	fault creeping, stable-substable extentions of ruptures, unstable rupturing before impending quakes, intense movements of fluid and thermal materials		unstable faultings	large-scale regional stress adjustments

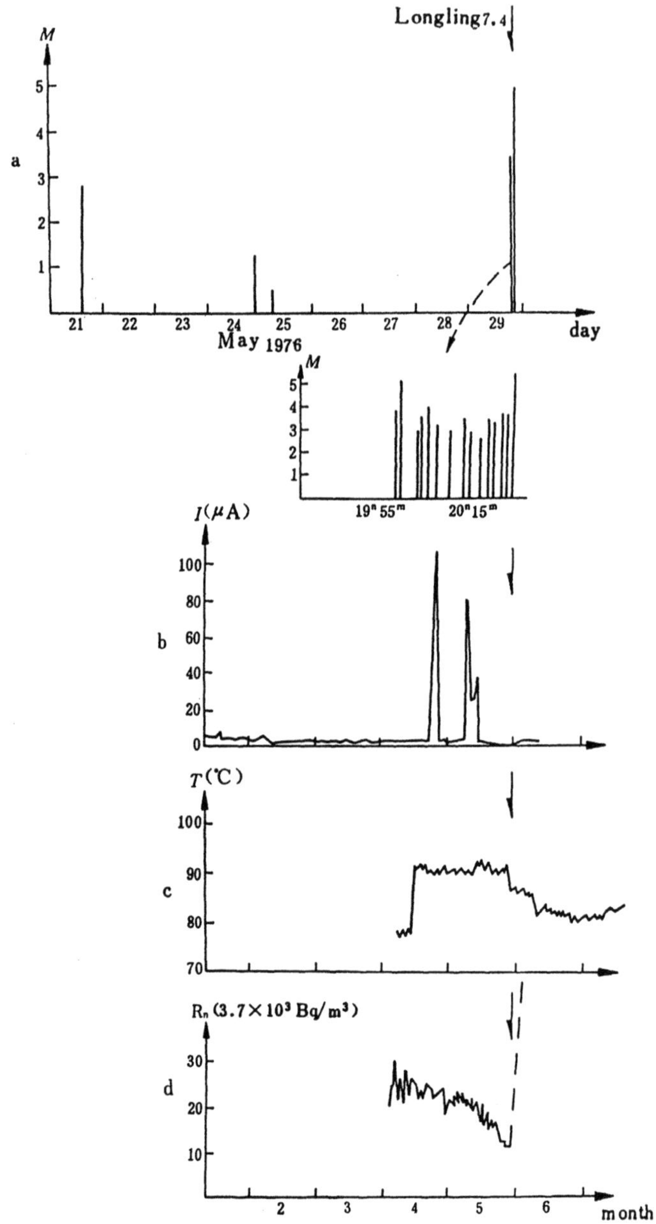

Fig. 10.9. Sudden anomalous variations observed several days before the Longling earthquake; *a* foreshocks; *b* telluric current variations at No.4 Middle School in Tengchong ($\Delta=40$ km); *c* water temperature of spring in Balazhang ($\Delta=10$ km); *d* radon content changes in spring in Balazhang

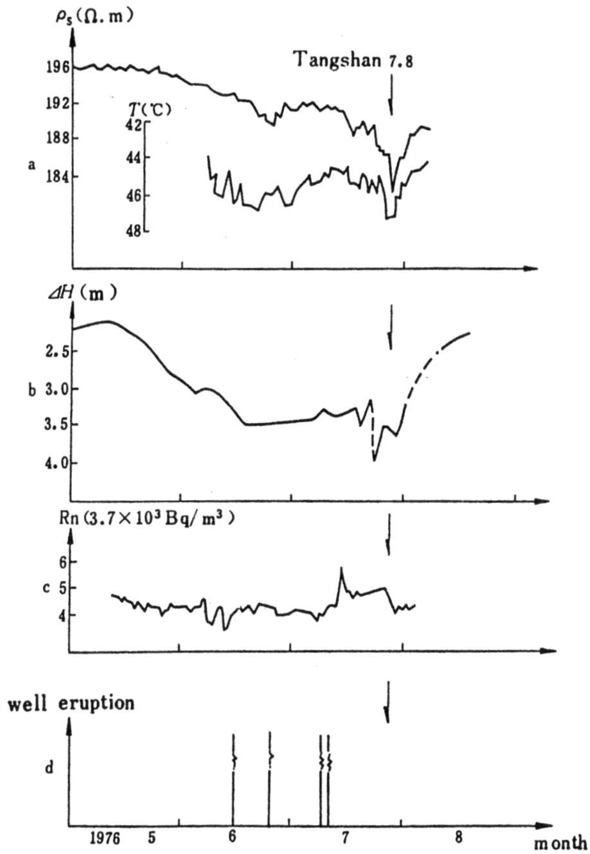

Fig. 10.10. Sudden anomalous variations observed several days before the Tangshan earthquake; *a* resistivity and ground–water temperature at Changli (Δ= 70 km); *b* water level of well at Boguzhuan (Δ= 40 km); *c* radon content changes at Langfang (Δ=130km); *d* well ejection of a wasted oil well at Qingxian (Δ= 150 km)

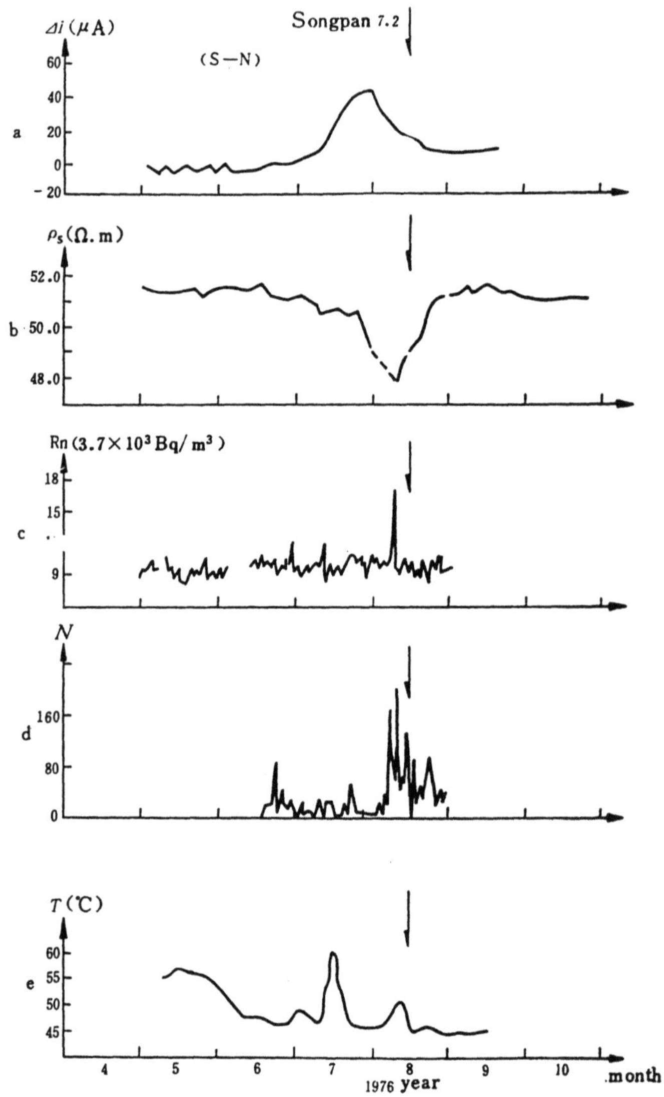

Fig.10.11. Sudden anomalous variations observed several days before the Songpan earthquake; *a* telluric current at the Kangding Middle School (Δ= 350 km); *b* resistivity at Wudu (Δ= 100 km); *c* radon content at Guzan (Δ= 350 km); *d* frequency of macroscopic anomalies in Songpan–Longmenshan region; *e* water temperature of spring at Litang (Δ= 350 km)

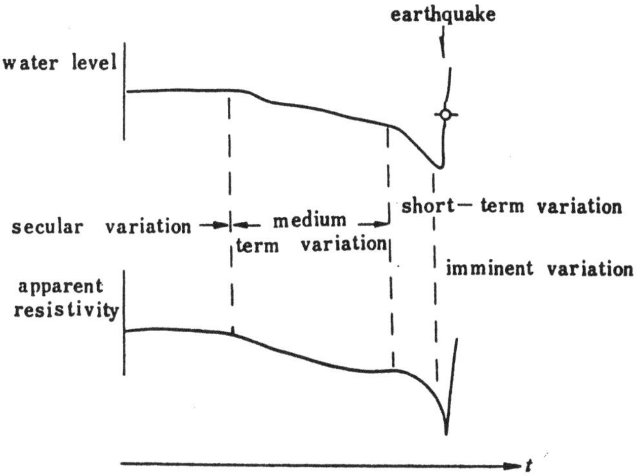

Fig.10.12. Diagram for the whole process of anomalous variations before a large earthquake

10.2 Common Features of the Spatial Distribution of Anomalous Phenomena

The time processes of long–, mid–, short–term, and imminent anomalous variations before large earthquakes and their characteristics are analyzed above. In this section, mainly the spatial distributions of anomalies and the characteristic of anomaly migrations are dealt with. As in many cases, it is difficult to clearly distinguish the short–term from imminent phases, both of them, usually termed short and imminent anomalies, are often discussed together. With the present level of research and observations, it is impossible to perfectly describe the features of the spatial distribution of anomalies. This is because there must be sufficiently dense seismic networks set up in accordance with unified requirements, strictly standardized observation systems and methods for various stations, and a profound understanding of the crustal structures and physical properties of media in the regions under observation. In fact, these conditions cannot really be provided yet. Therefore, based on presently available data, we can only make very preliminary and rough analyses of the spatial distribution of anomalies, and discuss in particular the characteristics of spatial distributions (mainly by analyzing the data of regional seismicity) of four large earthquakes (Tangshan, Haicheng, Longling and Songpan) which occurred from 1975 to 1976.

10.2.1 Tangshan Earthquake

1) Spatial distribution of mid–term anomalies.

Figure 10.13 gives the spatial distribution of various mid–term anomalies before the Tangshan earthquake. The following points are shown:

a. Spatial migration of earthquakes with M > 4. From 1966 to August 1972, the earthquakes predominantly took place in the area west of Beijing and 200 km from Tangshan; from September 1972 to July 1976, regional seismic activities were remarkably concentrated in Tangshan and its vicinity, while seismicity tended to be quiet in outside areas, hence implying a process of migration of earthquakes toward the future mainshock epicenter. While examining the features of spatial distributions of much larger–scale regional seismicity (see Chap. 9), it was discovered that the large–scale regional seismicity moved torward the latitude at which the Tangshan earthquake was located.

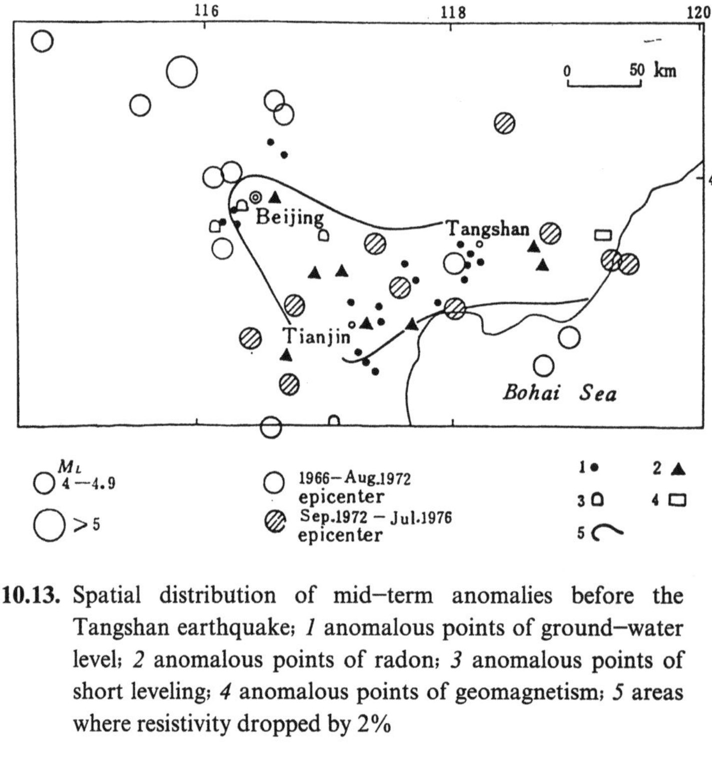

Fig. 10.13. Spatial distribution of mid–term anomalies before the Tangshan earthquake; *1* anomalous points of ground–water level; *2* anomalous points of radon; *3* anomalous points of short leveling; *4* anomalous points of geomagnetism; *5* areas where resistivity dropped by 2%

b. In Fig. 10.13, the distributions of some mid–term anomalies including apparent resistivity, fault displacement, ground–water level, water radon content, and geomagnetism are also given, thus illustrating that various mid–term anomalies were approximately spread over a range of 200 km in and around Tangshan. This is consistent with the area where the local seismicity concentrated after September 1972. These mid–term anomalies began mostly to appear from the late half of 1972 and roughly coincided with the beginning of the concentration process of seismic activity.

c. The spatial distribution of mid–term anomalies seemed to be controlled tectonically. The seismicity, contour lines of apparent resistivity and other anomalies were predominantly distributed in NE and NW orientations, roughly associated with two sets of tectonic frameworks in NE and NW (see Chap. 3).

d. Certain anomalies observed at some stations for apparent resistivity and gravity also showed a great variation near by the epicenter and a gradual decrease off it (see Chap. 5).

2) Spatial distribution of short–term and imminent earthquake anomalies.

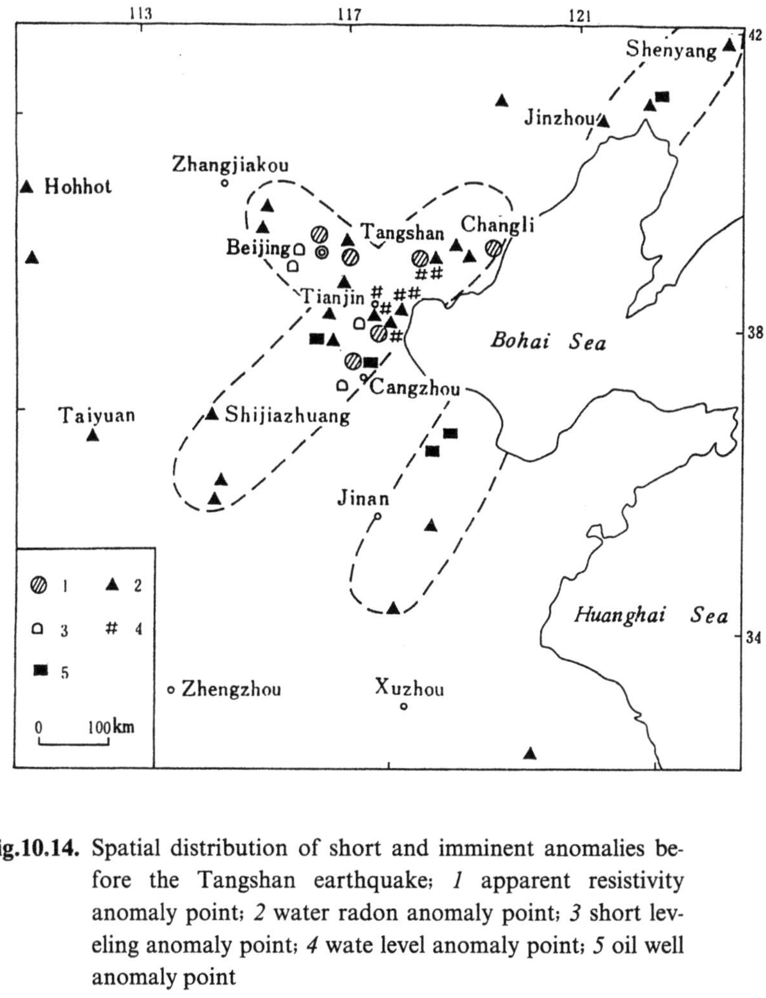

Fig.10.14. Spatial distribution of short and imminent anomalies before the Tangshan earthquake; *1* apparent resistivity anomaly point; *2* water radon anomaly point; *3* short leveling anomaly point; *4* wate level anomaly point; *5* oil well anomaly point

Figure 10.14 gives the spatial distribution of short–term and imminent anomalies before the Tangshan earthquake, including a variety of anomalies, such as apparent resistivity, radon concentration, water– level, oil–well changes, and short levelling. At some stations, the tendency anomalies accelerated or recovered within 2 or 3 months before the Tangshan earthquake, while at the great majority of stations various abrupt and rapid anomalies were observed in 1 or 2, or 10 days and more before the impending mainshock. The three points made in Figure 10.14 are as follows:

a. The short–term and imminent earthquake anomalies were widely scattered in a region much larger than that for the mid–term anomalies. As mentioned in Chap. 9, the short–term and imminent anomalies were even observed about 500 or 600 km away.

b. The short and imminent earthquake anomalies were distributed inhomogeneously. On the one hand, the general spatial distribution of anomalous points was inhomogeneous, for example there were anomalies at some points far away in one direction, and no anomalies in areas closer to the epicenter in other directions; on the other hand, only a part of the observation points and items showed short–term and imminent earthquake anomalies in one and the same region with multiple observation points and items.

c. The anomalies were relatively concentrated in the epicentral area and surroundings, the anomalies in amplitude being larger than those in outside areas.

In addition,some macroscopic anomalies like ground–water level and animal behaviour in the epicentral area before the Tangshan earthquake appeared rather late, and the macroscopic anomalies seemed to migrate toward the epicenter from outside (see Chap. 9).

10.2.2 Haicheng Earthquake

1) Spatial distribution of mid–term anomalies.

One or two years before the Haicheng earthquake, the regional seismicity was relatively concentrated within two bands in NE and NW orientations (Chap. 4); at the same time, the main mid–term anomalies , such as radon content, fault displacement, and ground tilt were also distributed within the same two bands and their vicinities. The characteristics of spatial distribution of mid–term anomalies before the Haicheng earthquake were very similar to those listed under 1) and 2) of Sect. 10.2.1 for the Tangshan earthquake.

2) Spatial distribution of short–term and imminent anomalies.

The distribution of short–term and imminent anomalies before the Haicheng earthquake was, besides the orientation, also characterized by a notable time–dependent migration of anomalies toward the epicenter (see Fig. 9.10). In general, the anomalies over a rather large–scale region migrated toward the epicenter along the WNW and NNE tectonic zones. The time–space relationship of the migrations is shown in Fig. 10.7.

10.2.3 Longling Earthquake

1) Spatial distribution of mid–term anomalies.

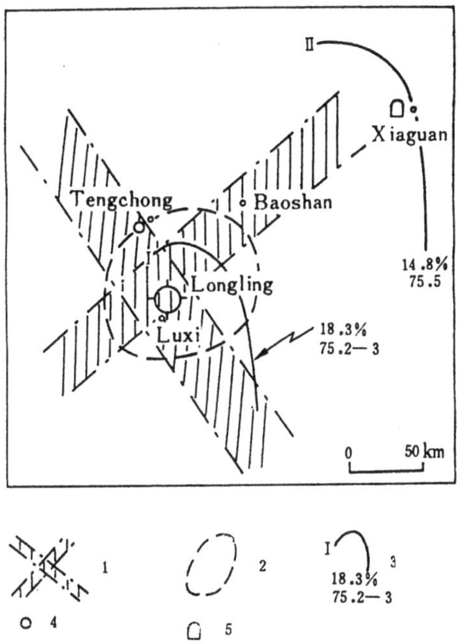

Fig. 10.15. Spatial distribution of mid–term anomalies before the Longling earthquake; *1* band–like distribution of seismicity; *2* anomalous wave velocity area; *3* anomalous radon content area, numerals represent anomaly quantity (%) and the time (year and month); *4* anomalous apparent resistivity point; *5* anomalous baseline point

Figure 10.15 shows the distribution of some main anomalies before the Longling earthquake.From this figure, it is seen that 3 or 4 years before the mainshock the regional seismicity increased in frequency, plotting a band–like distribution, and at the same time the seismic wave velocity ratio dropped near the epicenter, and some mid–term anomalies like water radon content increase were observed 1 year or more before the mainshock. Furthermore,variations were observed in apparent resistivity at Tengchong and in base line at Xiaguan. Generally speaking, the mid–term anomalies were mainly distributed over a range of 100 or 200 km in and around the epicenter.

2) Spatial distribution of short–term and imminent earthquake anomalies.

Chen Lide et al. (1979) suggested that the short and imminent anomalies were predominantly spread over the high–intensity area of tens of kilometers around the epicenter.

In addition, the short and imminent anomalies before the Longling earthquake were spread converging toward the epicenter with time. One or two months before the mainshock, the ground–water anomalies and others were spread over an area of 60–70 km (Longling–Tengchong–Baoshan area) several days to several hours prior to the mainshock, the anomalous phenomena occurring in the great majority of wells and springs as large–amplitude increases in water–level and flow, changes in water colour and odour, muddled water and so on and were mainly concentrated within a range of 30–40 km in and around the epicenter.

10.2.4 Songpan Earthquake

1) Spatial distribution of mid–term anomalies.

Three or four years before the Songpan earthquake, the regional seismicity was distributed in a NE band,forming a seismic gap for the forthcoming mainshock. The mainshock was located at the end of the band, on the inner side of the gap. At the same time, a great number of anomalies were observed, such as faulting intensification, a drop in seismic wave velocity, radon content increase, apparent resistivity decrease, electro–inductional stress variation, and so on, which were mostly distributed in a regional seismic band and its surroundings along a NE direction around the epicenter.

2) Spatial distribution of short–term and imminent earthquake anomalies.

Figure 10.16 shows the distribution and migration of short and imminent macroscopic anomalies before the Songpan earthquake. Starting about the end of May 1976, the macroscopic anomalies increased remarkably, constituting three high–tide periods and migrated gradually toward the epicenter. The first high–tide began with June 20 to 23, during which the anomalies were centered in the Dayi–Qionglai–Tianquan–Baoxing region, south end of the buried Qionglai–Mianzhu fault of Longmenshan piedmont; the second high–tide started with July 19 to 23, during which the anomalies spread over the Guanxian–Pengxian–Shifang–Mianzhu region in the central Longmenshan fracture zone; and the third high–tide appeared from August 8 to 11, during which the number, kinds, amplitude, and distributive range of the anomalies were far greater than during the first two high–tides, with all anomalies concentrated closely around the forthcoming epicenter, as shown in Fig. 10.17. The migration rate was about several kilometers per day (see Chap. 9).

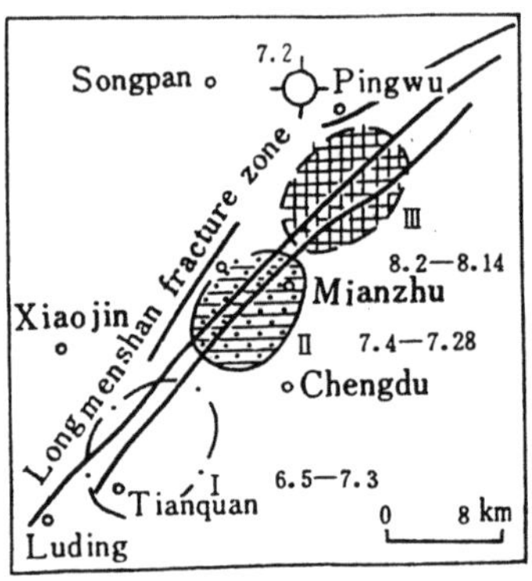

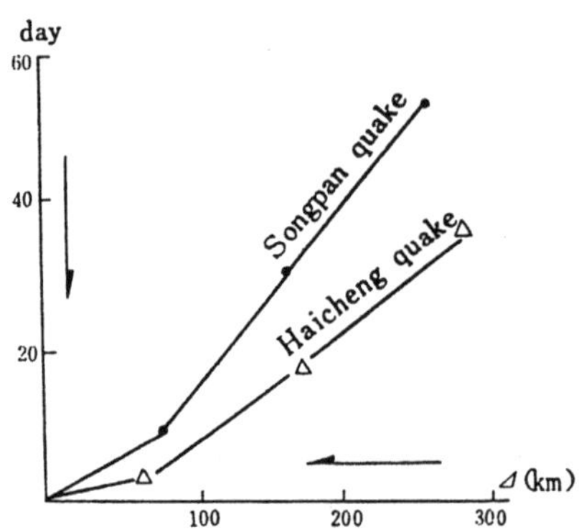

Fig.10.16. Spatial distribution of short and imminent anomalies before the Songpan earthquake

Fig.10.17. Migrations of short and imminent macroscopic anomalies toward the epicenters before the Songpan and Haicheng earthquakes

10.2.5 Common features of the spatial distribution of anomalies.

The aforesaid mainly described the spatial distribution of anomalies before the Tangshan, Haicheng, Longling and Songpan earthquakes. Because of the insufficient mumber of stations and their uneven distribution with different observational conditions, the data described above can hardly avoid the inclusions of some false phenomena. However, a measure of preliminary understanding may still be gained from the above said data.

1) Most of the intermediate–term anomalies appeared to be concentrated within a range of 100 to 200 km near the forthcoming epicenters, of which some were distributed in an inhomogèneous pattern, with some bands and scattered points of anomalies along certain tectonic zones. The spatial distribution range for the short–term and imminent anomalies were larger than for the mid–term anomalies.

2) all of the medium–term or the short–term and imminent anomalies were associated with the regional tectonic faults, exhibiting marked directions of their distributions.

3) At the early period of the medium–term anomalies the seismic activities very often occurred outside of the epicentral regions. When the regional seismicity migrated into the forthcoming epicenters, or approached them, other medium–term anomalies followed one after another in seismicity–concentrated areas.

4) The short–term and imminent anomalies seemed to move toward the forthcoming epicenters from outside regions.

10.3 Mechanical Analysis of Common Features of Anomalous Variations

The foregoing two sections summarized some common features of temporal and spatial processes of various anomalies. These anomalies were mainly caused by the variations in the mechanical states of the media of various crustal layers. In the following, the temporal and spatial characteristics of variations of these anomalies will be studied, based on experimental laboratory results of rock samples and the analyses of the mechanical and physical states of the crust.

10.3.1 Comparison Between Laboratory Experiments of Rock Samples and the Variational Process of Earthquake Anomalies

Brace (1968) reported a set of experimental curves for variations of wave velocity, resistivity, volumetric strain and microcracks of granitic rock samples with changes of loading pressure. Mogi (1968) also reported the developmental process of microcracks in granitic rocks under pressure. Their results indicated that when the acting stress reached 50% of the rock rupture strength, the physical properties of rock samples would be significantly changed, including decrease of

P wave velocity, drop of resistivity of saturated rocks(at the orientation of parallel maximum compressive pressure) by an order of magnitude, acceleration of frequency and concentration of microcracks, and increase of volumetric strain. Luo Guangwei et al. have made laboratory experiments to study rock ruptures and radon content variations, suggesting that the growth of microcracks of rock samples at high stress stage would be the primary cause for the increase of radon content. In addition, the experiments with granitic rock samples of low permeability also showed that pore pressure would be decreased at a high stress stage. Therefore, according to the experimental results, in the course of stress loading, some anomalies may appear, such as decrease of wave velocity, increase of frequency of small shocks, drop of b value, diminution of apparent resistivity, increase of radon content, fall of water–level, and acceleration of deformation rate. These results are given in Fig.10.18 a, and the common anomalies observed through different approaches before the nine large earthquakes are also shown in Fig.10.18b. By comparing the two figsures, it is seen that the variational patterns of anomalies observed before earthquakes are generally consistent with the experimental results. It may be thus suggested that the pre–seismic anomalies perhaps reflected some variational processes like fissure development and fluid flow as mediums in and around source regions under high stress during seismogenic processes. The mid–term anomaly stage might correspond to the phase of elastic strain buildup, concomitant with the increase of stress and strain and accelerations of seismicity and regional crustal deformation within a considerable large–scale region. Some particular highly stressed areas were deformed non–elastically allowing microcracks to grow and develop and underground water to flow, thus leading to the various mid–term anomalies. With the speeding up of the processes for growth and interaction of microcracks, some locked segments of the fault zone were softened in strength, causing pre–displacement of the fault zone and rapid variations of local stress and strain. Therefore, the mid–term anomalies were typically accelerated or decreased (reversed), hence showing characteristics of the short–term anomaly phase. The development of pre–displacements, the linking up, opening and closing of microcracks during critical rupture must be accompanied with rapid movement of underground flow, and the intense motion of subsurface materials might also produce more complicated physical and chemical processes,so causing sudden variations in water, gas, oil and geotemperature, as well as imminent anomalies like geosound, earth light, electricity, magnetism and animal behaviour. During and after the large earthquakes, the prompt dislocations of crustal rock bodies and the rapid release of stresses within source regions would surely lead to large–scale regional stress adjustments, causing post–seismic effects within a large–scale region.

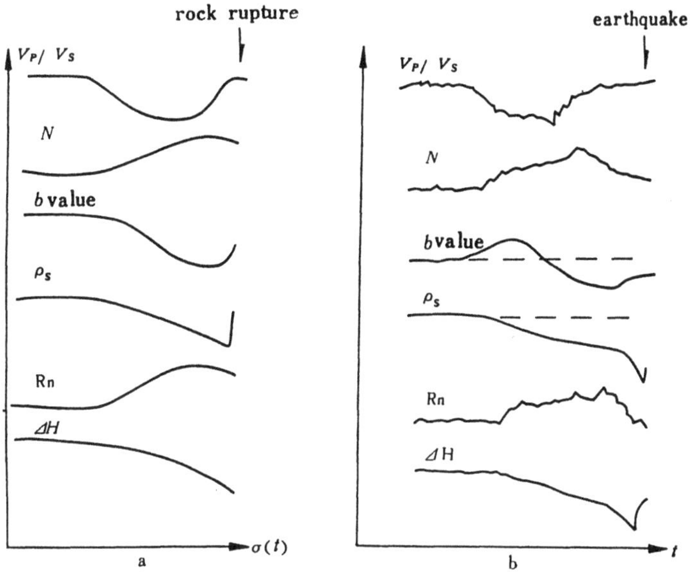

Fig.10.18. *a.b.* Comparison between mid–term anomalous variations before large earthquakes and laboratory experiments; *a* laboratory results; *b* observed results; Vp / Vs wave velocity ratio; *N* earthquake frequency (microrupture); *b* b value; ρ_s resistivity; Rn radon content; Δ *H* water level

Table 10.2. Probing depth of various observation means

serial number	observation means	depth
1	seismicity	5–20km
2	wave velocity ratio	whole crust
3	gravity	0–tens of km
4	magnetotelluric sounding	0–tens of km
5	apparent resistivity	0–1 km
6	geomagnetism	0–tens of km
7	groundwater level	0–several km
8	radon content	0–several km
9	crustal deformation	shallow crustal depth
10	fault displacement	shallow depth
11	electro–induction stress	0– 1 km
12	meteorology	ground and low altitude space

10.3.2 Profile Environments of Anomalous Variations

Currently, multiple means of observation have been utilized to probe and reveal medium states at different crustal depths(or upper mantle). The recorded anomalous variations seemed to be related with processes within crustal environments in different depths.

1) Probing depth of various observation means.

Table 10.2 shows the currently used observation means in earthquake precursory research, and their possible probing depth or responding depth. From this table it is seen that seismicity, wave velocity, gravity, magnetotelluric sounding, and geomagnetism possibly have the capability of reaching a greater depth, and might directly reflect changes in the nature of the medium at a depth equivalent to source depth. The ground–water level, radon content, apparent resistivity, crustal deformation, and electro–inductional stress measurements can survey a shallower depth only reflecting some variations at a depth close to surface and several kilometers beneath the surface.

2) Analyses on profile environments of anomalous variations.

Let us divide the earth crust and the parts above and below it roughly into four parts: (1) the lower crust and upper mantle; (2) the middle crust; (3) the upper crust; and (4) the low–altitude air space. Since the mechanical physiochemical regimes are different for these four profiles, the anomalies along these profiles may deviate from each other.

First of all, the middle–crust mainly denotes the earthquake–prone layer (competent layer). The role of the competent layer in overall crustal deformation was described in Chap. 3, so it should be a principal element in causing various anomalous changes. In fact, this has been demonstrated by correlations between seismicity and various anomalies.

Compared with non–seismic areas, the thickness of the subcrustal basaltic layer within seismic regions is greater, with upwarping upper mantle and low resistivity layer, hence showing that there might be rather peculiar environments at the lower–crust level beneath the seismic regions. There might exist partially melting materials,areas of local heat anomaly, and dynamic action causing upwelling of heated materials. As a result, the local anomalous areas of gravity and geomagnetism have been generated. Such variations of local states are not separated, and their development and disappearence could not be performed within a short period of time. Correlations between such local states and the competent layer should be taken into account. If the competent layer was buckled upwardly, causing its axial part to disjoin the rock stratum, a chance is provided for the plastic materials below to fill in, so that the upper mantle would partly be heaved up. The local variations of the competent layer would be certainly developed in case the upper mantle is uplifted.

The medium states of the upper crust are rather complicated, with their variations indirectly reflected by variations of the earthquake energy layer within the middle crust. However, the actual variations of the competent layer would be very often magnified since the upper crust is rich in fluids in various states. The variations in resistivity, wave velocity, crustal deformation, and fault weakening associated with the fluid flow, together with the local geomagnetic changes caused by the fluid flow in fissures, would eminently indicate slight strain in the competent layer. Meanwhile, the fluid flow may have an indirect influence on various sudden anomalies in the low–altitude space.

The above analyses made on the state of different crustal layers contribute to understanding the origin of different anomalies and their correlations. But the studies performed are yet preliminary and tentative. Nevertheless, it is suggested that comprehensive analyses on the profiles of anomalous changes still offer a worthy approach in research for studying the nature and original mechanism of precursory anomalies.

10.3.3 Mechanical Analyses of Anomalous Variations

Figure 10.18 shows that the anomalous variations are the direct result of the changes in the mechanical state of rocks (stress– strain). The following will discuss possible mechanisms for various anomalous variations, based on the physical, chemical and thermomechanical conditions of the earth crust:

1) Anomalous seismicity.

Before large shocks, the increased frequency of moderate and small earthquakes, the drop of b values , and the migrations and band–like concentrations of seismic activities may be associated with the process in which microcracks within the middle–crustal competent layer developed, interacted and connected (rock strength weakening) as the regional stress field gradually intensified.

2) Gravity anomalies.

The theoretical studies (see Chap. 5) of the gravity anomalies before and after the Haicheng, Tangshan and Longling earthquakes have suggested that the gravity variations may be due to volumetric strain and mass migrations within the crustal layers, particularly the fluid flow, including magmatic materials wedging into the middle crust.

3) Geomagnetic field anomalies.

There are mainly long–period and induced geomagnetic variations for geomagnetic anomalies before large earthquakes. Under crustal conditions, the piezomagnetic and thermomagnetic effects are possible main mechanisms for secular variation of geomagnetism.

4) Resistivity anomalies.

The decreases of shallow–depth apparent resistivity may be associated with the intensified strain, fissure growth, and fluid injection within the shallow crustal media. As fissures were generated near the bottom of the competent layer during structural deformation of the crust before large earthquakes, the fluid within the saturated low–resistivity layer was compelled to penetrate into the competent layer, leading to the drop of resistivity of the lower competent layer within the middle crust.

5) Deformation and stress anomalies.

The observations of crustal deformation and stress which directly recorded the amplitude of regional strain and fault displacement, and fault creeping (deep or shallow) before large earthquakes, made a great contribution to the strain field.

6) Ground–water (gas) anomalies.

The fall of ground–water level and the increases of radon content within the shallow crust before large earthquakes seemed to be mainly related with the fact that developments and connections of cracks within the water–saturated layer brought about ground–water movement, and rupture of cavities with higher radon content, causing the escape and diffusion of radon.

7) Sudden anomalies of animal behaviour and physical meteorology in low–altitude space.

There appeared a wide variety of anomalies prior to impending mainshocks, mainly including animal behaviour, electromagnetic, geosound, earth light, atmospheric and heat anomalies. Possibly, before impending mainshocks, the preseismic slip occurred on related faults and in the epicentral areas, causing rapid connection of cracks within rock bodies, into which the subcrustal fluid and gases penetrated swiftly, or issued fast,and even erupted out onto the surface. Efflux phenomena or electrical charge absorptions were thus produced, forming an anomalous low–altitude electrostatic field. All this might be an important reason for abnormal animal behaviour and physico–meteorological phenomena.

10.4 Differences in Time–Space Distributions of Anomalous Phenomena

The above three sections have summarized and discussed the common features of time–space processes of anomalies precursory to the nine large earthquakes, proposing general concepts for classifying the entire seismogenic process, instead of a unified model suitable for each type of earthquake, which it is impossible to establish at present. In fact, there were significant differences in precursory anomalies for the nine large earthquakes besides these common features. Therefore, emphasizing and studying the diversity of precursors would help in improving prediction quality in practice. Only by analyzing dialectically both the generality and the individuality of these precursors, can our understanding of seismogenic processes be deepened.

Darwin once said that there were no two leaves to be found in the world which were absolutely identical. Similarly, it is hard to say that two earthquake sequences in nature are in every respect the same. However, it is possible to classify some earthquake sequences according to their basic characteristics, so as to emphasize significant differences among the various types.

10.4.1 Differences Among Anomalous Variations

First of all, let us sum up some significant differences among anomalous variations for the nine large earthquakes:

1) Major differences in regional seismic frequency a year before mainshocks.

As mentioned above, there existed usually a gradual increase of regional seismicity several years before large earthquakes. However,the seismic frequency changed a year before large earthquakes. For instance, the frequency of seismic activity was highest 1 year before the Haicheng mainshock, while it decreased 1 year before the Tangshan earthquake, characterized by pre–seismic quiescence.

2) Differences in characteristic migrations of regional seismicity.

There was a common phenomenon in that regional seismic activities approached the forthcoming epicenters from outside before the mainshocks. However, for some earthquakes, like the Haicheng and Longling ones, the pre–seismic migrations went on until the mainshocks, and the seismic activities moved into the future epicentral area, while the moderate and small earthquake activities stopped several months to a year before the Tangshan and Songpan earthquakes, and they did not spread toward the epicentral areas.

3) Differences in direct foreshock activities.

Among the nine great earthquakes, only before the Xingtai, Haicheng and Longling ones were direct foreshock activities observed (see Chap. 4).

4) Deviations of seismic wave velocity in anomalous forms.

Some large earthquakes in Southwest China occurred either just when the wave velocity curves began to rise or soon after they had risen, while the Haicheng and Tangshen earthquakes in North China occurred a considerably long time(waiting period) after the wave velocity curves had began to rise, with the waiting period amounting to 13 months for the Tangshan earthquake. In addition, there were major deviations in the time lengths of the anomalous processes of wave velocity.

5) Different forms of gravity variations.

Gravity decreased by hundreds of microgals near the epicenter (relative to the outside area) 2 or 3 years before the Haicheng earthquake, but the gravity value increased relatively by more than 100 microgals before the Tangshan earthquake.

6) Major differences in macroscopic anomalies.

There are different types of these: intensity, appearence time, spatial distribution and migrations of macroscopic anomalies before separate shocks. For example, it was remarkable that the macroscopic anomalies appeared 2 months before the Haicheng and Songpan earthquakes, with their gradual migrations toward the epicenters. However, they were known to approach very late and close in the case of the Tangshan earthquake: 1 or 2 days prior to the mainshock, and, though not with the same intensity, before the Haicheng and Songpan earthquakes.

7) Regional differences in ground–water level anomalies.

Before several large earthquakes in North China, the ground–water level tended to decrease during a lengthy period, with relatively distinct stages in the anomalies, while it tended to fall in a shorter period, with uncertain stage division of anomalies, before some large earthquakes in Southwest China.

8) Regional discrepancies in deformation characteristics during earthquakes.

As stated in Chap. 6, the ratio between maximum descent and maximum uplift during the large earthquakes in North China was 5:1–7:1, but it was 1.5:1–4:1 for shocks in Southwest China. The most significant coseismic vertical deformation can be observed near the mainshock epicenters in North China, but in Southwest China it occurred at the end of a fault dislocation segment. Meanwhile, the ratio of coseismic horizontal deformation to vertical deformation in North China was 2:1–3:1; whereas in Southwest China it was 9:1–12:1. In addition, some earthquakes in North China were accompanied by strike–slip faulting with normal faulting component, and those in Southwest China displayed strike–slip faulting with reverse–faulting component.

9) Different features of aftershock activities.

The aftershock frequency of some large earthquakes in North China attenuated more slowly than in Southwest China, with a long duration of strong aftershock activity, and high b and a values for the aftershock series.

10.4.2 Analyses of Differences Between Anomalies

All of the nine great earthquakes are tectonic ones at a shallow–depth within the continental crust, and the seismogenic conditions for them are generally alike. This makes it possible to derive a unified insight into general seismogenic processes. The major differences presented during a seismogenic process may be due to the following three important elements, as we have suggested: (1) regional differences of earthquakes; (2) earthquake types; and (3) historical conditions for a seismogenic preparation.

1) Regional differences of earthquakes.

As discussed in Chap. 3, since there were obvious differences in the physical properties of crustal rocks, crustal layering, tectonic frameworks and regional dynamic environments in both the North China seismic region and the Southwest China Sichuan–Yunnan seismic region, there were different rupture modes, sequence types, aftershock attenuations, and anomalies of pre–seismic seismicity patterns, crustal deformation and ground–water level as well. In brief, the Sichuan–Yunnan seismic region is a combined body with thick–crust–compressed deformation areas and fault blocks cut by large regional faults of strike–slip nature, intensely acted upon by regionally marginal dynamic forces. Within this seismic region the large earthquakes occurred mostly at fracture zones, and the anomaly features of crustal deformation and ground–water level were mainly associated with the shallow–depth regime of strike–slip fault zones. Whereas in the North China seismic region the crustal thickness was small, and the crust was deformed due to volumetrical compression, extension and contortion, cut by numerous large or small fault zones, significantly acted upon by regional marginal dynamic and subcrustal thermo–dynamic forces. Within this seismic region, the large earthquakes occurred mainly at secondary or side–tectonic intersections related to large regional faults, or in small fault blocks, instead of being right within large fault zones. The crustal deformation and ground–water level anomalies seemed to be related with widespread recent accumulations and the deformation of water–saturated crust.

2) Earthquake types.

Research on earthquake types covers a wide range, involving various aspects of classification studies in accordance with earthquake origin, seismic intensity, focal depth, earthquake sequence, focal mechanism, source medium and tectonics. Different types of earthquakes directly influence the behaviour of precursory anomalies. For instance, the tectonic types of earthquakes would determine the spatial distribution patterns of precursory anomalies; and the homogeneity of source mediums might be associated with the occurrences of foreshock activity. In this book, great importance will be attached to discussing the effects of tectonic types of earthquakes on a multiplicity of seismogenic processes (see Chap. 11).

3) Historical conditions for a seismogenic preparation.

We can imagine that the present seismicity is only a short episode in the long history of crust movement for a given hypocentral region, and that the seismogenic processes of earthquakes are based on prequake crustal stress and state of the medium. Therefore, it cannot be expected that two large earthquakes with the same magnitude but different preparation conditions should have similar features of anomalous variations. In other words, the differences of anomalies may also be related to the initial stress state of a seismogenic process.

CHAPTER 11

Sources of Tectonic Earthquakes

A tectonic earthquake process develops within the earth crust. In principle, geophysical techniques available today cannot be used to directly survey an earthquake process within the crustal source region. Therefore, currently studies on source processes as well as theoretical assumptions have been made mainly based on surface and shallow–depth measurements and the experimental results from small–scale rock samples in the laboratory. In the previous chapters, the data on seismicity, crustal deformation, telluric current, geomagnetism, gravity, ground–water level, hydrochemical composition of ground–water and so on obtained before and after the nine large earthquakes were all recorded at the earth surface. The variations of these parameters may be associated with seismogenic processes within the crust, but in some of them, such as seismicity, geomagnetism and gravity, the variations may be caused by subcrustal processes. Field observations and some laboratory experiments make it necessary to carry out theoretical studies so as to acquire knowledge on crustal seismogenic processes and characteristics.

Some seismologists might think that at present great importance should be attached to accumulating more observation data. No doubt, a great deal of accurate data obtained may form a solid foundation for theoretical research. However, a relatively reasonable theory must exist which corresponds to each developmental stage of a science, and which is, furthermore, very conducive to its further advancement. This has been vividly proved by the developmental history of natural sciences. After the 1906 San Francisco earthquake, Reid (1910) proposed the "elastic rebound hypothesis of earthquake occurrence" in accordance with the data of crustal deformation measurements.Some theses of his theory have been adopted by many seismologists and maintained by them even today.

The purpose of this chapter is not to describe the history and contents of studies of focal physics, and details of different views points, but to depict and discuss some aspects of focal physics of tectonic earthquakes, with special emphasis on the important role of crustal structures in seismogenic processes, in combination with the features of these processes in the case of the nine large earthquakes.

On the strength of the data mentioned in the previous chapters, the following fundamental understandings of physical processes of these earthquakes may be gained: (1) These are shallow intracontinental tectonic earthquakes (dominant in 10–30 km); (2) A majority of these earthquakes are due to stick–slip movements on the pre–existing fracture planes within the crust; (3) Various geophysical phenomena, seismogeological background, coseismic crustal deformation and aftershocks-present in the source region and surroundings can be classified as of different types. Therefore, it is impossible for us to propose a unified focal physics model applicable to all the nine large earthquakes. The discussions below will be particularly on the crustal environments, fault friction state, condition and mode of stress concentration (fault plane locking) and release (fault plane unlocking) in source regions, as well as the tectonic origin of differences in seismogenic processes, while research on the classification of seismogenic processes will be also emphasized.

11.1 Source Physics of Tectonic Earthquakes

The whole process of a tectonic earthquake consists in that during the action of regional tectonic stresses and shearing crustal deformation stresses are concentrated at or near some particular localities (locking segments with relatively high strength) of fault planes within crustal mediums, fault planes rupture and displace rapidly, a major rupture develops and terminates, and stresses are then readjusted and tend to be balanced within and around hypocentral areas. Therefore, source physics is naturally divided into three main components: (1) physics of earthquake preparation; (2) physics of earthquake generation; (3) physics of earthquake aftermath.

In general, source physics involves rather broad aspects for studies, as of crustal environments (chemical composition of crust, pressure status, temperature distribution, fluid flow as well as mechanical, thermal, magnetic, electrical and structural properties of rocks), and studies of conditions for earthquake preparation; research on modes of crustal movement and deformation features of different layers and blocks with inhomogeneous strength, and their relations with energy storage in source regions (stress concentration); studies on the history of microcrack developments within source regions and the accompanying geophysical phenomena during seismogenic processes; studies on dynamic and kinematic aspects of rapid ruptures of faults, and investigation of problems (field observations, experimental and theoretical research) of the starting and terminating conditions of faulting and aftershocks.

Source physics is an important component in earthquake prediction research, playing a significant role in the comprehensive understanding of the nature of various anomalous phenomena and their connections. Since the above–mentioned numerous scientific branches cannot be extensively dealt with in this chapter, the focal environments and structures of tectonic earthquakes, including modes of fault locking (stress concentration) and unlocking (stress release, faulting) and their relations to precursory anomalies are discussed mainly on the basis of the analyses of seismogeological backgrounds and seismogenic processes of the nine large earthquakes, mentioned in Chap. 3.

11.1.1 Implications of Seismic Sources

The implications of seismic sources vary greatly as seismological science advances, being related perhaps to the research subjects at different periods.

1) Seismic wave (point) source. Usually in seismometry, data on the travel–time of various phases of seismic waves are used to locate the position of

the seismic wave source, which is regarded as "point source", and understood as the starting point of an earthquake rupture.

2) Fault plane dislocation source. As Knopoff (1958) suggested, the total energy released by shock is proportional to the product of squared depth and length of fault planes, that is, a seismic source is related to the scales of earthquake fault planes.

3) Source body. The source is taken as a volume, in which seismic energy is considered to be in store. Benioff (1951b) suggested that before an earthquake, the rock strength limit can be reached everywhere within a source body, and during an earthquake the media underwent a rupture everywhere within the source body. Kostrof (1975) defined a seismic source as an irreversible non–elastic deformation area during an earthquake, including major fault and associated secondary fault. Therefore, the aftershock area is often considered as a source area. Bullen (1965) thought that though the fractured region was very extensive at the time of a great earthquake, a large amount of energy was released by the earthquake from a limited area, much smaller than the rupture zone. Such a limited area is called seismic source area. Fu Chengyi (1976) indicated that the energy released during an earthquake was naturally confined to where the fault was dislocated, but the range affected by the energy concentration was much bigger, and a considerably large part of crustal media was in a state of accelerated stress accumulations before an impending earthquake. Fu Chengyi termed this part of crustal media as "seismogenic zone" or "inflammation zone", which is much larger than the aftershock area. However, we suggested that the concrete role of crustal structures within a source region in a process of seismic energy storage should be also emphasized,namely, the source area is composed of several segments with relatively high strength (locked segments) of a fault, hence being prone to energy accumulations, and the energy results from deformations of the media within a large–scale region.

11.1.2 Two Types of Modes of Source Ruptures for Tectonic Earthquakes

The two types of basic source ruptures for the nine large earthquakes have been mentioned in Chap. 3.

1) Sudden shearing of pre–existing crustal fractures.

The nine large earthquakes mainly featured such a type of rupture. There might be large historical earthquakes at these original fractures, such as the Xianshuihe fault zone at which the Luhuo earthquake occurred. This mode of rupture corresponds to the so–called " stick–slip" motion in mechanics. "Stick–slip" is an important phenomenon in frictional sliding. In 1966, Brace et al. proposed that " stick–slip" might be a possible mechanism for a shallow tectonic earthquake. They considered the section, or rupture plane, or sawcut in some direction as a field fault, and they carried out stick–slip experiments on rock specimens with section. The rock samples were first experimented upon under a confining pressure of P = 2.14 kb, and then F was added at uniaxial direction, with variations of axial displacement recorded. When F was loaded to a certain extent, the opposing sides of a fault would be dislocated abruptly, with stress decreased. If the sample was stressed again, a similar sudden slip would be repeated.

2) Rupture events in perfect crustal rock mass.

The Longling earthquake may belong to this type (see Chap. 3). Brace (1966) also made brittle fracture experiments on granite rock samples without section. The stress–strain processes were similar for both perfect rock samples and those with section, with only slightly larger stress drop for the former.

We thought that the shallow crustal earthquakes in Mainland China were basically the outcome of reactivation of pre–existing faults,and even though there were no sufficiently large fractures within the rock mass for the Longling earthquake, there still existed some weakened planes (rupture zones)(see Chap.3).

In addition, for the movements of rock walls with fractures there was a "stable slip" motion, which seemed rather slow and homogeneous, similar to the creeping of natural faults, besides the discontinuous rapid "stick–slip" dislocation (Stesky 1978).

11.1.3 Crustal Stick–slip Source Environments

In physics of seismic source by Guo Zengjian and others (1979), it is indicated that there are nine aspects for researching crustal source environments (chemical compositions, pressure, temperature and water content of crust, mechanical, thermal and magnetic and electrical properties of crust, and structures of crust). As for the mechanical property of rocks, the rupture strength of perfect rocks or the frictional strength of a rock section may be critical elements for seismic sources to endure huge compresive stresses, and some factors of the above–mentioned crustal environments have considerable effects on the strength of rocks.

1) Crustal temperature and pressure and rock strength.

Based on experimental results of some rock samples and theoretical considerations, Guo ZengJian and others (1979) gave the temperature, confining pressure, and rock strength in different crustal depths, as shown in Table 11.1. It is

Table 11.1.

Depth(km)	Confining Pressure(kb)	temperature (° C)	Granite strength(kb)
0	0	0	1.5
5	1.3	200	5.95
10	2.7	350	7.50
20	5.5	550	7.35
30	8.5	700	6.6
40	11.5	850	4.8
50	15.0	980	2.6
60	18.5	1100	0

indicated that the crustal temperature, pressure and rock strength are increased with depth, but the rock strength is gradually attenuated below a depth of 20 km.

2) Crustal "stick–slip" and "stable–slip" conditions.

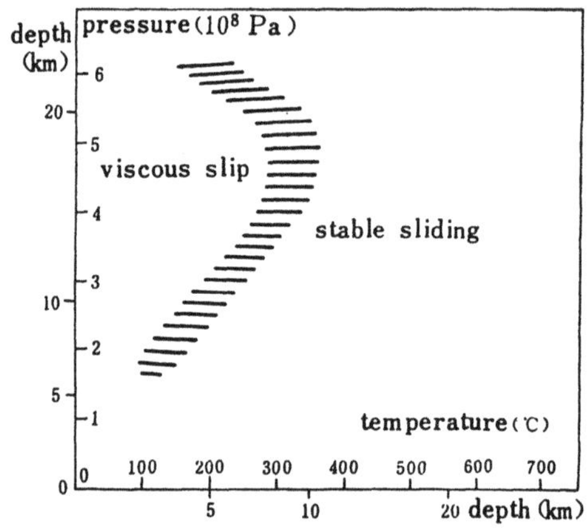

Fig.11.1. Influences of temperature and pressure on rock fractures

The conditions for stick–slip and stable–slip along fractures in the rock mass are very complicated, and the slip characteristics are associated in an intricate form with the presence of pressure (or normal compression), temperature, mineral composition, fault gouge, and water (or other fluid) and the geometry of fractures. Generally speaking, stick–slip motions easily occur under conditions of high confining pressure, low temperature, strongly brittle rocks (like quartz and feldspar), low asperity, and without fault gouge. By contrast, stable–slip motions are apt to happen if there exist a low confining pressure, low strength and especially high temperature. Stesky (1974) gave the range in temperature and pressure coordinates, within which stick–slip and stable–slip are prone to occur, based on the results of stick–slip and stable–slip experiments with rocks, as shown in Fig. 11.1. From this Figure, it can be seen that the motions are stable under high temperature and low pressure, and they are abrupt and rapid (stick–slip) under high pressure and low temperature. If the temperature and pressure coordinates in Fig. 11.1 were transformed into the crustal thickness according to the data in Table 11.1, then the upper crust (0–5 km) and the lower crust (greater than 15 km) are prone to stable–slip, and the middle crust (5–15 km) to stick–slip motions, even though, and this is very tentative, a certain qualitative concept can be acquired, that is, the middle crust possesses conditions for generating earthquakes. In other words, the middle crust acts as a part conducive to accumulations of seismic energy and occurrences of stick–slip motions (earthquake competent layer) (see Chap. 3). The data on focal depths of the nine great earthquakes show that these occurred just in the middle crust.

From the analyses on vertical crustal profiles, it is suggested that the middle crust is one of the most advantageous environments for producing stick–slip earthquakes. Furthermore, it should be noted that there exist certain differences in some factors which affect rock strength, such as temperature, water content and porosity in the middle crust. Even along the same large fault zone in the middle crust, some segments would be advantageous for generating stable–slip, if the hot mantle materials upwelled from these segments, or these segments were water saturated due to strong porosity caused by ruptures. Therefore, the rapid stick–slip motion at the pre–existing fractures in the middle crust (earthquake competent layer), provoked by regional tectonic stresses, is a particular expression of the tectonic movements of the crust, accompanied by aseismic stable–slip motions. In fact, shallow–depth earthquakes in the mid–crustal layer are distributed unevenly both in China and the world at large. In the sense of earthquake preparation, one of the prerequisites for stick–slip earthquakes along these segments is the stable–slip motions at the neighbouring segments. For example, before the Tangshan earthquake, the stable–slip motions might have occurred in the upper and lower parts of the earthquake–related fracture (see Chap. 6), thus, providing energy for producing fast stick–slip dislocation in the mid–part of the fracture.

11.1.4 Friction on Fracture Surface

In order to understand the mechanism of rapid shearing ruptures of pre–existing faults within the crust, it is very important to study the friction mechanism and other problems on the fracture surface of rocks.

1) Frictional strength of the fracture surface.

The frictional strength of rocks implies the minimum shear stress needed on pre–existing fractures to slide under given conditions of confining pressure, positive pressure, temperature, pore pressure and loading speed.

a. Generally speaking, frictional strength would be increased if the time of static contact of the fractures surface is a lasting one. Dieterich (1978) has made experiments on the relation between static frictional coefficient ($\mu_s = \tau / \sigma$) and static contact time upon rock samples, with the results satisfying the following expression:

$$\mu_S = \mu_0 + A\log(Bt+1), \tag{11.1}$$

where μ_s represents frictional coefficient in static contact time t (unit in second); μ_0, A and B are constants ($\mu_0 = 0.6$–0.8, $A = 0.01$–0.02, $B = 1.0$–2.0).

b. For the relation between the frictional strength and effective normal stress Stesky (1978) synthesized the experimental results of frictional strength on sandstone and granite and 13 kinds of rocks with joints, sawcuts or fractures under different temperature, loading rate and water–saturation conditions, indicating that within a range of temperature less than 400℃, the frictional strength τ was mainly proportional to the effective normal stress σ_n as shown by:

$$\tau = \sigma_n \mathrm{tg}(\varphi + i), \tag{11.2}$$

where φ denotes resistance of frictional slip in contacts of asperities of a fracture;

i,dilatancy angle, and tgi = $\Delta h / \Delta s$(h, normal displacement; and s,shear displacement). Usually, $\sigma_n = P - p$ (P, normal stress at fault plane, and p pore pressure), hence illustrating that frictional strength would be decreased as pore pressure augmented, even if the normal stress at a fracture remained unchanged.

Byerlee (1978) gave an empiric relation between τ and σ_n. If the normal stress $\sigma_n < 2$kb,

$$\tau = 0.85\sigma_n. \tag{11.3}$$

and if $2\text{kb} \leqslant \sigma_n \leqslant 20\text{kb}$,

$$\tau = 0.5 + 0.6\sigma_n. \tag{11.4}$$

c. The frictional strength basically seemed to be independent of temperature, in case the temperature was not close to that for medium softening.

d. The frictional strength would be decreased if there existed fault gouge amidst sliding surfaces (Byerlee 1978).

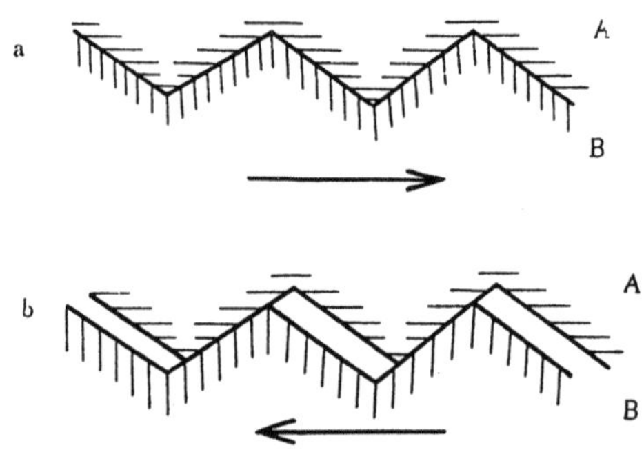

Fig. 11.2. Concavo–convex sawcut contact of a fracture surface

2) Modes of fracture surface contacts.

For realistic fracture surface contacts it is impossible to be ideally flat conjuctions. Preliminarily, there are two typical modes for fracture surface contacts:

a. Concavo–convex sawcut contact.

As shown in Fig. 11.2a, if shear stress acts on the two walls of a fracture, the A wall starts to move relatively along the sawcut oblique surface (as shown in Fig. 11.2b), and finally gets into a neighbouring sawcut slit. If the friction of the sawcut is very strong, great shearing force needs to accumulate to destroy the protruding sawcut enabling the two fracture walls to slide against each other.

b. Contact of the convex parts.

As shown in Fig. 11.3, since the area for contact of the fracture walls is fairly small, friction of the contact parts becomes so strong as "welding".

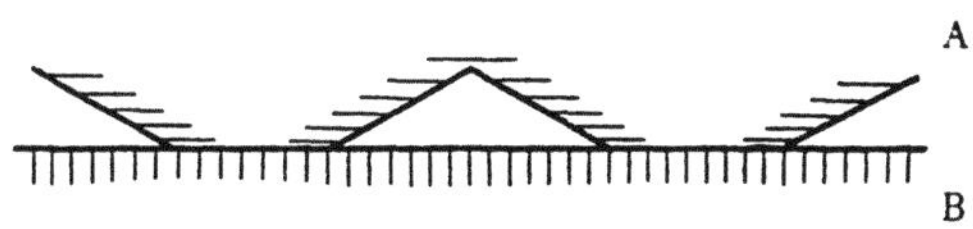

Fig. 11.3. Contact of the protruding parts of a fracture surface

11.1.5 Precursory Slip on Fracture Surface

Before the stick–slip motion of a pre–existing fracture surface on a rock sample, the precursory slip is very often observed, while during the precursory slip some properties (like pore pressure) of a rock sample would be changed.

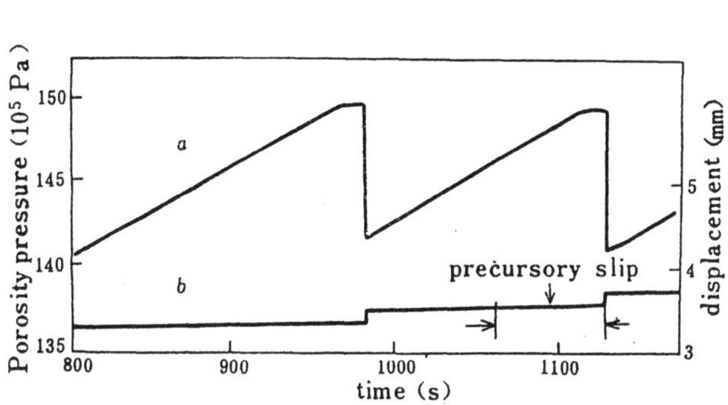

Fig. 11.4. The deformation and stick–slip curves of the sandstone sample under 500 b effective pressure; *a* pore pressure– time curve; *b* displacement–time curve

Figure 11.4 gives the experimental curves of precursory slip and pore pressure variations of the sandstone sample (Lagan 1978). The segment between the two arrows in this figure indicates precursory slip, and during this period the pore pressure increases. However, immediately prior to the stick–slipping, pore pressure diminishes or remains constant. This might be either because of increases of porosity due to ruptures of a fracture surface, or the opening of the preexisting fracture, or both happening at the same time. The porosity increase forces the pore pressure to expand. During the principal sliding, the pore further widens, and pore pressure drops rapidly.

In addition, the laboratory experiments made by Byerlee et al. showed that the stable–slip distances diminished intensively with the increases of confining pressure, but independently of strain rate, before the stick–slip motion at a fracture of granite. The softening of contact surfaces (corresponding to plastic deformation) might also be one of the reasons for precursory slip.

Before the stick–slip motion, the stresses are concentrated at the ends of "locked" segments of fracture surface, and the precursory slip may start to propagate from the ends to the center. The observed macroscopic precursory anomalies before some large earthquakes (the Haicheng and Songpan earthquakes) were distributed along the forthcoming earthquake–generating faults toward the epicenters from the outside (see Chap. 10). Furthermore, 1 or 2 months before the Tangshan earthquake, a fairly large–amplitude precursory slip along the fault was observed at the Ninghe deformation station located at the southern end of the earthquake–generating fault (see Chap. 6).

11.2 "Locking" of Source Fracture Planes and Stress Distribution

In the above section, the middle–crust competent layer for earthquake preparation has been described. Since there existed differences in mechanics and physics of the media, some segments were prone to stable–slip displacements, whereas the others were subject to strong friction and became so–called "locked" segments, due to regional tectonic stress action, even along the same fracture zone. The stable–slip movements of nearby segments as well as the deformation process of surrounding media would be contributive to stress concentration at "locked" segments. If further more one takes into account the complexity of crustal structure, a seismogenic process may be under way not only at a fracture, but also within a seismogenic body with multiple fractures as main structures.

We will now make qualitative analyses of the "locking" modes of single or multiple fractures within the crust, and the distribution patterns of stress and strain at or around the locking surfaces. The following sections will analyze the condition and manner of releases and of stress "unlocking" of fractures.

11.2.1 "Locking" Types of Fractures

1) Single Fracture.

Figure 11.5 shows the locking segments at a single fracture. Figure 11.5a presents a locking model of a fracture. The shadowed area represents the locking area and, of course, part of the unlocking segments may be contained in the locking areas. In fact, a single fracture is not always straight and flat in form, but also inflected or with abrupt bends (step–like valve) in some portions. The dimension of the inflexion and bend is much smaller than the length of a fracture, and the step–like valves may be also understood as a kind of locking (Fig.11.6). Such single fracture with locking segments can be used to simulate the seismogeological background for large strike–slip earthquakes, like the Tonghai and Luhuo ones.

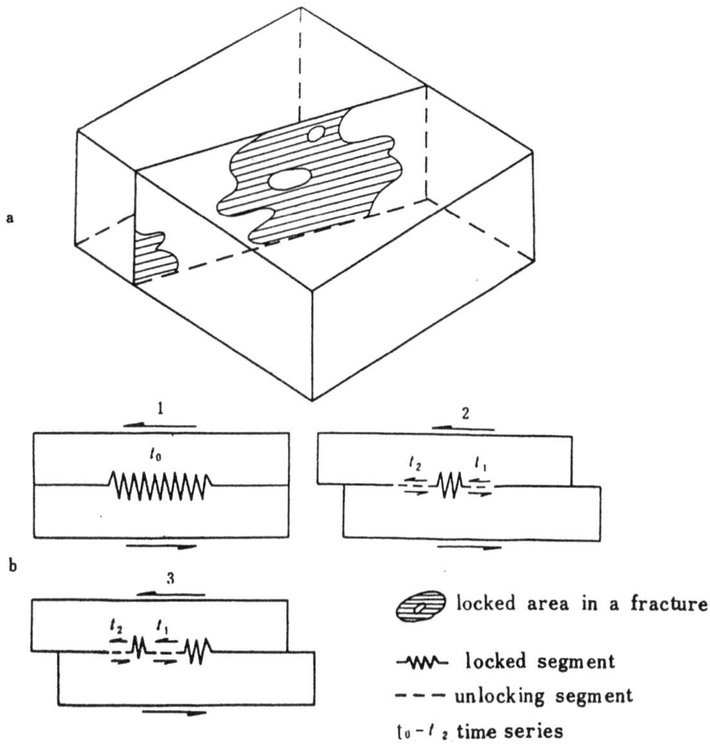

Fig. 11.5. *a.b.* Locking and unlocking of a single fracture surface; *a* Locking model; *b* unlocking mode; *1* diagram before locking (t_0); *2* unlocking starting at the two ends of a locking segment (t_1); *3* unlocking at the central part of a locking segment (t_2)

a

b

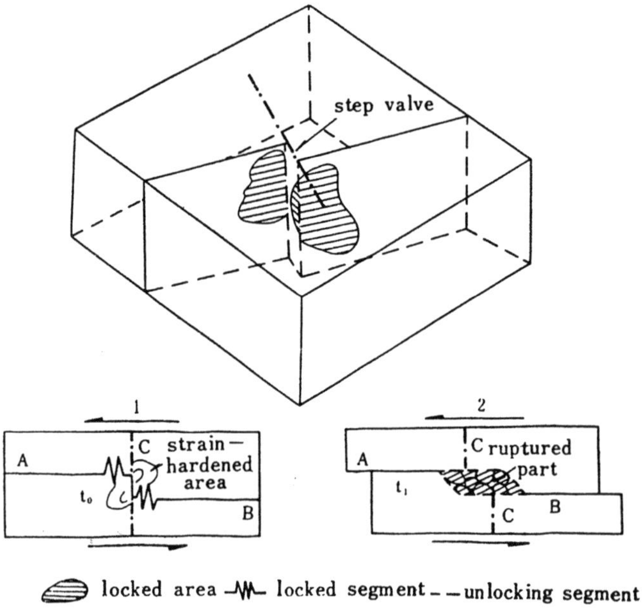

Fig.11.6. *a.b.* Locking and unlocking of a single fracture with step–like valve; *a* locking model; *b* unlocking modes: *1* locking segment diagram before unlocking (t_0); *2* unlocking (t_1)

2) Two Crossed Fractures.

Figure 11.7 shows the situation of two straight crossed fractures, with locking segments perhaps near to the intersection part.

Figure 11.8 presents a combination of Fig. 11.5 and Fig. 11.6. The earthquake–generating structures for the Haicheng and Yongshan earthquakes seemed to be similar to the situation shown in Fig. 11.7 or Fig.11.8.

3) Crossed and jointed multiple fractures.

Figure 11.9 indicates a complicated framework consisting of four fractures, with a set of crossed fractures inside, behaving like the seismogeological background of the Tangshan earthquake.

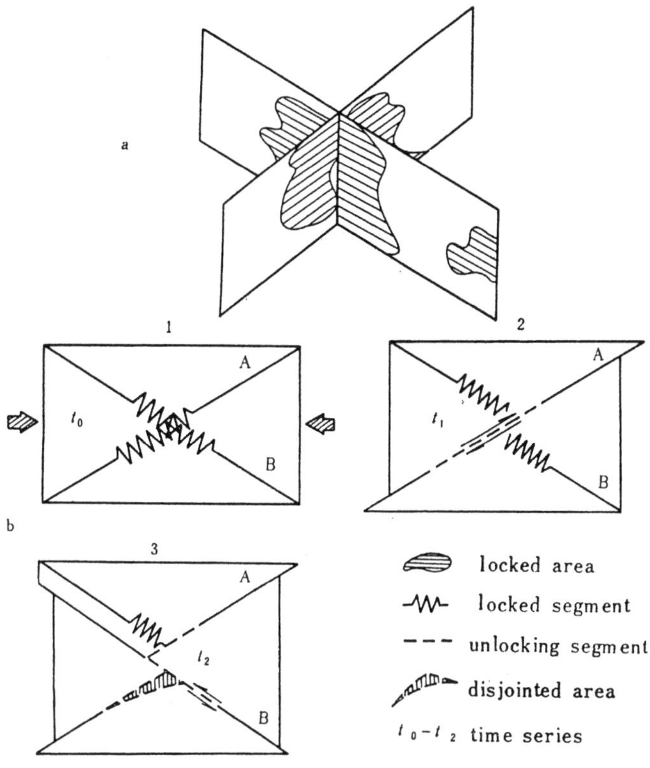

Fig.11.7.*a.b*. Locking and unlocking of two crossed fractures; *a* locking model; *b* unlocking modes; *1* before unlocking (t_0); *2* unlocking of fracture A (t_1); *3* unlocking of fracture B (t_2)

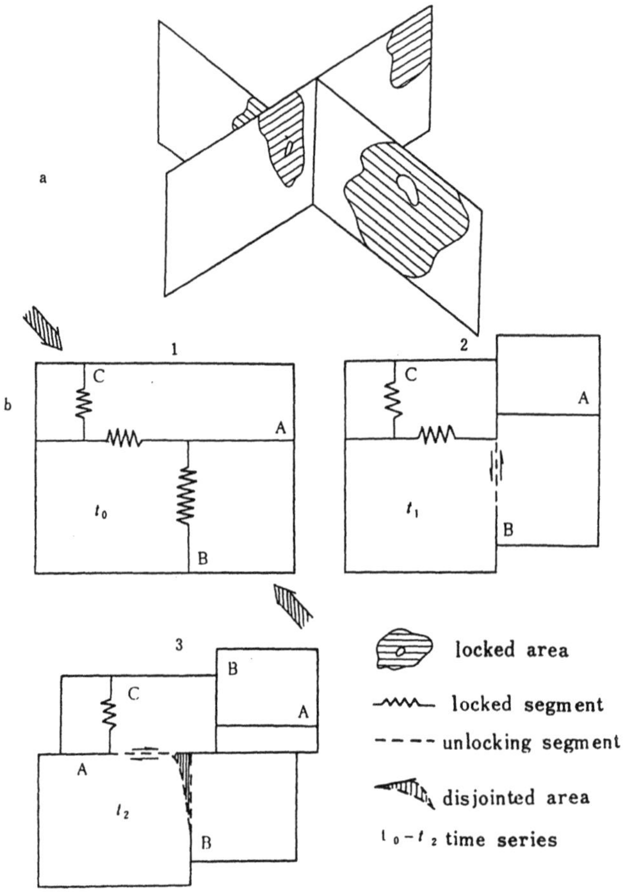

Fig. 11.8.*a.b.* Locking and unlocking of two crossed fractures; *a* Locking model; *b* unlocking modes; *1* before unlocking (t_0); *2* unlocking of fracture A (t_1); *3* unlocking of fracture A (t_2)

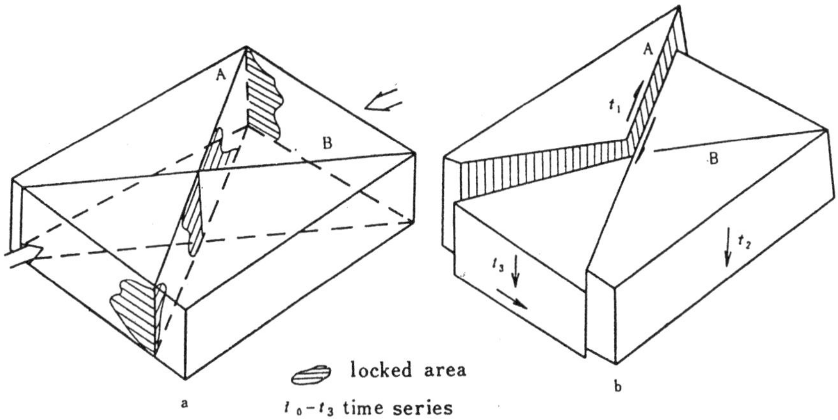

Fig. 11.9.*a.b.* Locking and unlocking of multiple fractures; *a.* model; *b* unlocking modes; t_1 unlocking of fracture A; t_2 rising and falling movements of the walls of fracture A; t_3 descending or lateral motions of small fault block due to disjunction

The above–mentioned fractures may either be the boundaries of crustal blocks, or the fractures within the blocks.

11.2.2 Stress–Strain Distributions

It seems impossible at present to offer an explanation of the stress–strain distribution within a crustal rock mass cut vertically and horizontally by faults under the action of regional tectonic stresses. A solution can be obtained only by assuming ideal crustal media, or an approximate solution for complicated situations can be derived by using the finite element method. Meanwhile, modelling of photoelastic tests in this regard can be carried out in laboratories. We shall take some results as an example to demonstrate the characteristic stress–strain distribution at locking segments of fractures:

1) Photoelastic tests.

Pan Qiuye et al. (1979) made analyses on the stress field of rock samples with typical cracks by means of the photoelastic method.

Figure 11.10 a and b show the results of photoelastic tests on stress distributions in rocks with a multitude of parallel and intersected cracks under uniaxial pressure. Some segments of these cracks are locking segments. The numerals in the figure represent the stress levels.

The test results indicate that:

a. The extent of stress concentration at the locking segments is related with the total length of cracks and the size of the locking segments. If the crack length is similar, and the locking segments are longer, the stress build–up would become stronger.

b. The ends, inflexion and intersected points are stress–concentrated localities.

2) Theoretical studies.

Niu Zhiren et al. (1976) established the mathematical model of one–dimensional earthquake preparation (Fig. 11.11) based on the combination model provded by Guo Zengjian et al. (1973). Figure 11.11 shows the locking elements $(-H/2, -L/2)$ and $(L/2, H/2)$ with stress accumulation $(-L/2$ and $L/2)$ and the sliding element with stress adjustment on the boundary $(y=0)$ of two blocks. The arrows in the figure represent the driving modes of the blocks.

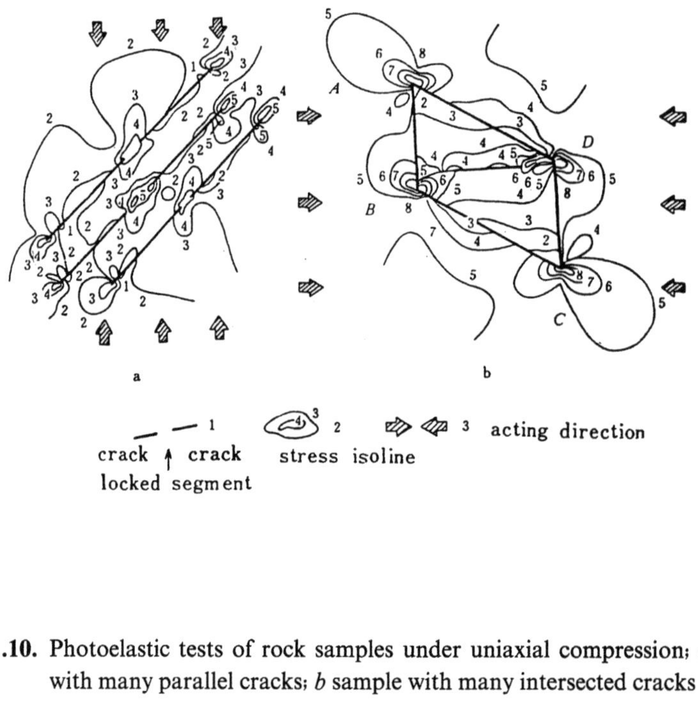

Fig. 11.10. Photoelastic tests of rock samples under uniaxial compression; *a* sample with many parallel cracks; *b* sample with many intersected cracks

In the model given by Niu Zhiren et al., the locking and sliding elements were simulated separately by the strengthened elastoplastic material and the Newtonian viscous body to derive the distributions of normal stress σ and shear stress of τ the model during elastic deformation:

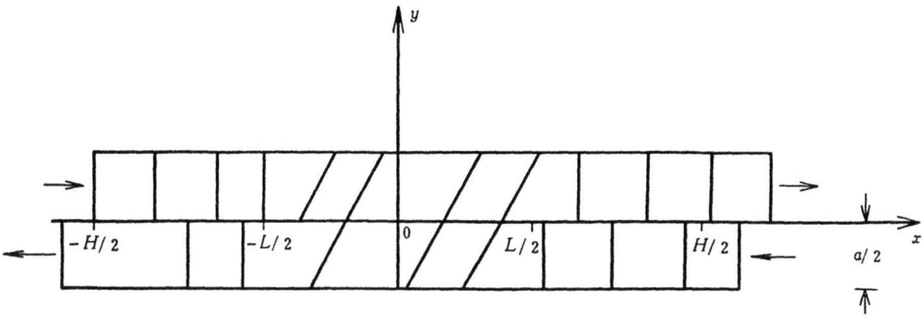

Fig. 11.11. Seismogenic model of one–dimensional source

$$\sigma(x, t)=\begin{cases}\dfrac{4V_0\eta}{a^2}\left(x-\dfrac{x}{|x|}\cdot\dfrac{L}{2}\right)+p_1(t)\dfrac{x}{|x|}, & |x|>\dfrac{L}{2}\\ p_1(t)\dfrac{\mathrm{sh}\left(\sqrt{\dfrac{K}{E_0}}\cdot\dfrac{2x}{a}\right)}{\mathrm{sh}\left(\sqrt{\dfrac{K}{E_0}}\cdot\dfrac{L}{a}\right)}, & |x|\leqslant\dfrac{L}{2}\end{cases} \tag{11.5}$$

$$\tau(x, t)=\begin{cases}\dfrac{2\eta}{a}\left[V_0-\dfrac{dp_1(t)}{dt}\cdot\dfrac{\dfrac{H}{2}-|x|}{E}\right], & |x|>\dfrac{L}{2}\\ \sqrt{\dfrac{K}{E_0}}\,p_1(t)\dfrac{\mathrm{ch}\left(\sqrt{\dfrac{K}{E_0}}\cdot\dfrac{2x}{a}\right)}{\mathrm{sh}\left(\sqrt{\dfrac{K}{E_0}}\cdot\dfrac{L}{a}\right)}, & |x|\leqslant\dfrac{L}{2}\end{cases} \tag{11.6}$$

$$p_1(t)=\frac{V_0t-\dfrac{2V_0\eta}{a^2E}\left(\dfrac{H-L}{2}\right)^2}{\dfrac{H-L}{2E}+\dfrac{a}{2\sqrt{KE_0}}\mathrm{cth}\left(\sqrt{\dfrac{K}{E_0}}\cdot\dfrac{L}{a}\right)},$$

where, p_1 (t) : normal stress at L / 2 at time t during elastic deformation.

E_0,K: Young's modulus and shearing resistance coefficient before yielding of the locking element.

V_0: driving velocity.

η: viscous coefficient of the adjusted sliding element.

Equations (11.5) and (11.6) show that either the normal stress or the shear stress at the ends (± L / 2) of the locking elements with stress accumulation reached the maximum value. The normal stress was mainly distributed near the ends of the locking element, while the shear stress was transferred at the same time toward the center of the element. This result is not in contradiction with the result of the above photoelastic tests. In Niu's paper, the solutions for stress and strain during elastoplastic deformation stage and shearing rupture stage were also given.

11.3 "Unlocking" Conditions for Source Fuacture Planes

Under the action of regional tectonic stresses, if the shear stress acting on a locking segment of a fracture reached the strength of static friction, the locking segment would be dislocated rapidly, thus releasing energy, that is the unlocking of fracture planes. The conditions influencing the unlocking may be very complicated before the impending unlocking of a seismogenic fracture. In addition to the increase of shear stress at a fracture, the decrease of strength of static friction at the locking segment of a fracture, and the variations of the state of the medium around the fracture, will also play an important role.

11.3.1 Intensification of Shear Stress at a Fracture Plane

With the increase of regional tectonic stress level, if the shear stress acting on a fracture had been also built up, the fracture would be unlocked quickly when the shear stress level reached the static friction strength of the fracture.

11.3.2 Attenuation of Friction Strength of a Fracture

There may be many reasons for the decreases of friction strength of a fracture:

1) Pore pressure increased.

As Eq. (11.2) indicates, when pore fluid existed in rocks, the friction strength of fault planes would be in direct proportion to the effective normal stress ($\sigma_n = P - p$). If the pore pressure P increased faster than the normal pressure p during compression, the friction strength would be made to attenuate.

If compression of the pore water in rocks was more difficult than of the rock framework, the water could help the rock framework to endure compression, while pore pressure increased during compression of the two sides of a fracture. If the pore water became vaporized due to other factors, pore pressure would also be expanded. These situations can make the friction strength of a fracture weaken (Guo Zengjian et al. 1979).

2) Fluid wedging.

If remarkable pressure differences exist between some neighbouring locations in the crust, the fluid may wedge itself in where the pressure is small. If the fluid became squeezed somewhere into the locking segment of a fracture, the friction of the fracture would be greatly reduced as the fluid failed to support the shear stress. Meanwhile, if the squeezing fluid acts as a kind of solution which at a certain temperature is prone to chemical reaction, the rupture would be induced to grow and expand slowly.

3) Microfracture growth.

It is well known that if in the pressurizing test on rocks the acting force reached one half of their rupture strength, the microcracks began to develop, and at the same time they seemed to be concentrated in form of a band or cluster near the fracture plane, prior to the imminent macrorupture. This phenomenon not only illustrates that the shear stress intensified along this direction, but it also indicates that the stresses at the forthcoming fracture and in its vicinity had approached the rupture strength of rocks.

In addition, the dehydration effects of crustal rocks under high temperature and high pressure, and the heat at a fracture produced by precursory sliding, will also reduce the rock strength or the friction strength of fault planes.

The anomalous variations observed before the nine great earthquakes have all demonstrated the existence of the above unlocking conditions. For example, the band–like distributions of small quakes at or around the forthcoming earthquake–generating fault before a large shock, and the penetration of high–pressure water in the deep low–resistivity layer into cracks of a source region, are important factors for speeding up the occurrence of large earthquakes.

11.3.3 "Abdication" and Bending Disjointedness Within a Source Region

The combination model of earthquake preparation given by Guo Zengjian et al. (1979) emphasized that before a large earthquake the border of the opposing sides of a source fracture which is in the frocess of dislocating should give a seat ("abdication") to dislocated rocks, and create conditions for stopping the propagation of fault dislocation. Such a border is called the adjusted element. The adjusted element may be the creeping fault segment or the geometrically irregular area with low rock strength (such as ruptured area, plastic area, thermal anomaly area and water–saturated area at intersections of fractures). The adjusted element is an indivisible part of the combination model. One of its effects is to make shear stresses concentrated at two ends of the locked stress–accumulated element, hence being contributive to the unlocking of a locked segment.

In profile, the crustal is layered and dissected by faults of different scales, the crustal rocks being flexured under the action of great horizontal tectonic stresses. If there are differences in bending degree for various layered rocks, the rock layers would be "disjointed" due to discontinuous deformation (Fig. 11.12a), and if the friction of various segments of fractures which dissected the layers is not homogeneous, disjointed displacement would also take place (Fig .11.12b). The disjointedness of rock layers may be the best abdicating condition for source dislocation. After the great Tangshan and Xingtai earthquakes, the subsidence areas observed in the epicentral regions (see Chap. 6) might illustrate that there existed perhaps disjointed abdication in or around the source areas before the mainshocks.

11.3.4 Triggering Effect of External Factors

In a seismogenic source area, when the shear stress at a locked fault segment reached a critical state of the friction strength, and the external factors conducive to unlocking dislocation of the fault occurred, shocks would be triggered and unlocking accelerated, even under small external action. The short–period variations of lunar and solar tidal forces, atmospheric pressure, inertial force of the Earth's rotation, and polar wandering may produce such a triggering effect.

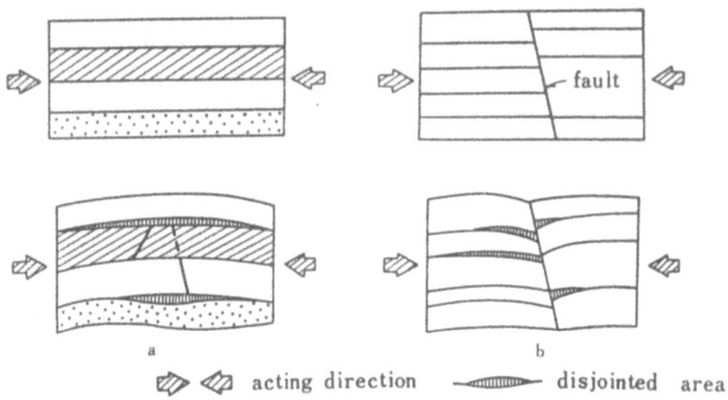

Fig. 11.12. Disjointedness of rock layers; *a* axial disjointedness; *b* disjointedness near faults

Many earthquakes seemed to take place during or around full moon and new moon. For example, the Xingtai earthquake occurred on March 1 of the lunar calendar year, and the Tangshan earthquake took place on July 2 of the lunar calendar year. The triggering effects of atmospheric pressure in the nine great earthquakes were noted in Chap. 8. As for the relation of variations in the Earth's rotation rate with earthquake occurrences, this will be discussed in the last section of this chapter.

11.4 "Unlocking" Types of Source Fracture Planes and Aftermath

When the locked element (stress concentration) of a source fracture gets ready for unlocking, the unlocking of the latter will begin at a place of maximum shear stress, or of the most intense temporary decrease of static friction strength, or near a full "abdicated" fault segment. Once the rapid displacements due to fracture unlocking has taken place, the stresses over a fracture would be promptly redistributed, and the stress–strain state of the surrounding media and fractures would be strongly disturbed and quickly changed. After main ruptures (unlocking), the mode of repeated or secondary ruptures (unlocking) are closely related with the seismotectonic pattern and the redistribution of stresses, thus affecting directly a series of geophysical aftermath phenomena, like aftershock activity and others.

We shall attempt to describe qualitatively the unlocking process under different locked conditions, as follows.

11.4.1 Unlocking Dislocation of a Single Fracture

In case of a locking of asingle fracture (see Sect. 11.2), the stresses are concentrated near the ends of the locked segment and, as a result, the unlocking will start from an end (see Fig. 11.5b–1), or from a local part where the static friction strength is decreased at the locked segment (Fig. 11.5b–2). The initial unlocking may not be able to make the whole locked surface relaxed, and some residual parts of the locked segment will remain locked. Furthermore, during a directional unlocking process, the unlocking will proceed over a highly strengthened locked segment. Once the unlocking front reaches a low–stress area (like plastic, high temperature and ruptured areas), the dislocation will stop. With such unlocking manners and processes, the reason why the mainshocks were located at the extremes of source areas, or near the central parts of source areas, and why the late strong aftershocks occurred possibly at some residual locked segments (aftershock activity gaps) can be understood.

11.4.2 Unlocking Dislocation of Two Crossed Fractures

As for the locking types of two crossed and intersected fractures (Fig. 11.7 and 11.8), their intersections might be locked parts of fractures. The unlocking will start firstly due to displacement of one of the fractures, then the principal dislocation will excite the unlocking of another fracture. Figure 11.7b shows such an unlocking process. The unlocking is initiated at fracture A, and after it has been set off, the great momentum of the rapid motion of the rock walls is apt to promote the unlocking of fracture B. This situation seems to arise readily in the case of two non–orthogonally crossed fractures. Assuming that the original locking of fracture B was weak, a series of small earthquakes would take place along fracture B. Such phenomena were clearly observed in the Xingtai, Yongshan and Haicheng earthquake sequences. If the energy accumulated at the original locked segment of fracture B was not released because of the unlocking of fracture A, major earthquakes may perhaps also occur at fracture B.

During the fast and complex unlocking process, a temporary "disjointed" state may be generated, causing so deep crustal materials to well up (Fig. 11.7b–3). In addition, the block which was loosened in the course of the unlocking process might be turned (Fig. 11.8b).

11.4.3 Unlocking Dislocations of Multiple Crossed Fractures

Figure 11.9 shows the case of multiple crossed fractures. Assuming that fracture A was first unlocked, the unlocking process would be more complicated, as there are multiple locked parts, and small blocks surrounded by multiple fractures would be set in motion in all directions so as to shake off the locking besides the above–mentioned two unlocking modes (single fracture and two crossed fractures). If there were disjointed segments at the blocks, it would be quite possible for the blocks to descend.In the Tangshan and Xingtai earthquakes, distinct vertical motions accompanied the significant horizontal movements. Such phenomena appear to be related with the existence of disjointedness. What is more, the horizontal and vertical motions might occur successively at different times.

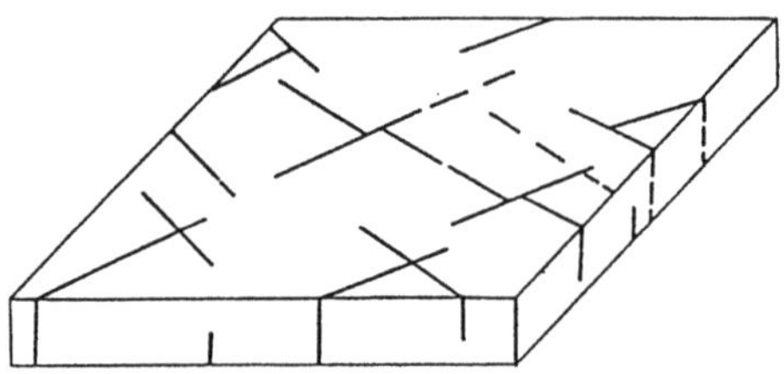

Fig. 11.13. Net–like distribution of rupture zones in a perfect rock body

We would like to mention yet another kind of tectonic framework of earthquake preparation, namely net–like distribution of rupture zones in a perfect rock body, as shown in Fig. 11.13. These rupture zones act as weak zones in the rock body, and under certain conditions seismic events would take place due to fast dislocation of rocks on both sides of these rupture zones. The Longling earthquake (M = 7.3 and 7.4) has been considered to be the consequence of shearing slip of such rupture zones in a perfect granite rock body, as well as the result of two intense dislocations at the two crossed rupture zones (see Chap. 3).

11.5 Classifications of Seismic Sources

In the above sections, some aspects have been discussed concerning crustal environments, types of fracture connections, and the conditions and manners for the locking of fractures (stress concentration) and for the unlocking of fractures (stress release) in seismogenic processes of shallow continental tectonic earthquakes. As there exist clear differences in crustal environments, state of the medium, tectonic framework and acting modes of force for various regions, and as these are not homogeneous even in a same region, it follows that differences exist in the seismogenic processes of earthquakes, and it is necessary to emphasize the need for the study and classification of the characteristics of focal processes. Assuming that the property of the medium in the crust (including structures) and the mode of tectonic and regional actions are the two major elements which decide focal processes, it is a justified approach to classify seismic sources based on these two elements and the related time–space variation features of stress–strain processes in the source.

11.5.1 Classification Based on Source Medium Properties

Mogi (1962) found from laboratory rock experiments that the acoustic emission process can be divided into three types for samples with different grades of inhomogeneity under stable rate of loading: (1) in inhomogeneous samples, numerous microruptures are produced before principal ruptures; (2) in very homogeneous samples, no microruptures are recorded before principal ruptures; (3) in very inhomogeneous samples, a great number of microruptures are generated, without a principal one. The experimental results can be regarded as a basis for classifying time–series features of natural earthquakes. According to the above three characteristic microrupture series in rock experiments, the earthquake sequences can be divided into the mainshock, isolated, and swarm types (see Chap. 2).

11.5.2 Classification Based on Seismotectonics

Ma Zongjin et.al. (1980c) emphasized that the joining and crossing modes of active fault zones and the manner of their motions should serve as bases for seismotectonic classification, based on studies of the seismogeological backgrounds of the nine great earthquakes. They suggested three types of seismotectonics for almost vertical earthquake–generating faults, namely: single–fracture strike–silp faulting; couple–fracture alternated faulting; and multiple–fracture overlapped faulting. In Sects. 11.2,3 and 4 of this chapter, the stress–strain concentration and release processes for these seismotectonic types have been discussed, and the tectonic classifications of the nine great earthquakes given.

11.5.3 Classification Based on the Manner of Regional Stress Actions

There exist discrepancies in direction, magnitude and grade for different regional tectonic stress actions.

For example, the focal mechanism solutions obtained from properties of P wave onsets were used to infer the direction of the principal stress of regional tectonic stress actions, and so the regional features of recent tectonic stress fields in or around China could be also recognized (Yan Jiaquan et al. 1979). The results show that the regional stress field in South China is characterized by the WNW horizontal compressive stresses; the North China region was dominated by the NNE horizontal compressive stresses; and in most parts of the Qinghai–Xizang plateau to the west Mongolia, the SN or NE horizontal compressive stresses are in action. The corresponding classifications of the nine great earthquakes are consistent with the above features of regional stress fields.

The influence of variable loadings and different strain grades on the rupture process of rocks can be observed in laboratory experiments (Chen Yong 1977).

11.5.4 Classification Based on the Characteristics of Precursory and Aftermath Anomalies

The differences in the properties of earthquake precursors and aftermath indicated that there were differences in the variation of medium and stress states during a seismogenic process. For earthquake prediction research it is therefore very important to classify the features for both precursors and aftermath effect. In the first eight chapters of this book attempts at classifying observation data about the processes of the nine great earthquakes have been discussed..

11.6 Seismic Energy

In the first section of this chapter we mentioned that the energy supplied and accumulated for earthquake preparation originated from large–scale regional tectonic deformation movements. Seismogenic processes should then be connected with the mode of crustal tectonic movements. Actually, this involves the energy problem of crustal movements too. Still, this is an extremely complicated problem requiring wide–ranging discussion. Here, we shall outline several potential sources for earthquake energy, mainly based on data concerning seismisity and others.

11.6.1 Plate Movement

The focal mechanism solutions of the majority of earthquakes in China have revealed that the recent crustal tectonic movements there are dominated by horizontal motions, and the directions of compressive stresses in regional tectonic stress fields are closely associated with the westward movement of the Pacific Plate and the northeastward movement of the Indian Plate, pushing against continental China (Shi Zhenlian et al. 1973, Yan Jiaqian et al. 1979). Furthermore, certain regular time–space correlations between seismicity in North China and in the Japanese trench (Wu Jiayi et al. 1979) have indicated that the earthquakes occurring in North China may be controlled by the underthrusting of the Japan Sea Plate. Therefore, many researchers have suggested that one of the sources for seismic energy in China may originate from the pushing force during movements of the oceanic plates.

11.6.2 Subcrustal Material Motions

The spatial distribution of some earthquakes are perhaps coincident with high–gravity gradient areas. This might signify that the upwelling of high–temperature and high–density materials of the upper mantle into the crust, accompanied by thermal stress, may provide a kind of earthquake energy. In addition, local mantle convection below the crust may also produce shear stress on the crustal basement, hence causing stable–slipping of subcrustal faults and setting off the earthquake preparation process at locked areas within the middle crust. We (Chang Guomin et al. 1981) discussed some precursory anomalies before some of the nine great earthquakes in the light of the above–mentioned idea.

11.6.3 Non–uniform Earth Rotation

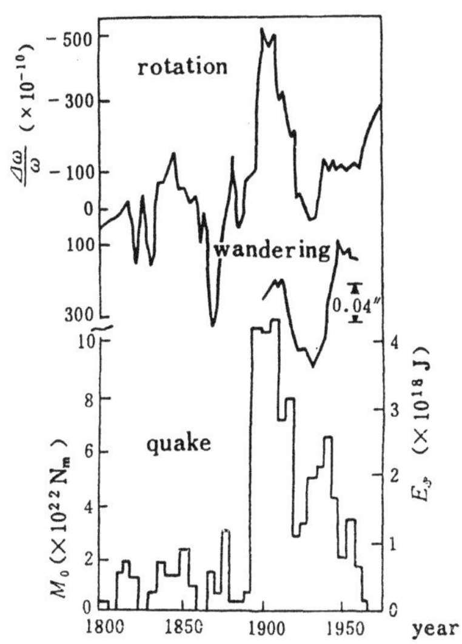

Fig. 11.14. Relation between global seismicity and non–uniform Earth rotation

There are two aspects in non–uniform earth rotation, i.e., variations of angular velocity and polar wandering, and the inertia and the polar drift forces produced thereby will act on the Earth, thus providing energy for crustal tectonic movements. Numerous papers about the relations between time–space distributions of global seismicity and non–uniform earth rotations have already been published. For example, Anderson (1974) gave a diagram of the correlation between global seismic intensity (shown by 5–yearly average values of seismic moment $M_0$and energy E_s) and variations of the angular velocity of the Earth's rotation($\Delta\omega / \omega$) and the amplitude changes of polar wandering (Fig.11.14). This figure shows that when the seismicity level increased, the rate of Earth rotation would be decreased. In addition, while studying the relations between the spatial distribution of shallow earthquakes in China (earthquakes of Ms$\geqslant$6.0 in East China, and of Ms$\geqslant$7.0 in West China) and the variations of angular velocity of the Earth's rotation, we discovered that the epicentral distributions of earthquakes, which occurred in years when the Earth's rotation rate increased or when the rate decreased, were separated by some longitude and latitude lines (Fu Zhenxiang 1981). The existence of the above correlations enable us to stress the importance of the non–uniform Earth's rotation in dynamic studies of seismogenic processes. Ma Zongjin et al.(1980b) analyzed some regularities of seismotectonics in the Eurasian continental seismic system, pointing out that the regularities could be explained more clearly by taking into account the dynamic factors of the Earth rotation more clear than by the theory of plate tectonics. Anderson (1974) definitely indicated that the Earth's rotation makes for a large amount of elastic energy being stored in the upper mantle and crust, and a small variation of the rotational parameters (velocity and polar wandering) might trigger off global seismicity.

However, either plate movement or mantle convection can be considered as a kind of motion in the interior of the rotating Earth. It is necessary to make detailed analyses as to which is has the leading or the secondary role in crust tectonic movements and the seismogenic processes of earthquakes, that is, to determine which driving force is the guiding factor for the actual behaviour of crustal movements in different regions and during different periods.

CHAPTER 12

Earthquake Field

As described above, a certain relationship exists between strong seismic activities in a large–scale region, and the various anomalies observed before and after large earthquakes may be possible expressions of crustal movements and material motions of the mantle, so that it is necessary to study certain problems concerning crustal structures and stress states distributed due to regional tectonic forces, formations of seismic sources, and effects of seismogenic preparation and generation on the development of regional stress fields. This chapter will firstly deal with the conception of the earthquake field, and depict both research methods for studying crustal stress fields and major research results, then go on to propose some mechanical models of crustal structures and medium properties for the mechanical analysis of the earthquake field, and finally, give the results from analyzing earthquake fields in North and Southwest China by the use of the finite element method, and discuss relations between variations in the earthquake field, regional seismicity and precursory observations, with a view to offering some explanations for the large–scale regional precursory anomalies as seen from the angle of stress field.

12.1 Concept of the Earthquake Field

12.1.1 Single–source Field

The concepts of source and field have been widely employed in seismological research. Based on the assumption that the mediums are homogeneous, isotropical and fully elastic, the theory of elastic wave motion field has been established. This theory has principally emphasized the leading role of sources, that is, discussions restricted only made to the distribution of stresses or displacements at various parts of the field caused by the forces acting on, or dislocations of, a seismic source (namely, the so–called hypocentral stress field or displacement field). These have been expounded in detail in the seismological theory and the theory of fields. In studying the earthquake prediction problem for explaining various precursory phenomena observed before earthquakes, such a conception has been very often quoted, and a series of seismogenic models have been thus proposed, including the elasto–plastic hardening model, the dilatancy model, the inclusion theory, the cracks–connected theory, the microrupture pre–displacement model, the pursuance model, and so on. For details of these models, please refer to chapter 2 of Physics of Seismic Sources coauthored by Guo Zengjian and others (1979). These models all discuss the influence of variations in rock properties, development, growth and mutual connections of cracks, and fluid flow in seismic sources on the mediums around them, suggesting that a variety of anomalies observed before earthquakes originated from seismic sources, and were also constrained by focal processes. In these models only the influences of a single source were discussed. We have called such spatial distributions of the influences on the surrounding medium produced by variations in rock properties or the mechanical state of a single seismic source, the single–source field.

Studies of the single–source field have involved the rupture process of a source, and the signle–source field can be used to explain a part of the observed anomalous phenomena. No doubt, this is an important aspect for research in earthquake prediction. However, regarding the relation between a field and a source, it must be emphasized that the sources are assumed to be homogeneous. These models are put forward based mostly on the results of experiments with standard rocks, or on the constitutive relations of idealized rock elasticity and elastoplasticity.

12.1.2 Multiple–source Field

While using the single–source field theory discussing the nine great earthquakes, we encountered difficulties in two aspects: (1) The distribution area is very large for the anomalous points observed before and after the large earthquakes, it being beyond the scope of what is described in some models. (2) Strong earthquakes occurred successively in both North and Southwest China, with very small distances between these earthquake areas, so the areas affected before and after these earthquakes very often overlapped. This problem remains yet untouched in the study of the single–source field.

There are the following interpretations for the anomalies in a large–scale region:

1) The inflammation hypothesis (Fu Chenyi 1976), in which Professor Fu noted that "... before an earthquake occurred, a considerable part of earth mediums was under a state of accelerated stress build–up, and this part of the mediums was the seismogenic area, the inflamed area. Due to rock deformation, material migration and other forms of material motion in the seismogenic area, the mediums of the upper earth layer was changed on a large–scale in nature. Such variations might involve changes in mechanical properties, electric properties, magnetism, thermal properties and other physical and chemical aspects. Various earthquake precursors are just reflecting these variations". This hypothesis makes use of material migration and motion to explain precursory anomalous phenomena in a large–scale range.

2) The viewpoint of "block movement" and "common zone movement". Guo et al. (1979) suggested that when a large earthquake was in progress on a boundary of crustal blocks, the entire crustal block might be motivated as pre–seismic slip took place, hence, there could be anomalous expressions on other boundaries of the block, and the larger the block was, the farther away, the anomalies appeared. Meanwhile, it was thought that various segments of the same fracture zone might be in motion simultaneously. The appearance of anomalies in some segments implied that this fracture zone was commonly under in-

tense stress action and a large earthquake might occur at any other segment of the same zone.

This idea employed the boundary effects of block movement to account for the large–scale regional anomalies.

3) The idea of multiple (stress–concentration) point field (Ma Zongjin et al. 1980a), which suggested that the existence of inhomogeneous and discontinuous crustal media should be firstly taken into consideration in studying seismogenic processes and cluster development of strong earthquakes. Let us regard the crust as a multi–layered elastic plate body consisting of some discontinuous fault blocks. Acted upon by larger–scale regional dynamic forces, the fault blocks themselves are strained and relatively twisting motion occurrs along fault zones between the fault blocks. The stresses are thus concentrated at the locked segments of the twisting zones and the extremes of the fault zones. As a result, the fracture expand, creep, become connected or dislocated further, showing the uneven development of many stress–concentration points. Some of these points release their strain energy with the occurrence of moderate and small earthquakes or fault creeping, while some gradually develop into the largest stress concentration point (area), with support from the entire regional stress field, and in turn the evolution of this point will play a feedback role in the whole region. Such mutual influences and constraints of these multiple stress concentration points will decide the seismogenic preparation of earthquakes in various areas, as well as the development of anomalies in the fields. We have termed the stress–strain field commonly affected by multiple points, the multiple source field.

This idea essentially emphasized the predominant role of the field, instead of the source, which takes shape gradually in the process of stress–strain developments in the field.

12.1.3 Earthquake Field

On the basis of the above–mentioned research work, we have advanced further, evolving the concept of the earthquake field.

The stress field of the multiple–source field includes the basic stress field and the variable stress field. The basic stress field represents the stress distribution within the crust caused by the larger–scale regional dynamic forces (such as plate compression, earth rotation, mantle material convection, and so on). This stress field decides the basic state of crustal stresses, and tends to be rather stable for a considerably long time. The variable stress field involves two parts, one is produced by weak action of the external forces outside the crust, and the other is caused by variations of medium property and stress state in a region, and by mutual effects of various parts there. The variable stress field overlaps the basic stress field, with the magnitudes and directions of its various points changing frequently due to various factors. This varying stress field plays a major role in seismogenic preparation and generation of an earthquake. Therefore, we have simply considered such variable stress–strain field in the crust subjected to the action of a larger–scale regional stress as an earthquake field. The features and evolutions of earthquake fields are not only associated with crustal structures and the basic stress field, but are also related to motion of the mantle material. The following will lay emphasis on describing research methods and results for basic crustal stress fields pertinent to the nine great earthquakes, and discuss models of crustal structures closely related to the characteristics of earthquake fields.

12.2Methods for Studying Basic Crustal Stress Fields of the North China and Southwest China Regions , and Principal Research Results

12.2.1 Direct Measurements of Earth Stresses

Measurements of the currently existing stress state of the crust are usually called absolute measurements of earth stress values by the methods of stress recovery, stress release, and hydrofracturing. At present, the borehole coring stress release method has been mostly used in China.

Li Fangquan et al. (1979) performed stress release work in a shallow depth of 30 m in the North China region, obtaining the following main results:

1) The orientations of the maximum principal stresses at various points in the northern part of North China run roughly NWW or nearly EW. For the Tancheng–Lujiang fault zone in the eastern part of North China, the stress pattern is rather complicated close to the fault zone, but at points far away from the zone the directions of the principal maximum stresses are similar to those in the northern part of North China, still running NWW or nearly EW.

2) Measurements in the epicentral areas of the large earthquakes showed that the stress values in these areas were lower than those in the outside areas after the earthquakes. For example, the stress values in the epicentral areas of the Haicheng and Tangshan earthquakes were 92 $kg/cm^2$and 25 kg/cm^2, while those in the outside lying Dashiqiao (30 km or more away from the Haicheng epicenter) and Luanxian (about 50 km from the Tangshan epicenter) areas were 166 $kg/cm^2$and 58 kg/cm^2, respectively. Such was the case too in the Longling earthquake area in Southwest China. This was the result of stresses released in the epicentral areas after the earthquakes.

Although the direct stress measurement methods can provide the magnitude and orientation of stresses, the depth measured by these methods is very shallow, and these methods are much restricted by subcrustal rock conditions. Furthermore, the question how to use shallow–depth measurements to determine deeper stress values still remains unsolved.

12.2.2 Tectonic Mechanical Analysis

Based on the data about neotectonic movement, geomorphology, focal mechanisms, earthquake rupture zones and geodesy, the basic crustal stress state can be given comprehensively by using the method of tectonic mechanical analysis. The axis of maximum compressive stress and of tensional stress obtained from focal mechanism solutions provide information on stress variations during a seismogenic process, as their orientation is different from that of the overall action of regional tectonic forces. But it is usually suggested that the predominant direction of focal mechanisms for various earthquakes in a given region might be close to that of the action of regional tectonic forces. Therefore focal mechanism solutions can be still used to derive the basic crustal stress state in a given region.

The overall characteristics of the crustal tectonic stress fields in Mainland China from the Neogene up to the present, arrived at by using the above method, are basically consistent, with manifest continuity It is revealed that the orientation of the principal compressive stresses and the mechanical properties of major tectonic zones in various regions are rather stable. There was a significant variation in the crustal tectonic stress field in North China during the middle Tertiary, but since then the recent tectonic stress field basically remains unchanged, with the maximum compressive stress axis running NEE, and the minimum compressive stress axis NNW, horizontally. The tectonic stress field in Southwest China is relatively complicated, displaying subregional differences. The preferred orientation of the principal compressive stress axis in the Anninghe fault and Xiaojiang fault and to the east of it is NWW (280–300 °); in the Songpan area, including the Xianshuihe fault zone and east of it, and north of the Longmenshan fault, the predominant azimuth of the principal compressive stress axis lies NEE–NE (50–80 °); in the west Sichuan region north of the rhombic block of the west Sichuan–Yunnan region, the principal compressive stress axis is directed NNE–NE. In the northern part of the rhombic block from north to south the principal compressive stress axis changes its NE strike into a nearly EW orientation, and then to NW, and in the south rhombic block area the preferred orientation of the principal compressive stress axis is NNW (325–355 °); and in the Simao and Puer areas it is NNW (340–360 °).

By using this method, only the directions of the basic crustal stress fields and the orientations of the fault motions can be approximately derived.

12.2.3 Physical Simulation Method

With this method, certain modelling materials, such as photoelastic material, gelatin or soil with some carved–in cracks are used to simulate discontinuous structures of media to fit a certain region, and for some models various materials are used to simulate differences in lithological character, with stress or displacement acting on the boundaries, so that the internal deformations and stress distributions of the materials used can be observed.

Guo Shifeng et al. (1981) modelled the stress field for North China, suggesting that the region of maximum shear stress concentration obtained with loading in an EW direction was in fairly good agreement with the distribution of $M > 6$ earthquakes. This revealed that the present–day tectonic stress field of North China has a nearly EW orientation.

The results of photoelastic experiments in simulating the Southwest China region have indicated that this region has been subjected to a joint attack by three plates (Pacific plate, Indian plate and Philippine Sea plate), with the Indian plate acting predominantly. As a result, the Southwest China region is under intense NW–orientated compression which is diminishing southward from the north, and the "rhombic block" (namely the Kangding, Ganzi, Tonghai, Xiaguan areas and their surroundings) is intruding south–southeastwardly. This is an overall feature of the present–day tectonic stress field in this region.

Photoelastic experiments can reconstruct a relative stress concentration area, besides the general outline of a stress field. Presently, since mostly homogeneous modelling materials are used, this method is mainly adopted to study basic crustal stress fields.

12.2.4 Mathematical Simulation Method

This means that the mathematical analysis method of media continua mechanics is directly used to calculate a stress field. In 1945, Professor Li Siguang used the elastic plane strain method for computing the stress field of E–shaped structures. Since the finite element method has been introduced to analyze properties of crustal media, there have been considerable developments in the analysis of regional stress fields with the finite element method in recent years in China. By using this method, the structure body can be regarded as a combined body consisting of finite, simple and discrete elements with connections at the nodes. For each discrete element, a displacement model is to be established, and the rigidity and loading characteristics of each discrete element are derived according to the displacement mode. Finally, an overall analysis of the structure is made. This method simplifies the complicated problem, so that the displacement of various points within each element can be approximately described by some simple functions, and the whole problem can be represented by algebraic equations, instead of using a partial differential equation for solving the complex boundary conditions. By using the structural mechanics method, from the characteristics of the entire finite element system, consisting of the features of separated elements, a set of balance equations for the displacements of nodal points can be obtained, for so deriving displacement solutions, and computing stress values of the separated elements. For applications of the finite element method in geological problems, we refer to the book Elementary Solid Mechanics (Wang Ren et al. 1979).

The finite element method, which has been developed from the solution of linear problems to that of non–linear ones, and from the analysis of two–dimensional stress to that of three–dimensional stress, has furnished us with favourable conditions for the further study of the basic stress field and the variable stress field of a structurally complicated and heterogeneous body. However, little work has been done on computations of three–dimensional stress fields with the finite element method, due to insufficient understanding so far of the nature of subcrustal structures.

Wang Ren et al. (1987) used the finite element method of plane strain for an ideal elastoplastic body to simulate the earthquake migrations in North China since 1966, showing the influence of an event on the stress field over all the region, and revealing the problems of the variable stress field after earthquakes.

Wang Suyun et al. (1987) used the finite element method of plane elastic strain to model the recent tectonic stress fields in China, suggesting that the largest forces acting on the tectonic stress fields in China have originated from the

Indian plate, and are about twice as large as those of the Pacific and Philippine Sea plates.

Luo Huanyan et al. (1983) used the three–dimensional finite element method to simulate the stress state of a geological body with faults and soft layers, stating the amount of stress concentration and the effects of median principal stress on the maximum shear stress at the ends of a fault.

The above–mentioned studies have mostly dealt with the basic stress field of the regional crust or the stress distribution in certain structure and medium conditions. No doubt, these researches have provided a basis for studying crustal tectonic movement. Nevertheless, for earthquake prediction it is still necessary to understand the variable stress field which may overlap the basic stress field, that is, to study variations of the earthquake field in the process of seismogenic preparation and generation of an earthquake.

12.3 Mechanical Models of Crust and Evolution of the Earthquake Field

12.3.1 Structural Models of Crust

1) Structural framework.

In order to analyze the features of an earthquake field, first of all it is desirable to know the structures and medium state of the crust. Structurally, there exist many discontinuous bedded boundaries, and the discontinuous character is more apparent if the crust is dissected by vertical and oblique faults. The regional tectonic outlines in both North China and Southwest China, metioned in Chap. 3, show that there are discontinuous features in lateral (horizontal) crustal structures, and the epicenters are very often located at compound parts of some structures, forming a plane–netted pattern of seismotectonic framework. If we take further the vertical (depth direction) heterogeneity of the crust and the distribution of hypocenters at depth into consideration, the tectonic framework associated with earthquakes may be regarded as a complex stereo framework.

The inhomogeneity of crustal media has been studied already in geophysics. The upper crustal rocks seem to be brittle in nature, but below a depth of 20 km of the lower crust, the rocks seem to be clearly plastic and viscous, due to the increase of confining pressure and temperature. In addition to this vertically layered crust, the crustal media are also different laterally and vertically. From the differences in rupturing velocity, the pre– and post–earthquake quality factors of the media, and the b–values in recurrence rate curves obtained from seismic waves for the nine great earthquakes, it is indicated that there are clear differences in medium character among focal regions as well as between focal regions and surroundings. Therefore, the structural models of crust can be represented by a heterogeneity and discontinuous multiple–layered elastic platy body which is laterally–blocked, and vertically–layered, as shown in Fig. 12.1. In this figure, A and B denote two sets of tectonic zones separately, C_1 and C_2 represent earthquake–prone boundary and intermediate layers, which are developed to different extents, and L_1, $L_2$and L_3 show the layers with different physical properties in the Earth's crust.

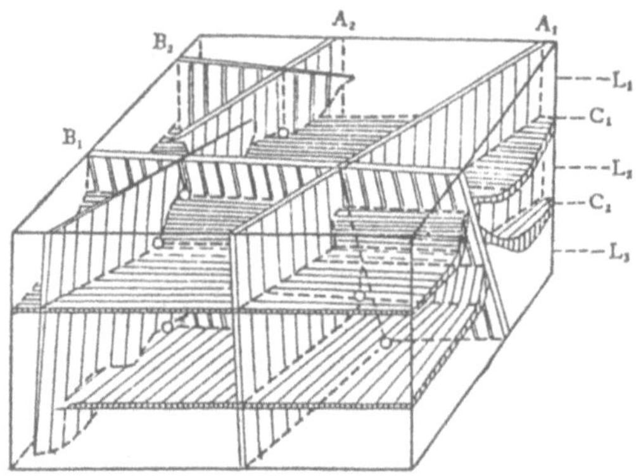

Fig. 12.1. Diagram of a crust structural model

The above–mentioned heterogeneious and discontinuous platy body overlies the mantle. Due to the thrusting force acting on the large regional tectonic margins, uneven shifts of mantle materials, and the action of the inertia forces produced by non–uniform rotations of the Earth, the platy body is compelled to move and deform, hence causing earthquakes in some particular portions.

2) Tectonic elements of the discontinuous platy body mainly consist of the following:

a. Fracture zone, which acts as a boundary for small fault blocks within the platy body or the marginal zone of the platy body. When the acting stress is small, with good cementation of a fracture (large friction coefficient), and the stress is less than friction strength of rocks in this zone, the rock mass satisfies the elastic stress–strain relation, while the stresses are highly concentrated in a discontinuity nearby, with main expression being shearing deformation. Therefore, we have termed this zone "torsional deformation zone", from an angle of the strain.

b. Creeping segment. At some segments of a fault zone, where the stress exceeds friction strength, or the friction coefficient of fault planes is reduced due to the effects of temperature and fluid flow, slipping occurs. The creeping segment acts releasing and trainsmitting stresses, and may also provide conditions for the intrusion of subcrustal materials.

c. Locked segment. There are some particular parts in a fracture zone, such as non–dislocated fault locations, blocking parts or intersections of multiple faults where there is a large friction coefficient of fault planes and sliding does not occur easily. Here the locking state makes its first appearance and prevails during a certain period of time. The blocked segment is the part where the stresses are very highly concentrated, or highly strained in a torsional deformation zone. This segment, by its existence exerts an influence on the relative motion of fault blocks and the uneven distribution of stresses. Such segments may be candidates for the occurrence of potentially large earthquakes.

d. Inter–zone area. This is a small fault–block body confined by various fracture zones. The media within this body are relatively uniform with respect to the entire platy body. The torsional deformation zones are manifestations of non–uniform deformations of these small fault–block bodies at their margins. The elastic deformation of the inter–zone area provides energy for the locked segment to dislocate.

The model of a crustal platy body is composed of the above–mentioned four elements. Some variations in an earthquake field reflect the mutual influences and conversions of these elements under stronger regional dynamic action.

12.3.2 Consideration of Mechanical Models of the Crust

Several mechanical models of a realistic geological body have been proposed. In order to select a mechanical model appropriate for the study of earthquake preparation and generation, the existing models, and a mechanical model we propose, are outlined as follows.

a. Elasto–viscous model ——The Maxwell model.

This model is composed of spring element and viscous element in series (Fig. 12. 2a), which can indicate constant creeping of a continuously deformed object in case the stress remains unchanged, and relaxation of gradually decreasing stress with deformation kept stable. This model features rheological property, suitable not only for analyzing a very slow permanent deformation process within the crust, but also useful for characterizing mechanical properties of the mantle media.

b. Firmo–viscous model ——The Kelvin model.

This model implies that the elastic strain is not completed immediately under stress loading, but there is a lag in strain. This model consists of spring element and viscous element in parallel (Fig. 12 2b), adaptable to analyzing the damping attenuation for a vibrating object, the attenuation of seismic waves, and the aftershock series. This model features solid property, so it is also called the Kelvin solid model.

c. Ideal elastoplastic model.

This model consists of plastic element and spring element in series (Fig. 12.2c). When the stress is smaller than the elastic limit, the model shows elastic stress–strain character. When the stress reaches the elastic limit, large plastic strain may be produced, and if the stress is reduced, the deformation would recover in line with the law of elasticity. The shortcoming of this model is that the deformation will be infinite when the stress exceeds the elastic limit.

d. The Bingham substance ——a non–linear model.

This model connects the elements of the ideal elastoplastic model with the viscous elements in series (Fig. 12.2d). When the stress is smaller than the elastic limit, it is an elastic body, and when the stress exceeds the elastic limit, the permanent strain would be increased constantly with time. This model features an elastic limit, and it was thought that this model could better represent long–term and large–scale orogenic movement compared with the above models.

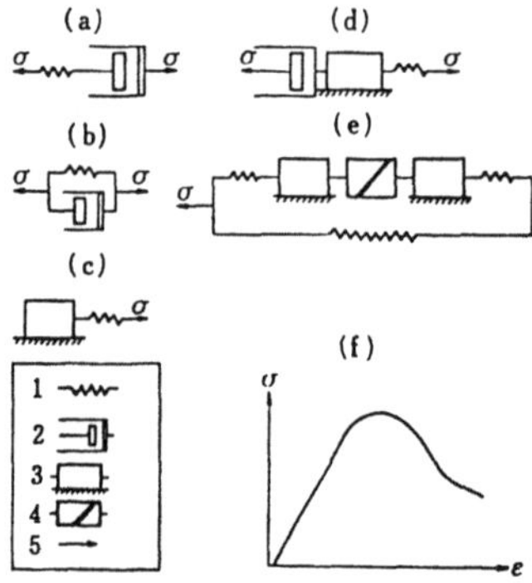

Fig. 12.2. Mechanical models of geological body; *a* viscoelastic model; *b* firmo–viscous model; *c* ideal elastoplastic model; *d* the Bingham model; *e* elastoplastic softening model; *f* stress–strain relation for plastic softening; *1* elastic element; *2* viscous element; *3* ideal plastic element; *4* plastic softening element; *5* loading direction

e. Elastoplastic softening model.

Based on the above–described models, we considered a mechanical model of crust appropriate for studying earthquake preparation and generation. Taking into account that for the above–mentioned tectonic elements of a platy body the inter–zone area is mainly a fairly homogeneous elastic deformation block, and the fracture zone is a bridge connecting elastically other elements (such as creeping segment and blocked segment) with the inter–zone area, we utilized elastic element E (spring) to represent the inter–zone area and fracture zone, an ideal plastic element to denote the creeping segment which deforms plastically when the stress exceeds the yield point, and then remains unchanged, and a plastic softening element to indicate the locked segment in which the media underwent the strain–strengthening and softening stages, and finally deformed plastically after the stress had exceeded the yield point. The strain–stress relation for the plastic softening element is shown in Fig. 12.2f. The model, suitable for studying the process of preparation and generation of large earthquakes, is composed of the three elements mentioned above (Fig. 12.2e). When the model body is stressed, first elastic deformation is produced. When the stress reaches a certain amount (beyond the yield point of the ideal plastic element), plastic deformation is caused locally, and meanwhile the transmitted stress adds more stress onto that acting on the plastic softening element. As the stress increases and exceeds the yield point of the element, strain hardening is brought about, and as the stress reaches the limit, the strain–softening element appears, featuring unstable nature. Finally, elastic recovery deformation and permanent deformation takes place in the element. The energy needed for deformation originated from the elastic strain energy of the entire model body. We regard the elastic recovery deformation and permanent deformation of the plastic softening element as an earthquake occurrence, and the process of strain hardening and strain softening of the element as a seismogenic process of quakes. The stress–strain relation for the plastic softening element seems to be similar to that of brittle rupture in experiments on rock samples. Therefore, the elastoplastic softening model can be used to describe a process of earthquake preparation and generation under regional stress actions.

12.3.3 Evolution of the Earthquake Field

As the elastic platy body is composed of the original fracture, the creeping segment and the locked segment, which are respectively represented by the aforesaid elastic element, ideal plastic element and plastic softening element, its basic stress field can be seen to remain unchanged in a short period in case the action of external force remains basically stable. However, the variable stress field will be produced if there are certain variations in the external forces or changes in nature of the media somewhere within the paty body. Figure 12.3 gives a plane diagram for evolutions of the earthquake field during a seismic process.

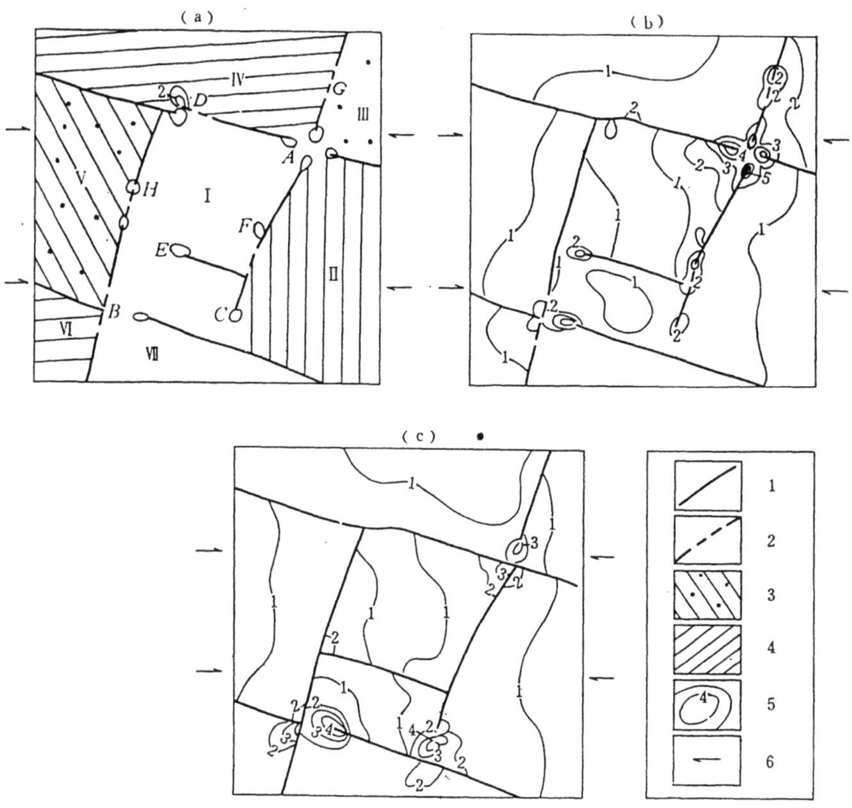

Fig. 12.3.*a–c*. Diagram of evolutional phases of an earthquake field; *1* fracture or tectonic zone; *2* locked segments of a fracture zone; *3* and *4* different fault block bodies; *5* degree of stress concentration; *6* large–scale regional stress actions

Figure 12.3a shows the starting phase of stress build–up, the existence of fault structures and tectonic points like A, B,··· G (either the unbroken parts, or compound fracture parts) within the crust, and the points with high stress concentration and the locality of relative stress concentration, such as D, in a given region.

Figure 12.3b indicates that due to the creeping of the locked segment, or to a moderate earthquake, the stresses accumulated at D are locally released, while a local area of stress concentration consisting of several small relatively densified stress–concentrated points is formed at A, since this point is a boundary of relative motion of the multiple fault blocks, and the combination of subcrustal fractures is rather complicated. When the stress at A exceeds the yield point, and the media undergo the strain–hardening and softening phases, it can be said that point A is just in a phase of seismogenic preparation, and during this phase the variations in medium property or in stress at A would bring to bear major influences upon the whole earthquake field.

As shown in Fig. 12.3c, when an unstable slip leads to large rupture at point A, the strain energy released at point A not only includes the energy accumulated at this point, but also contains the elastic strain energy associated with the deformation of the whole fault–block body. Point A is still an area with fairly concentrated stress, as it is constrained by incomplete rupture or the remaining connected sections. After point A has ruptured, the stress distribution over the entire region would be newly adjusted, with stress increased or diminished partially, possibly leading to new stress concentration points at other locked segments of fractures. After all of the locked segments within the whole region are dislocated, it can be thought that the strong seismicity within this fault–block body during this period is basically over. In this case, the earthquake field tends to be homogeneous, and the crust is simply featuring basic stress field. The anomalies observed before and after the earthquakes may be mostly expressions of stress variations in the earthquake field.

12.4 Results of Finite Element Analyses of the Earthquake Field

The methods for stress field research mentioned in Sect. 12.2 of this chapter can be in principle used for earthquake field studies. However, considering the inhomogeneity of crustal structures and media, it is more convenient to study variations of an earthquake field in two ways: the first is to use the data on recent regional seismicity patterns and precursory observations, and the second is to adopt the method of mathematical simulation. In the previous chapters, the observations of seismic activities and various precursors revealed intense variations in the regional stress fields both in North China and Southwest China in the recent decade and earlier. From the Bohai earthquake up to before and after the Tangshan earthquake, the variations in the distributions of anomalous points and the migration of moderate and small earthquakes before and after strong earthquakes, observed by using various methods, are all reflections of variations in the earthquake fields. In order to analyze the evolution of an earthquake field mechanically, and understand the influences of preparation and generation of multiple strong earthquakes on an earthquake field, we used the finite element method to simulate earthquake fields before and after some earthquakes in North China and Southwest China, so that some anomalous phenomena may be explained from the angle of stress variations.

12.4.1 Computation Method and Step

For simulating the variations of the earthquake field, we would have used the elastoplastic softening model mentioned above to carry out three–dimensional finite element analysis. However, due to inadequate understanding of the deep crustal structures, plus complex computation procedure and high cost for a three–dimensional finite element, we used the plane strain finite element program compiled by the Department of Mathematical Mechanics, Beijing University, in terms of the ideal elastoplastic body, to perform computations, so that a first approximation of the earthquake field can be described.

As mentioned above, the earthquake field is determined by tiny variations of external forces and changes in properties of the crustal media, and interactions of the two as well. But in the following calculations we only considered the latter, and assumed that the model of external force action remains unchanged temporarily. In accordance with a seismogenic process within a source area, because of increases of microfissures, the medium changes in nature, thus reducing the elastic modulus. So we used the medium–weakening method to model an earthquake. Using the shearing rupture criterion——the shear stress along the fault direction equal to the maximum friction sustained by the fault plane——the following relation

$$\tau = \mu \sigma_n \tag{12.1}$$

is adopted to simulate the occurrence of an earthquake.

For the computations to be performed, first the tectonic framework will be defined, the various elements to be computed divided, and the material properties of each element determined. By referring to the basic stress field defined by using other methods as described in Sect. 12.2, the direction and size of the external forces will be adjusted, thus making the stress field of a given tectonic framework consistent with the basic stress field; the directions of various fault motions must be also in agreement with the results obtained from the data on geology and crustal deformation. Then, the action mode of the external forces and the material natures of the various elements will be kept unchanged, and the elastic modulus of elements marking each seismic source will be revised, the high values being gradually decreased. By comparing the variations in the shear stress of various elements before and after the decrease, the effects of the variations in medium properties of a source region on the regional stress field will be derived. Finally, the frictional coefficients of elements on source fault planes will be changed, and the shear stress on fault planes altered so as to be less than the limit of static friction; as a result, fault planes will be dislocated. By adjusting the magnitudes of static friction coefficients of source fault planes, some main parameters of relative dislocations of fault planes —— such as dislocation orientation, dislocation amount, released strain energy and stress drop —— will be roughly identical with the results obtained from seismic waves and other data. By comparing the results of the two computations before and after the dislocation, some main earthquake parameters and their influences on the regional stress field may be obtained.

For expressing the influences of earthquake preparation and generation on a regional stress field, we can use the safety coefficient G:

$$G = 1 - \frac{\tau}{\mu \sigma_n}. \tag{12.2}$$

In comparing the results of two computations, if G is increased in a given element, it means that this element is fairly safe (usually corresponding to shear stress decreases). If G is decreased, it means that the danger of this element being close to failure increases (usually corresponding to shear stress increases).

12.4.2 Tectonic Frameworks and Material Parameters of Earthquake Fields in North and Southwest China

1) Determination of Tectonic Frameworks.

a. North China.

In order to simulate the evolution of the earthquake field from before the Bohai earthquake till after the Tangshan earthquake in the northern part of North China, we selected the tectonic framework shown in Fig. 12.4, based on the regional tectonic outline , as discussed in Chap. 3, by considering the distributions of anomalous points and small earthquake activities before and after the large earthquakes. This tectonic framework ranges from latitude of 37.5–40 ° N, and longitude of 116–123 ° E, with 467 elements and 253 nodal points. The main intercrossing form of this tectonic framework is roughly rhombic, the name of various faults within this framework are given in Chap. 3. As the northern boundaries of North China are affected by larger regional stresses, they are obviously active boundaries. Furthermore, as we discussed the problem of variation in stress field, we take the boundary conditions of external forces as such to be used in calculations. By referring to the research results of the basic stress field of North China, the maximum compressive stress is taken to lie within NE 70 –80 ° , and the minimum principal compressive stress is taken to be orientated in NW 10 – 20 ° .

b. Southwest China.

For the Southwest China region, the tectonic framework is chosen as shown in Fig. 12.9, mainly featuring small fault blocks. The names of various blocks and faults in this framework are given in Chap. 3. In this framework, with a range of latitude of 23–34 ° N, and longitude of 96–108 ° E, there are 563 elements and 305 nodal points. In considering the features of the basic stress field, and the direction of movements along these main fault zones, the maximum principal compressive stress is taken to be at NW 50 ° , and the minimum principal compressive stress at NE 40 ° ; zero displacement in EW direction is assumed for various points at the east boundary, and zero displacement in NS direction for various points at the south boundary.

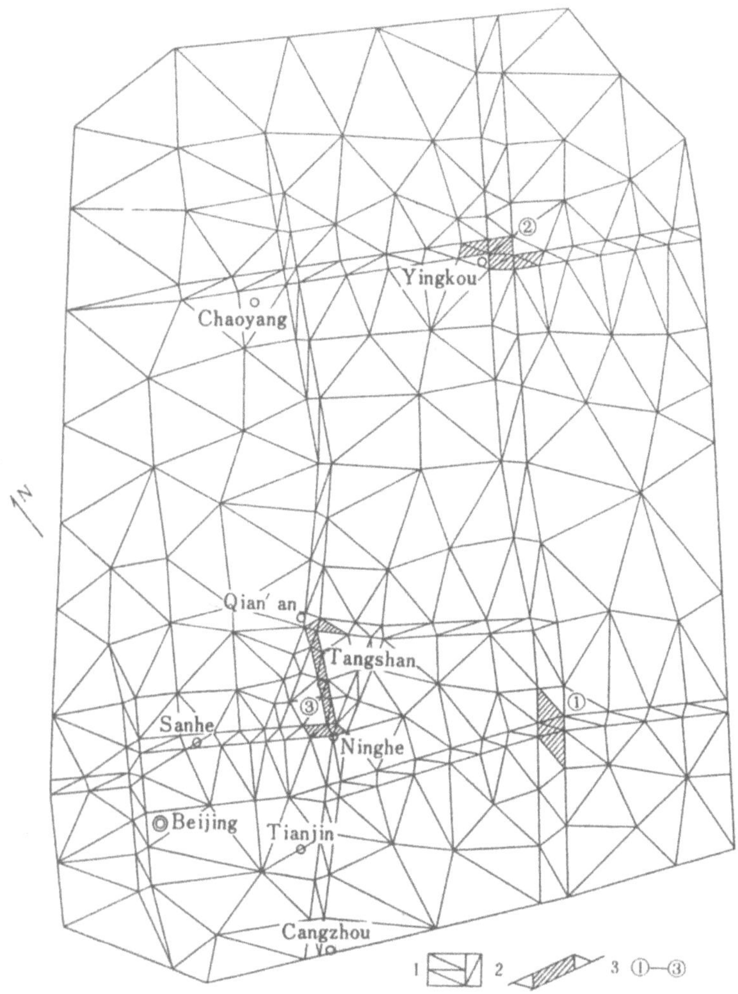

Fig. 12.4. Tectonic Framework in North China; *1* fault element; *2* fault element in strong earhtquake sources; *3* the number of earthquake areas

2) Choice of Material Mechanical Parameters.

a. North China.

For elements considered fracture–free (elastic material):

According to the P wave velocity, the elasticity modulus of material can be derived as follows

$$E = V_P^2 \rho(1+\lambda)(1-2\lambda)/(1-\lambda), \tag{12.3}$$

where V_p is P–wave volocity, about 5.9–6.0km / s, ρ, medium density, about 2.7g / cm^3; λ, Poisson ratio, about 0.25. So E = (8.3–8.5) × 10^5kg / cm^2.

For fractures (layered material):

In view of the results obained by Wunan(1980) while he studied the San Andreas fault, which showed faster propagation of P waves along the fault than along both sides of the fault, with a ratio of about 1.2–2.4, and by considering that we have taken the fault width to be about 10 km on average, wider than Wunan proposed in his paper, the average wave velocity ratio may be less than 1.2. When $V_p$along the fault zone is taken to be within 5.1–5.7 km / s, E should be (6–7.5) × 10^5kg / cm^2. As small seismic activities are very frequent at some faults (such as the Babaoshan and Cangdong faults) and obvious variations in fault displacements are measured, it is shown that the rock media of these faults are very ruptured, being unfavourable for energy accumulation. Therefore, a low value of elastic modulus is taken (6 × 10^5kg / cm^2).

Parameters of a source region.

As described above, research results of various models on seismogenic preparations have suggested that during the middle period of a seismogenic process the elastic modulus of the source is about one–third less than that of the surroundings with the cluster of microfissures. According to the differences in various source regions obtained from data on seismicity and others (like rupturing velocity of earthquakes, medium quality factors, b values and so on), the elastic modulus and Poisson ratio of media in various source regions will be adjusted to make the stress conditions and medium properties satisfy our requirements. Letting the magnitude of acting forces be −6 × 10^8Pa and −2 × 10^8Pa (equivalent to a depth of 15–20 km), and the direction of external forces in N70 ° E, the principal mechanical parameters of the three source regions can be obtained as shown in Table 12.1.

Table 12.1.

sources	elasticity modulus (kg/cm^2)	poisson ratio	limit static friction coefficient	shear stress background (bar)	
				average value	difference between maximum and minimum shear stresses
Bohai	6×10^5	0.22	0.56	960	290
Haicheng	5.5×10^5	0.22	0.40	1260	780
Tangshan	6.5×10^5	0.23	0.42	810	340

From Table 12.1 it can be seen that for the Haicheng earthquake the elastic modulus is the lowest, the stress background value is the highest, and the difference in shear stresses within various elements of the source region is the largest. If the rocks of the Haicheng source area were relatively ruptured, and the shear stress difference fairly large, the moderate and small earthquakes could easily occur before the mainshock. Whereas the Tangshan source rocks were of a relatively hard nature, with a fairly low stress value, and as E is supposed to be 9×10^5kg / cm^{2}for the Tangshan rhombic block, a hard block with weak zones is used to simulate the source region of the Tangshan earthquake. There are not many data about the Bohai source region, and considering that the source region is located in the area with relatively high geothermal flow values in North China, the modulus is assumed to be not very high, just within the values of the above–mentioned two source regions.

b. Southwest China.

The methods for selecting parameters for the mechanics of materials in Southwest China are similar to those used for North China. For the considered fracture–free elastic material elements, in addition to using the Eq. (12.3) to define average P wave velocity, the ages of rocks within various fault blocks are also taken into account, and this parameter is defined to be $(8.0–8.7)\times 10^5$kg / cm^2. For example, the Sichuan fault block and the Guizhou–Guangxi fault block are predominantly composed of ancient crystalline basements, so a slightly high elastic modulus is taken, about 8.7×10^5kg / cm^2, while the Sanjiang fold zone is mainly made of Mid–Cenozoic rocks, so the elastic modulus is proposed to be 8×10^5kg / cm^2. The mechanical parameters of various source regions are given in Table 12.2.

12.4.3 Computations of Earthquake Fields in North China and Southwest China

1) The North China earthquake field.

a. Fields effectted by seismogenic preparation of strong earthquakes.

Table 12.2.

Source region	Elasticity modulus (kg/cm^2)	Poisson ratio	Limit static friction coefficient	Average shear stress (bar)
Tonghai	5.5×10^5	0.21	0.34	520
Luhuo	6×10^5	0.22	0.30	770
Yongshan	5×10^5	0.21	0.36	680
Longling	7.5×10^5	0.23	0.61	1600
Songpan	6×10^5	0.22	0.67	1490

When the external force direction is taken to be N70 ° E, the maximum and minimum principal compressive stress values are -6×10^8Pa and -2×10^8Pa separately (the values will remain unchanged during computations), the material properties are mostly kept constant, and the elastic modulus for the source regions of the Bohai, Haicheng and Tangshan earthquakes, 8.5×10^5kg / cm^2(corresponding to a fault–free region), are reduced respectively to 6, 5.5, and

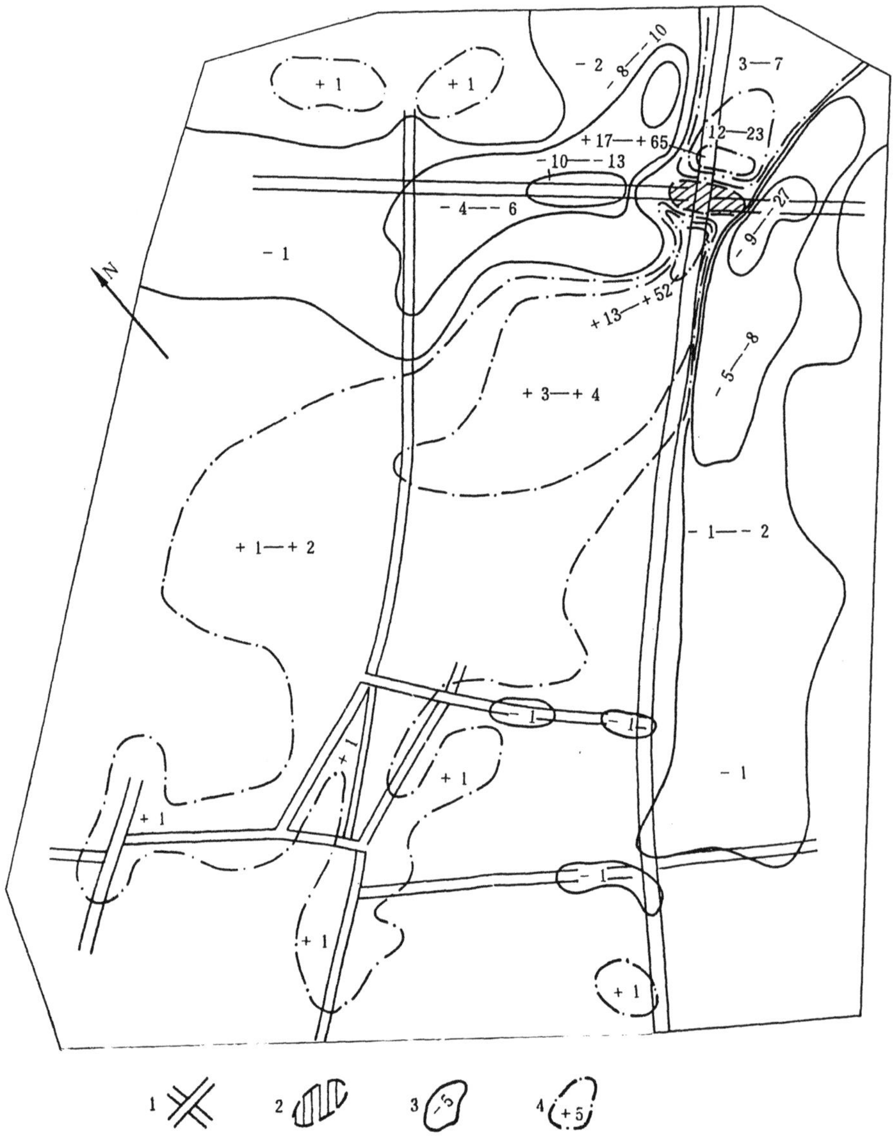

Fig. 12.5. The affected field of seismogenic preparation of the Haicheng earthquake (G value variation); *1* tectonic framework; 2 source region; *3* decreased G value and area; *4* increased G value and area (legend is same for Fig.12.6–Fig.12.9)

$6.5 \times 10^5 kg/cm^2$, with the differences in shear stresses within various elements derived before and after reduction, so that the influences of a seimogenic preparation on a regional stress field can be obtained. The variations of G values before and after the E values decreased within the Haicheng and Tangshan source regions are shown in Figs. 12.5 and 12.6. The numerical values given in the figures are 10^3times the values computed according to Eq. (12.2).

The computations show that after a reduction in the elasticity modulus of the Bohai source region, in addition to the obvious increase in shear stress value of extremes of the source region (maximum reaches 13×10^6Pa), the risks increased in the Bohai area of the north and south source region (value increased to about 3×10^6Pa); the risks also increased in the Tangshan, Tianjin and Babaoshan areas (value increased to $2–4 \times 10^5$Pa); the τ value increased to 1×10^5Pa in the Haicheng and Shengyang area in the northeast region; while in the Bohai area, east, northeast, west and southwest of the source region, safety increased (G value is positive).

Figure 12.5 shows the affected field of seismogenic preparation of the Haicheng earthquake. After the elasticity modulus of the source region decreased in the Liaoyang area north of the source region, the Chaoyang area and further south, northwest of it, and the Xiuyan and Dandong areas and the Liaodong peninsula, southeast of it, the risks increased, and τ increased to 7×10^5Pa on the average, with a maximum of 15×10^6Pa. While in the Benxi area, the Liaodong Bay and its west coast, in the east and west of the source region, the risks fell, with τ decreased to $2–11 \times 10^6$Pa, in the Beijing–Tianjin–Tangshan region the shear stress values are decreased to $1–2 \times 10^5$Pa on the average.

Figure 12.6 gives the field affected by seismogenic preparation of the Tangshan earthquake. As the elasticity modulus of the Tangshan source region reduced, in the northeast and southwest corners of the Tangshan block, the shear stress in particular increased apparently (to about $15–20 \times 10^5$Pa), and in the Zunhua and Yutian areas and the north section of the Babaoshan fault zone, to the north and northwest of the source region, shear streses increased to $6–50 \times 10^5$Pa.The risks also increased in the Bohai area, the Cangdong fault and at the central section of the Tanlu fault zone, to the south of the source region. In the Liaodong Bay, the Liaodong peninsula, to the east of the source region, and at the south section of the Babaoshan fault zone, to the west of the source region, the risks decreased, and shear stresses decreased to $6–30 \times 10^5$Pa.

The research results of the above–mentioned fields affected by seismogenic phases of strong earthquakes can be summed up as follows:

α. The degree of effects of the reduced elasticity modulus of a source region on the regional stress field is also associated with the structures of the source region, the focal scale and variations in the property of the media, in addition to the regional tectonic framework.

β. When the external forces acting on a given region are orientated NEE, the risks are usually increased in the north– south area or the northwest–southeast area of the source region, whereas the risks are decreased in the eastwest or northeast–southwest areas. However, the distribution patterns are not completely symmetrical, showing nonuniformity.

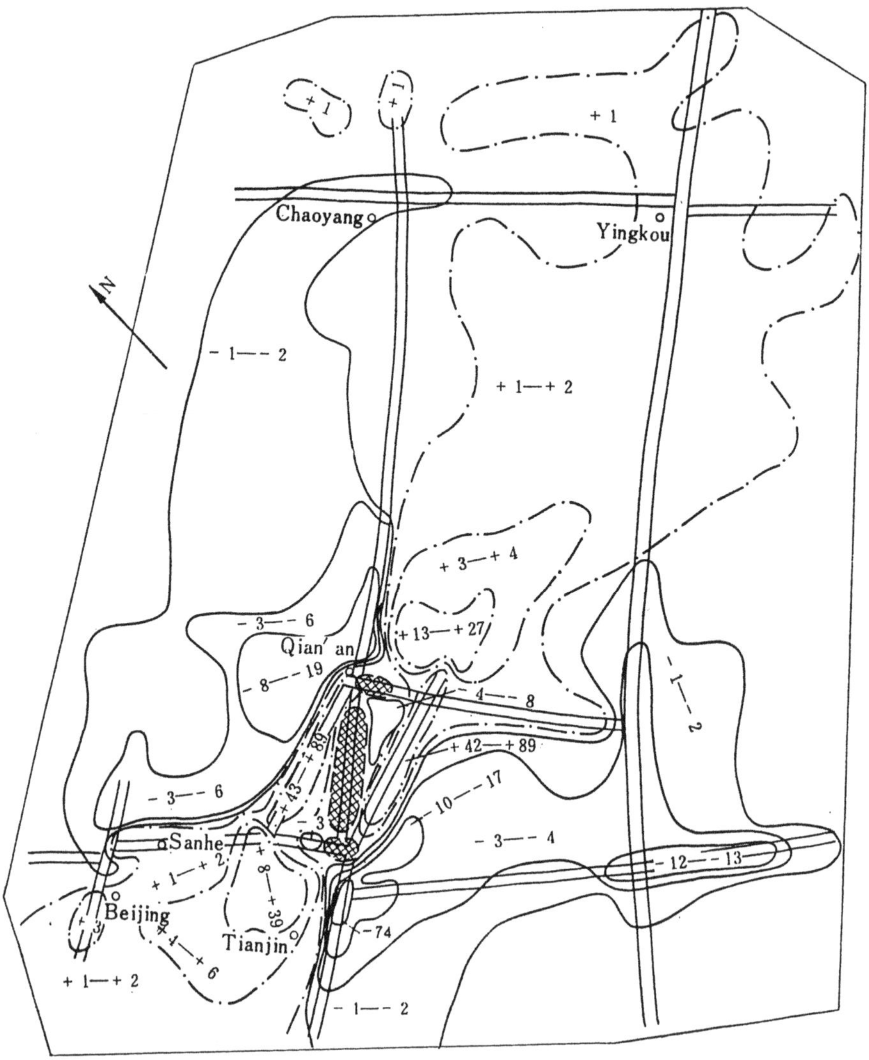

Fig. 12.6. The affected field of seismogenic preparation of the Tangshan earthquake

γ. The most affected areas are generally located at the extremes of a given source region. As the elasticity modulus is one–third lower here than in the surrounding areas, the affected range (shear stress variation of about 10×10^5Pa or more) will be one or two times larger than the focal dimension, and the most distant range of the affected area may be four to five times larger.

b. Fields affected by the occurrence of strong earthquakes.

Based on the above–described calculations, let the external forces and the material properties of various elements remain unchanged, with only the frictional coefficients of faults within each seismic source decreased, so that the relative dislocations (earthquakes) occur on fault planes. By comparing the results obtained before and after the dislocations, some main parameters during an earthquake (such as dislocation orientation, slip amount, and stress drop) and their effects on a regional stress field can be obtained. In this section, we particularly concern ourselves with the effects of these parameters on the regional stress field, and so the effects (G value variations) of an earthquake occurrence on a regional field are presented.

After the Bohai earthquake, the risks also increased in the northeast and southwest areas of the epicentral region, besides the apparent increase of shear stresses on faults near the extremes of the epicentral region (τ value increased to about 20×10^6Pa), while safety increased in the north (northwest) and south (southeast) areas (shear stresses diminished). Meanwhile, the risks increased in the Tangshan block and the north segment of the Cangdong fault, and the farthest part affected reached to the north of the Haicheng earthquake area.

After the dislocations on the fault planes of the Haicheng earthquake area, the risks increased in the NW tectonic zone from Dangdong to Chaoyang, the Tanlu fault zone, and the Benxi, Panshan and Beipiao areas, east of the epicentral region, whereas the risks decreased in the Taian area, north of the epicentral region, the Gaixian area, south of the epicentral region, and the Beijing–Tianjin–Bohai region (Fig. 12.7).

The occurrence of the Tangshan earthquake made the risks obviously increase along the Ninghe–Sanhe NW tectonic zone, the north segment of the Cangdong fault (near Tianjin), and in the Qianan–Xiangshan area (Fig. 12.8), with shear stresses increased to about 25×10^5Pa. The risks also increased from Qianan to Chaoyang and in the central part of the Tanlu fault zone (around the Bohai earthquake area), while the shear stress value decreased in Nanpu, south of the epicentral region, and in Zunhua, north of the epicentral region.

All in all, the above–mentioned affected fields associated with earthquake occurrences are similar to those associated with seismogenic phases of earthquakes. In addition to relations with the tectonic framework, and dislocation magnitude and direction, there are still some dissimilarities as follows:

α. The risks are generally increased along tectonic zones where dislocations occurred, or along tectonic zones across the dislocation, and they also extend very far along the tectonic zones in the direction of the dislocations. The affected fields are mainly "band–like".

β. When the acting forces run NEE, the risks are increased (τ value increased) in the eastern (northeast) and western (southwest) areas of a given dislocation source, while they are decreased in the south (southeast) and north (northwest) areas. The distribution patterns are not symmetrical, and vary with tectonic frameworks.

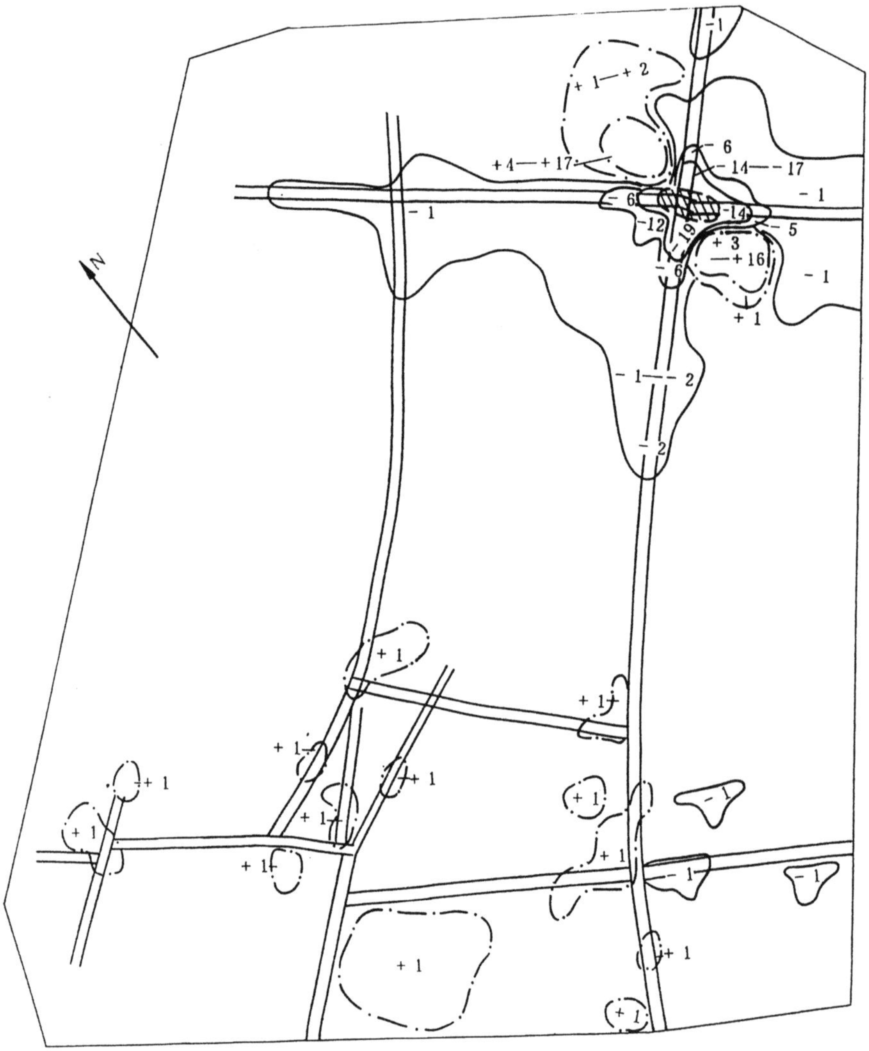

Fig. 12.7. The affected field of occurrence of the Haicheng earthquake

γ. Comparison between fields associated with earthquake preparations and those with earthquake occurrences indicates that the natures (implying increase or decrease of shear stresses) of some regions influenced by source regions before and after earthquakes are just opposite, that is, the area in which shear stresses are increased during a seismogenic preparation becomes an area with shear stresses decreased after a shock. For instance, in the Tianjin and Zunhua areas before and after the Tangshan earthquake, and in the Panshan and Gaixian areas before and after the Haicheng earthquake, the nature of the effects were just contrary.

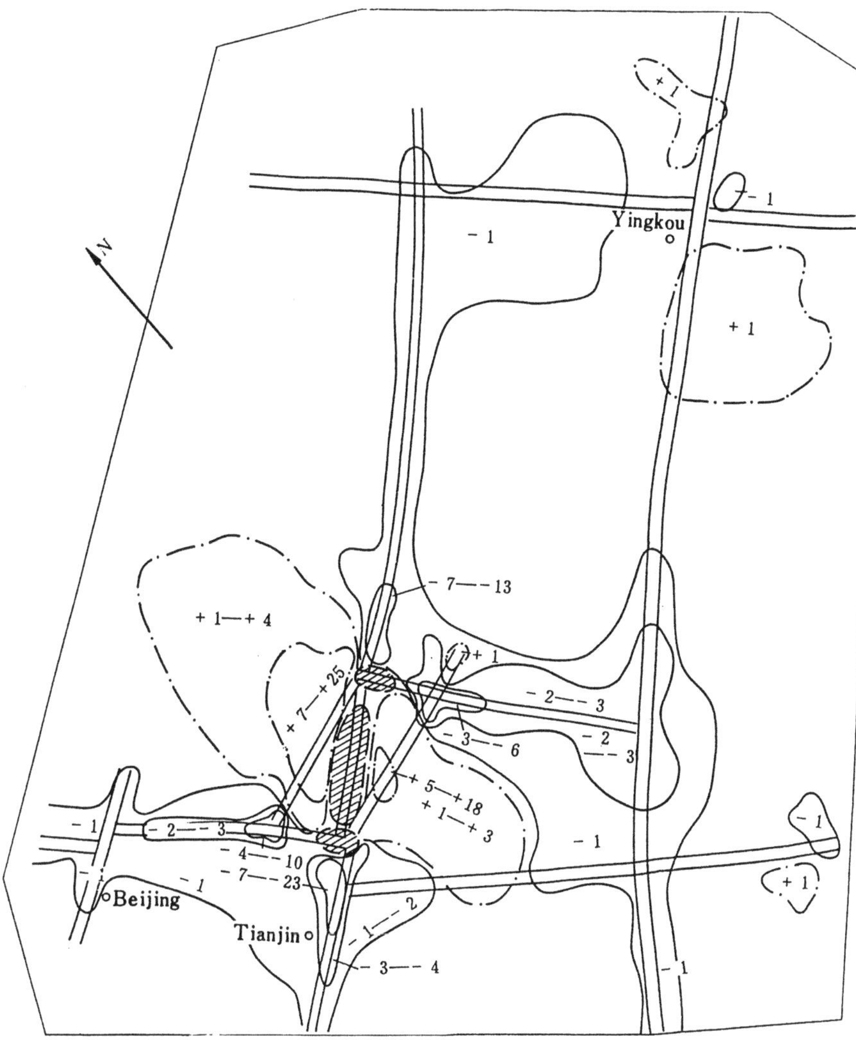

Fig. 12.8. The affected field of occurrence of the Tangshan earthquake

2) The Southwest China earthquake field.

For the Southwest China region, only the affected fields associated with seismogenic preparations of the above–mentioned five strong earthquakes, are computed. The results of two of them will be described here. If the elasticity modulus of the Yongshan earthquake source is reduced to 5×10^5kg / cm^2, the risks (shear stress increase) are remarkably increased in the Xichang–Qiaojia and Daguan–Xichang areas, and then in the Kangding–Mabian area. In addition, the risks are also increased in the Guanxian area, south segment of the Longmenshan fault zone, and in the Xiaguan–Honghe area of the Red River fault zone.

The affected field associated with the seismogenic preparation of the Longling earthquake is shown in Fig. 12.9. As the elasticity modulus of the Longling earthquake source is reduced to 7×10^5kg / cm^2, the risks are significantly increased in the Baoshan–Midu–Puer region, the Tonghai earthquake area, the Luxi–Ruili area, and then in the Xiaojiang fault, the Ganzi area and the Songpan area.

12.4.4 Relations Between Variations of the Earthquake Field and Pre– and Post–earthquake Anomalies

1) The North China region.

a. Seismic activity.

It was mentioned in Chap. 4 that before the Haicheng and Tangshan earthquakes, the earthquakes with $M>4$ were concentrated around the epicenters. The comparison between such a phenomenon and Figs. 12.5 and 12.6 suggests that the earthquakes of $M>4$ occurred in the areas of large G value variations in the fieds affected by seismogenic preparations of the Haicheng and Tangshan earthquakes, or very often in areas where G values were diminished (corresponding to increases of shear stresses). It is thus illustrated that due to increase of shear stresses moderate ruptures took place in areas for which it was difficult to sustain large stress loadings. After the Tangshan earthquake, the $M>4$ earthquake–densified zones were formed in the Ninghe–Sanhe and Qianan–Chaoyang areas. By comparing Fig. 9.2 and the field affected by the occurrence of the Tangshan earthquake (Fig. 12.8), it is indicated that the $M>4$ earthquake–concentrated zone was coincident with the G–values–reduced area, where the shear stresses were significantly increased with reference to those before the Tangshan earthquake(Zhang Yingzhen 1985). After the Tangshan earthquake, some so–called triggered earthquakes of M4–5 also occurred in the Baodi, Xiangshan and Chaoyang areas where the risks were obviously increased in the affected field (G–values–decreased areas).

b. Precursory observations.

The fairly obvious anomaly points before the Haicheng earthquake were mainly located in the vicinity of Jinxian, Liaodong peninsula, and the Dandong–Chaoyang NW tectonic zone. As the field affected by preparation of the Haicheng earthquake suggested, the areas of apparent G value variations (with values of 5–27) comparatively coincided with areas of concentrated anomaly points. In addition, there were slight influences from the preparation of the

Haicheng earthquake on the Beijing–Tianjin–Tangshan region (G value variation about 1), and shear stresses decreased to about 1×10^5Pa (Fig.12.5). However, the increase of shear stresses in the Beijing–Tianjin region associated with the preparation of the Tangshan earthquake was far greater than 1×10^5Pa, ranging from 6 to 5×10^5Pa. Therefore, it may be concluded that some precursory phenomena in the Beijing–Tianjin region from 1973 to the Tangshan earthquake occurrence were mainly caused by the preparation of the Tangshan earthquake.

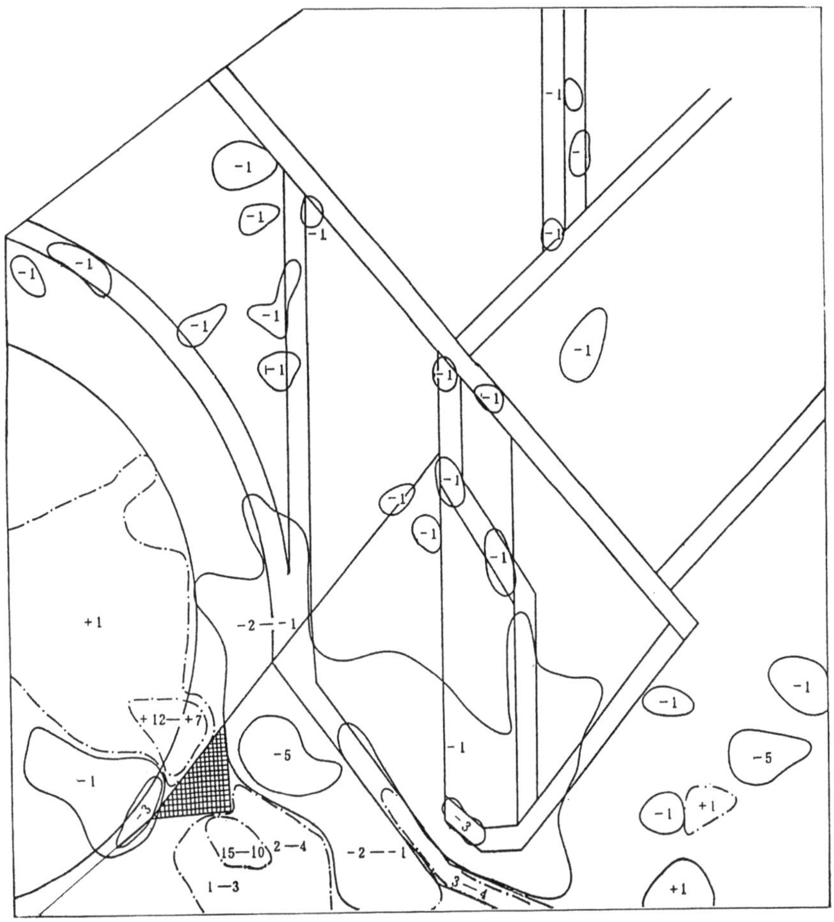

Fig. 12.9. The affected field of preparation of the Longling earthquake

Furthermore, the occurrence of the Haicheng earthquake also made the shear stresses diminish in the Beijing–Tianjin region (Fig. 12.7), hence it may be considered that the sudden recovery of earth resistivity at the Baodi station after the Haicheng earthquake discussed in Chap. 9 (Fig. 9.6a) may be due to the abrupt decrease of shear stresses in the Beijing–Tianjin region, associated with the occurrence of the Haicheng earthquake.

Before the Tangshan earthquake, distinct anomalous displacements along the fault were observed at the Dahuichang station at the north segment and the Nuikouyu station at the south segment of the Babaoshan fault zone in the Beijing region, but both stations showed the precisely opposite behaviour of forces acting on the two segments of the fault. From Fig. 12.6 it can be seen that, as the Babaoshan fault zone was affected by the preparation of the Tangshan earthquake, the shear stress increased in its north segment, while it diminished in its south segment. This might be one of the reasons why there were differences in the behaviour of variations at these two stations before the Tangshan earthquake. It is illustrated that the possible different effects of preparation or generation of one and the same earthquake on various segments of the same fault zone are chiefly dependent on the orientation of stations and source regions (seismic regions). For example, obvious anomalies before the Haicheng and Tangshan earthquakes were observed at the Jinxian station, Liaodong peninsula, but the mechanical behaviour of the anomalies for both earthquakes were different. This might be also due to the different nature of the effects of seismogenic preparations of these two earthqakes for the Jinxian area. By comparing Fig. 12.5 with Fig. 12.6, it is shown that the preparation of the Haicheng earthquake made the shear stresses increase, while that of the Tangshan earthquake made the shear stresses reduce.

Sudden reversal phenomena of pre– and post–earthquake anomalies were observed at numerous stations after some quakes, due perhaps to precisely opposite influences of preparation and generation of a given earthquake on some localities, that is, the area with shear stresses increased during the seismogenic preparation of an earthquake becomes one with shear stresses decreased after a shock.

2) Southwest China.

Relations between the evolutions of earthquake fields and the pre– and post–earthquake anomalies in Southwest China are similar to those in North China, and the areas with notable variations of the safety coefficient G before earthquakes are very often the areas of concentrated seismicity and anomaly points.

From the above–mentioned descriptions we can derive the following considerations about the variations of earthquake fields and the observations of precursory seismic activities:

a. The migration of regional seismic activity is closely related with the evolution of the earthquake field. Before a large shock the moderate earthquakes may be generally distributed in areas of major shear stress variations around the source region, and after a large shock they may be distributed over a band–like zone with increased shear stresses, and extended fairly far in the direction of the earthquake–generating fault.

b. The distribution of anomalous points in precursory observations and the mechanical behaviours reflected by the anomalies are dependent on the orientation of stations with respect to source regions, and the observation anomalies are the result of the comprehensive effects of preparations and generations of several earthquakes.

c. The medium–softening areas (that is the above–simulated areas in which strong shocks occurred) are essentially very sensitive in an earthquake field. For instance, the Bohai seismic area (Fig. 12.6) in the field affected by the preparation of the Tangshan earthquake, and the Tonghai seismic area in the field affected by the preparation of the Longling earthquake, are the areas with major G value variations far off the epicenters.

To summarize, in concluding this chapter, the following knowledge can be gained: (1) due to the laterally–blocked and vertically–layered crustal structures and media heterogeneity, and under the loading of larger–scale regional dynamic forces, the variations of pre– and post–seismic earthquake fields are rather complicated, extremely uneven in distribution patterns, and there are obviously stress–varying areas around a source region (or seismic region), whose effects are extended rather far along a tectonic zone, exhibiting stress–varying bands and points; (2) preparation and generation of a given earthquake seem to be associated with the distribution of crustal stress fields and the conditions causing softening of the medium; (3) not all stress–concentration areas would be earthquake–generating places, as thus depend on the variations of frictional coefficients of fault planes; and (4) the occurrence of a given earthquake will not always play a triggering role in generating the next earthquake in the same region, instead it might play a relaxing role, like the Haicheng earthquake, which caused shear stresses to decrease in the Tangshan source region. Therefore, the key to the regular occurrence of earthquakes in a given region lies in the extent of stress concentrations of various source regions and the conditions leading to variations in the nature of the medium.

The variations of crustal stress fiels before and after earthquakes are very complex, and the above–described modelling results can not be used to the explain all kinds of anomalous phenomena before and after a shock, owing to the simplification of crustal model and tectonic framework, plus insufficient knowledge of the mechanical parameters of crustal geology. Furthermore, in addition to the mechanical properties which can exhibit anomalies, there are still other physical and chemical properties bringing about anomalous variations. Therefore, the variations of an earthquake field can only be regarded as one of the approaches for interpreting large–scale regional anomalous phenomena.

CHAPTER 13

Earthquake Episodes

The seismic activities in the Sichuan–Yunnan and North China regions of continental China generally intensified in the period of 1966–1976, with strong earthquakes occurring in succession. Was this a particular period? Generally speaking, seismicity is relatively active within a given space in a given period of time, then becomes relatively quiet in another time period, and such fluctuations and alternations of intense and weak seismic activities are commonly existing phenomena. However, there seems to be a specific historic background for the concentration of strong earthquakes in the North China and Sichuan–Yunnan seismic regions within this particular time period. In this chapter, the generality and particularity of the phasic fluctuation of strong and weak seismicity are elucidated from long–range and large–scale regional earthquake behaviour in line with viewpoints on earthquake episodes, and the historic background, made up of successive occurrences, of the nine great earthquakes together with the origin and prediction of earthquake episodes are discussed.

13.1 Concept of the Earthquake Episode

Whether there exist any regularities governing seismicity distribution in time sequence has been a long–debated problem, about which advocates and adversaries of cyclicity have their own views. Nowadays, the majority of researchers emphasize the following facts: over the entire globe , or in a given space range, earthquakes are intense and frequent in a given time interval, but are weak and quiet in another interval, and there exist, though not very strict , time intervals between the fluctuation and alternation of strong and weak seismic activities. Benioff (1951b) divided earthquake activities in the first 50 years of this century into five time stages, suggesting that there existed active periods and relatively quiet periods for earthquakes. Mogi (1974) revealed that phenomena of alternation of strong and weak earthquakes occur in the world's chief seismic belts with a periodicity of about 20 years. The seismic activities in the North China and Sichuan–Yunnan regions since 1966 also displayed a seismically active time period, which Ma Zongjin (1987) deemed to be an expression of earthquake episode, and Li Qinzu et al. (1980) regarded as an earthquake activity in groups. Li Zhiqiang et al. (1980) made a specific study of the regional seismicity backgrounds for the clustering of strong earthquakes in North China in recent years. All in all, attention has been paid extensively to the phenomena of phasic fluctuation and alternation of strong and weak earthquakes.

A group of earthquakes, a cluster of earthquakes, or an episode of earthquakes are the notions for describing population earthquakes, as different from individual earthquakes. Studies on the problems of developmental regularity and the prediction of an earthquake population will provide many new research subjects. For example, how long a seismically active phase will usually last, and how is to be judged the initiation and termination of such a phase? Are there any spatial variations in occurrence of earthquake episodes? What is the relation between earthquake episodes? Is there any genetic relationship among multiple earthquakes within an earthquake episode? These problems are very important for prediction work, and also pertinent to theoretical studies of earthquake origins.

The reason why we introduced the concept of earthquake episode is illustrated below. In systematically sorting out the temporal distribution of earthquakes in China, we noticed that the phasic alternation of strong and weak earthquakes is a general phenomenon, and that the time lengths for variations in the fluctuation phenomena are different. From the historic records of about 2000 years, it is relatively practical and significant to classify them into three levels of fluctuation. The first level denotes a seismically active period of about 200 years with a relatively quiet interval of 100 years; the second level implies that there might exist several seismically active episodes with time lengths ranging from 10 to about 20 years, with relatively quiet episodes at intervals of 10–20 years, within a given seismically active period; the third level indicates that there may be several peak seismic stages with time lengths ranging from several months to a year in a given seismic episode. There might still exist a seismicity fluctuation phase longer than a "period" and shorter than a "stage", which cannot be identified definitely due to insufficient data.

The division of seismic period, episode and stage has been made in accordance with the division of era, period, epoch and stage in geochronology. The basic idea for such a division is to attribute a definite time in history to the earthquakes of a seismic period or a seismic episode. The division of levels is conducive to discussions on separate levels, and serves to avoid confusion in time scales, since the variational regularities of various levels may be associated with the action of different dynamic factors.

The successive occurrence of the nine great earthquakes can be regarded as a phenomenon of earthquake episode, so discussion of the seismic episode will receive special emphasis in this chapter. The implications from the study of an earthquake episode can be summarized in the following three points:

1) An earthquake episode is a seismically fluctuating and alternating phenomenon in a seismic period, with a time length of about 20 years.

2) Earthquakes in an episode are relatively concentrated in some regions or zones which are called the principal active regions or zones in this episode. The features of seismicity in different episodes are variable, showing a historical connection between the former and latter episodes. The development episode–after–episode seems to constitute a step or a rhythmical beat in the entire mechanical process of a seismic period. The mechanical relations of earthquake episodes are based on the evolution of seismic fields.

3) The episodic phenomena of seismic activity generally have a global significance, hence it is speculated that the rhythmical fluctuation of episodes is per-

haps related with the action of periodical dynamic factors of the Earth.

13.2 Episodic Phenomena of Seismicity in North China

In order to demonstrate the generality of earthquake episodes in the long history of seismicity and to study the relation between the evolutional process of a seismic episode and a seismic period, we shall emphasize the importance of some preliminary results of studies on the North China seismic region, with fairly complete historical earthquake data.

13.2.1 Division of Seismically Active Periods and Episodes

1) As the earthquake records of 1000 B.C. for the North China seismic region (North Latitude 34–41 ° , and East Longitude 109–123 °) are insufficient, only the relatively–concentrated earthquakes of M>6 of 1000 A.D. are separated roughly into four seismically active periods. The first might range from a certain time before 1000 B.C. to 1068 A.D. The second period is from 1209 to 1368 A.D. Beginning with the early Ming Dynaty (late fourteenth century), the Chinese imperial government decided that local county annals be established instead of compiling prefecture annals, so the calamities were recorded in fairly detailed form in different places and in different periods. The third period from 1484 to 1730 A.D and the fourth period from 1815 A.D. to the present seem to be documented more reliably, and these will from the main basis for studying episodic phenomena.

Table 13.1.

	Third period (1484–1730)		Fourth period (1851–)	
	active episode	quiescent episode	active episode	quiescent episode
1	1484–1487	1488–1496	1815–1820	1821–1828
2	1497–1506	1507–1521	1829–1835	1836–1854
3	1522–1538	1539–1577	1855–1862	1863–1879
4	1556–1569	1570–1577	1880–1893	1894–1908
5	1578–1588	1589–1613	1909–1923	1924–1928
6	1614–1627	1628–1639	1929–1948	1949–1965
7	1640–1627	1643–1663	1966–	
8	1640–1642	1684–1694		
9	1664–1683	1709–1719		
10	1695–1708 1720–1730			

2) Figure 13.1 shows earthquake episodes in the third and fourth periods. The principles for division of the episodes are: (1) the intervals with relatively concentrated earthquakes in the time series diagram of earthquakes with $M > 4\frac{3}{4}$ are defined as active earthquake episodes; and (2) there must be at least one

earthquake of M6 and the total number of earthquakes exceed four with intervals of less than 6 years within an episode.

From Fig.13.1 and Table 13.1, the following three points can be derived:

a. There was a quiet period of 85 years between the third and fourth period, so the relative independence of each period is clear. There were also relatively quiet periods between the second and third period, and the first and second period. It is thus ascertained that there existed successive phenomena of seismic activities for about 200 years in the North China seismic region. The general patterns of the temporal distribution of earthquakes were basically similar for the third and fourth period.

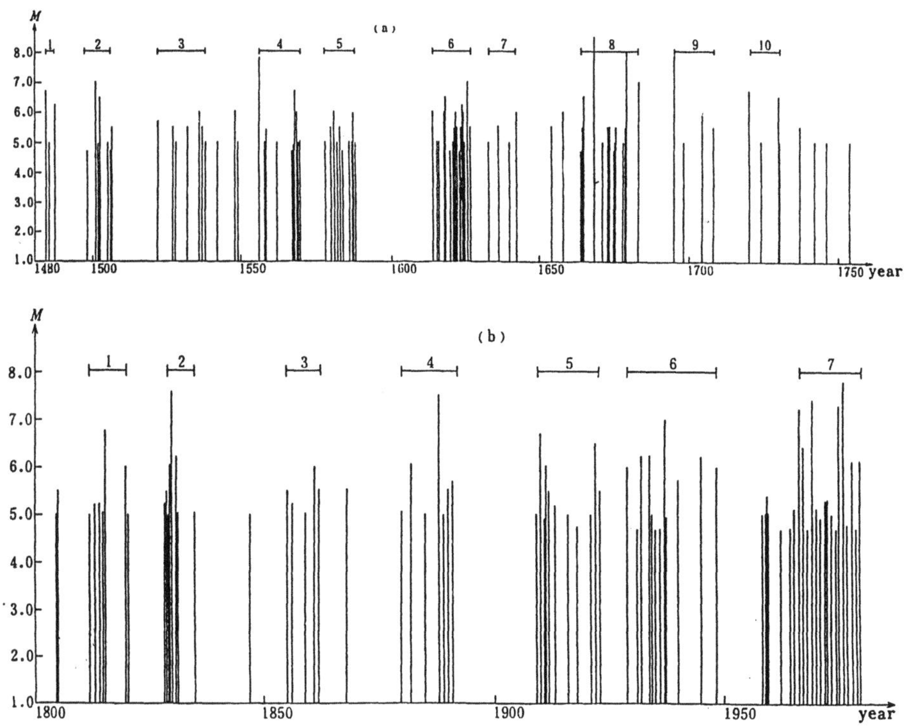

Fig. 13.1. The time series diagram of earthquakes with $M \geq 4.7$ in the North China seismic region; *a* the third active period (1484–1730); *b* the fourth active period (1815–)

b. Earthquake frequency and magnitude were fairly low in the first half of an active period including about five episodes, with individual ones of M7 or 8; while in the second half of the period, involving about five episodes, the frequency and magnitude of earthquakes were significantly increased, with the first several episodes characterized by high–frequency activity.

c. The lengths of seismic episodes were different from each other, usually lasting 10–20 years. The quiescent episodes had similar time lengths. Therefore, generally speaking, seismic activities feature rhythmical fluctuation and alternation of strong and weak earthquakes, with each fluctuation having a time length of about 20 years.

13.2.2 Spatial Distributions of Earthquake Episodes

First of all, let us look at the overall patterns of earthquakes with M6 in the third and fourth periods(Fig.13.2). The distribution of earthquakes during these two periods was in the form of interlocking circular patterns. The western boundary of the circular pattern for the third period is the Shanxi seismic belt of NE orientation, with very high earthquake frequency and epicenters aligned in an S–shaped line, and the eastern boundary is the NNE Tanlu fault zone. There was only one large earthquake during the third period at the Tanlu fault zone, but the M8.5 Juxian earthquake of 1668 affected a long stretch along the NNE fault zone. It is indicated that this fault zone is an important structure controlling the 1668 earthquake, and also a major seismotectonic boundary of the North China seismic region. The southern boundary of the circular pattern extends from Huaxian to Juxian, approximately along the line of latitude 35 ° N. In addition to the two shocks of M8 at both ends of the line, there were three earthquakes of M6 in the central part of it. The northern boundary of the circular pattern is the Huailai–Beijing–Bohai zone running NWW. The aforesaid four boundaries approximately made up a rhombic region in which there no earthquake of M over 6 occurred.

The strong earthquakes during the fourth period were also distributed in a circular pattern, but its main part moved eastwards for about 200 km in a horizontal direction. The western boundary of this pattern is the NE seismic zone in the North China Plain, along which the earthquakes were also roughly distributed in an S–shaped line. However, this line is still incomplete, and one or two stronger earthquakes may fill it up. The eastern boundary of this pattern is the nearly NS lying seismic zone of the Yellow Sea,while its northern boundary of it is not clear. If the Haicheng earthquake site was taken as an initial point at the north boundary, its limits should form a line running NWW along the tectonic zone starting from Dandong through Haicheng to Chaoyang. The southern boundary is not distinct either, having still some aseismic segments. If the southern boundary is assummed to take the line of latitude 35 ° N delimiting it in EW direction, and if the earthquake of M6 in south Anhui Province is taken to be an earthquake point at the southern end, the boundary would be running between NWW and nearly EW. This is still in need of discussion. Despite variations in the southern and northern boundaries, no influences are in evidence for determining the eastern and western boundaries, so it is established that the overall pattern for the fourth period has migrated eastward.

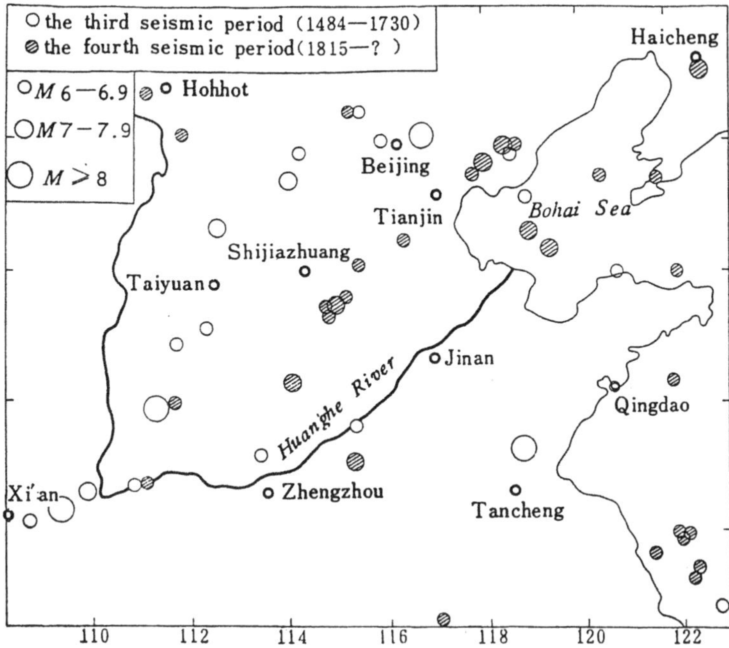

Fig.13.2. Epicentral distribution of earthquakes with M>6 during the third and fourth periods in North China

As the boundaries of two " circles" are seismic zones of the highest earthquake frequency, both arranged in a NE line, it is natural to have the impression that both are constrained by the NE structure. However, the space distribution of earthquakes with M over 4 in each episode and the details of the developmental process of each episode are studied, we would gain a new insight.

From Figure 13.3, it can be seen that in the first three episodes of the third period earthquake activities were mainly located at the southern and northern boundaries, at the south in particular, and the succession of seismicity, strong or weak, was aligned in a nearly EW direction. The earthquakes in the fourth and fifth episodes were arranged NE, exhibiting an S–shaped distribution, but the earthquakes in the fourth episode first occurred successively at the southern and northern extremes, and the ones in the fifth episode were concentrated in a NE direction, with a cluster of moderate and minor ones. The sixth episode has the highest earthquake frequency and vastest earthquake distribution, with the earthquakes still predominantly centered in a NE tectonic direction, and with successive occurrences of small earthquakes along the northwestern line. The seventh episode had rather low earthquake frequency and low magnitude. The eighth and ninth episodes were peak intervals for seismic energy releases, with three M8 (of which one was M8.5) earthquakes and one M7 earthquake. The three M over 8 earthquakes occurred separately at the corners of the earthquake "circle", i.e., the northwestern, southwestern, and southeastern, during the third active period, showing the overall activity of the fault block enveloped by the

(a)

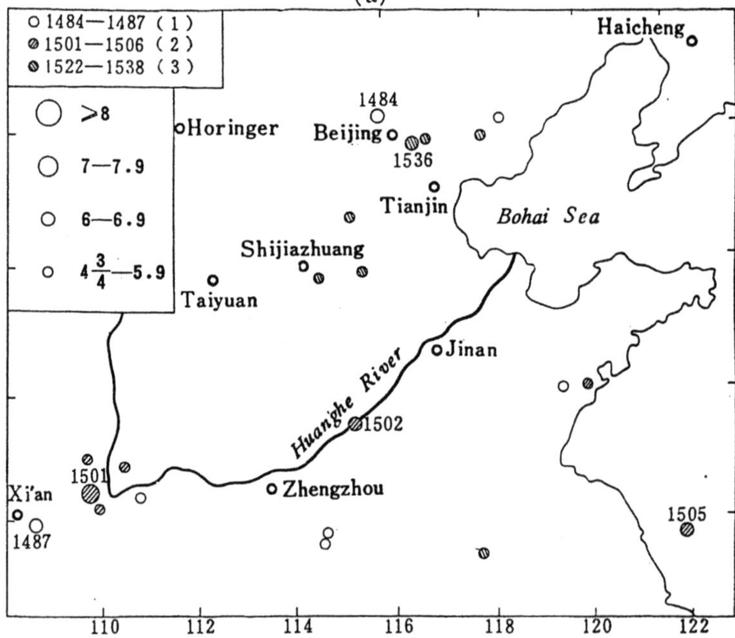

1484—1487 (1)
1501—1506 (2)
1522—1538 (3)
>8
7—7.9
6—6.9
$4\frac{3}{4}$—5.9
Haicheng
Horinger
Beijing
1484
1536
Tianjin
Bohai Sea
Shijiazhuang
Taiyuan
Jinan
Huanghe River
1502
1501
Xi'an
Zhengzhou
1487
1505
110
112
114
116
118
120
122
40
38
36
34

(b)

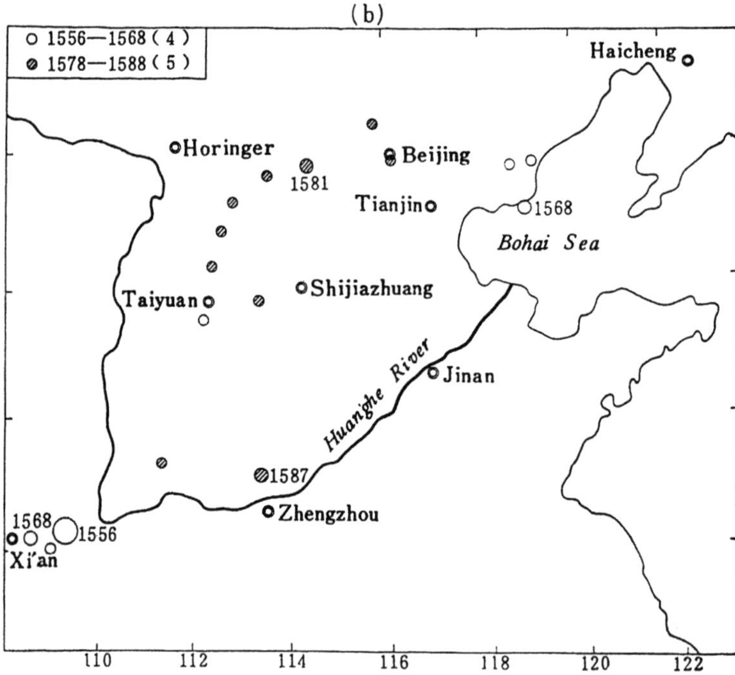

1556—1568 (4)
1578—1588 (5)
Haicheng
Horinger
Beijing
1581
Tianjin
1568
Bohai Sea
Taiyuan
Shijiazhuang
Jinan
Huanghe River
1587
Zhengzhou
1568
1556
Xi'an
110
112
114
116
118
120
122
40
38
36
34

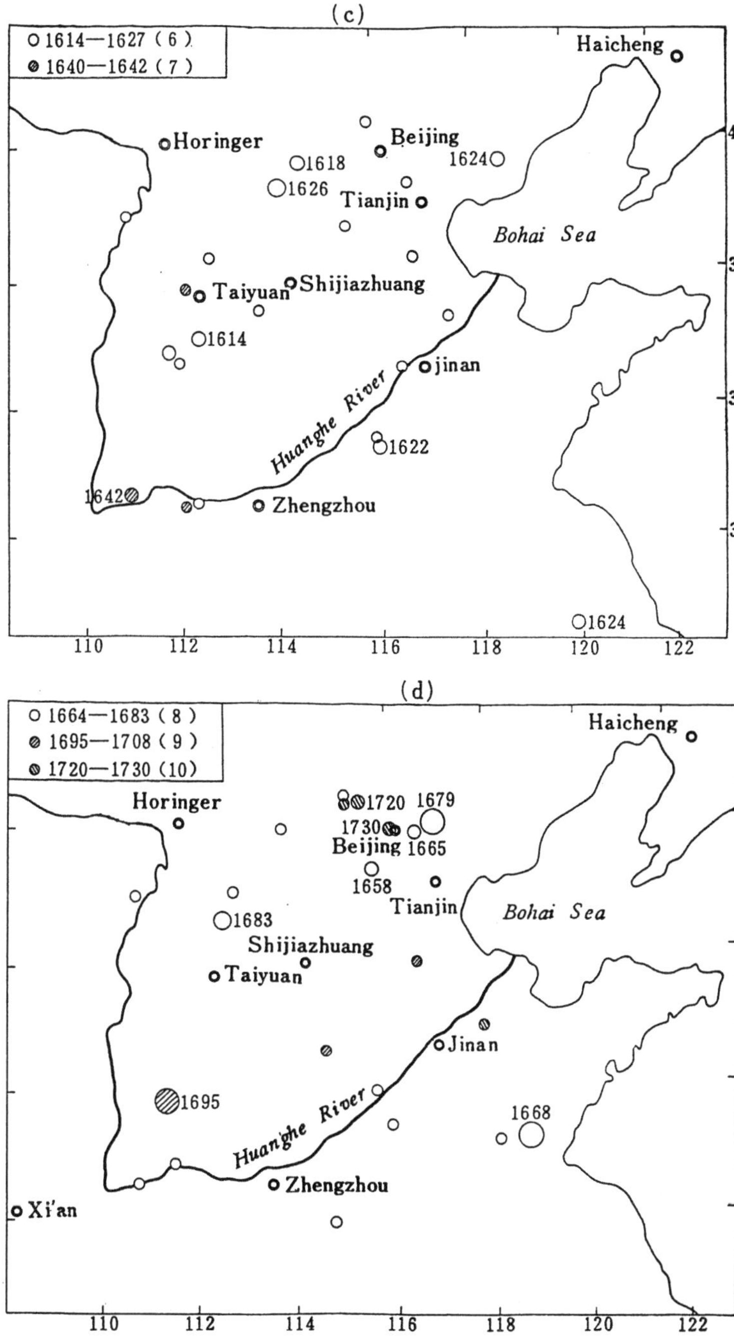

Fig. 13.3.*a–d.* Epicentral distribution of earthquakes in each episode during the third period in North China; *a* No. 1, 2 and 3 episodes; *b* No. 4 and 5 episodes; *c* No. 6 and 7 episodes; *d* No. 8, 9 and 10 episodes

" circle" . The tenth episode marked the end of the third period, without earthquakes of high frequency and magnitude.

The seismic activities in the southern part of the North China seismic region constituted episodes No. 1 and 2 of the fourth active period (Figure 13.4), generally arranged in two lines, with the southernmost runnning NWW and the other along latitude 36 ° N. The third episode had the earthquakes mainly concentrated in a segment of the northern boundary. The overall seismic activities in these three episodes seemed to be rather similar to those of the first three episodes during the third period. The earthquakes in the fourth episode were gathered in the northeastern part of the North China seismic region, with five earthquakes occurring at places similar to those of the earthquakes in the recent seventh episode. They are the 1880 Luanxian earthquake of M5.0, Hebei Province, the 1882 Shenxian earthquake of M6.0, Hebei Province, the 1885 Yingkou earthquake of M5.0, Liaoning Province, the 1888 Bohai earthquake of M7.5, and the 1893 Cangxian earthquake of M5.0, Hebei Province. The earthquakes in the fifth episode were small in magnitude, and scattered, showing northeasterly activity. The sixth episode had many earthquakes of M6, it being an episode with a high frequency of strong earthquakes, which were mainly located at the eastern boundary. These earthquakes were generally aligned in a nearly SN or NE direction, but the concrete succession of their occurrence still featured an east–west distribution, such as the 1939 Heze earthquake of M7 and the 1937 Yellow Sea earthquake of M6 which lay roughly in an EW direction, with the 1944 Dandong earthquake of M6, Liaoning Province and the 1945 Luanxian earthquake of M6, Hebei Province oriented approximately also in an EW direction. There had been a quiescent interval of about 20 years since 1948. Starting with 1966, the seventh episode saw consecutive occurrences of strong earthquakes. There were fourth earthquakes of M7 and one of M6 in this episode distributed in or around the Bohai area, but there were two phases of orderly occurrences of earthquakes. The first phase consisted of the 1966 Xingtai earthquake of M7.2, the 1967 Hejian earthquake of M6.3, and the 1969 Bohai earthquake of M7.4, all of which were centered near the latitudinal tectonic zone along latitude 38 ° N in central North China; the second phase had the 1975 Haicheng earthquake of M7.3, the 1976 Helingeer earthquake of M6.4, and the Tangshan earthquake of M7.8 near the latitudinal tectonic zone along Latitudes 40–41 ° N.

From the above descriptions it can be seen that the development of every episode during a period is very complicated. By comparing the third and fourth active periods, we shall discuss two related problems as follows: (1) the basically similar general development of the two periods; (2) the relation between the oblique (NE and NW) seismotectonic zone and the latitudinal seismotectonic zone.

(a)

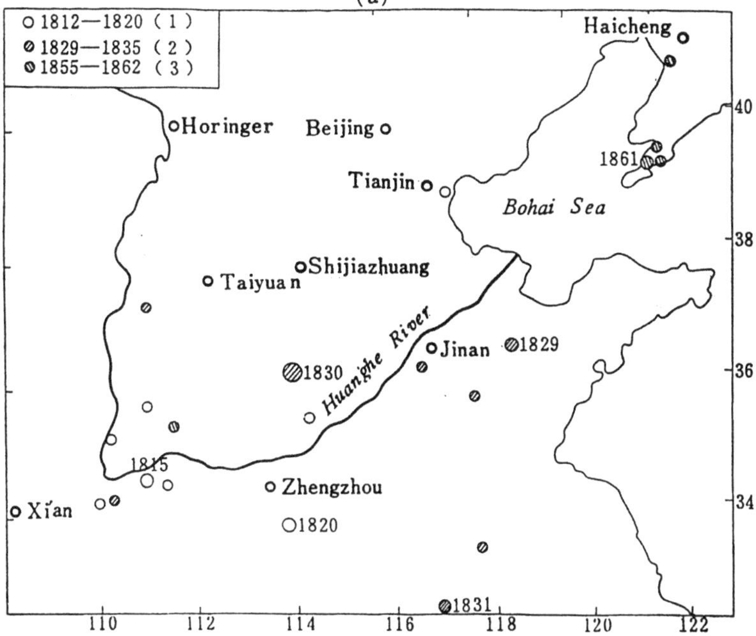
○ 1812—1820 (1)
⊘ 1829—1835 (2)
⊘ 1855—1862 (3)
Haicheng
Horinger
Beijing
Tianjin
1861
Bohai Sea
Shijiazhuang
Taiyuan
Huanghe River
Jinan
1829
1830
1815
Zhengzhou
Xi'an
1820
1831
110
112
114
116
118
120
122
40
38
36
34

(b)

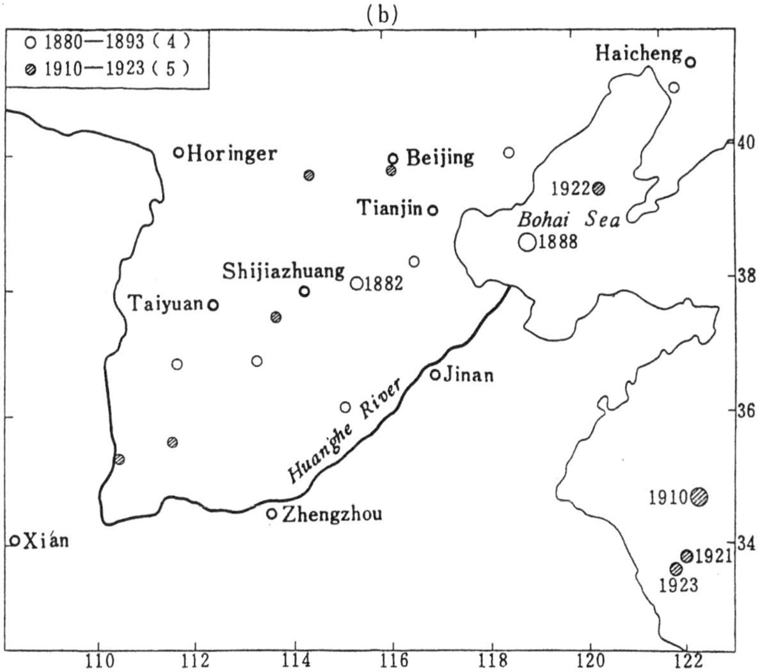
○ 1880—1893 (4)
⊘ 1910—1923 (5)
Haicheng
Horinger
Beijing
Tianjin
1922
Bohai Sea
1888
Shijiazhuang
1882
Taiyuan
Jinan
Huanghe River
Zhengzhou
1910
Xi'an
1921
1923
110
112
114
116
118
120
122
40
38
36
34

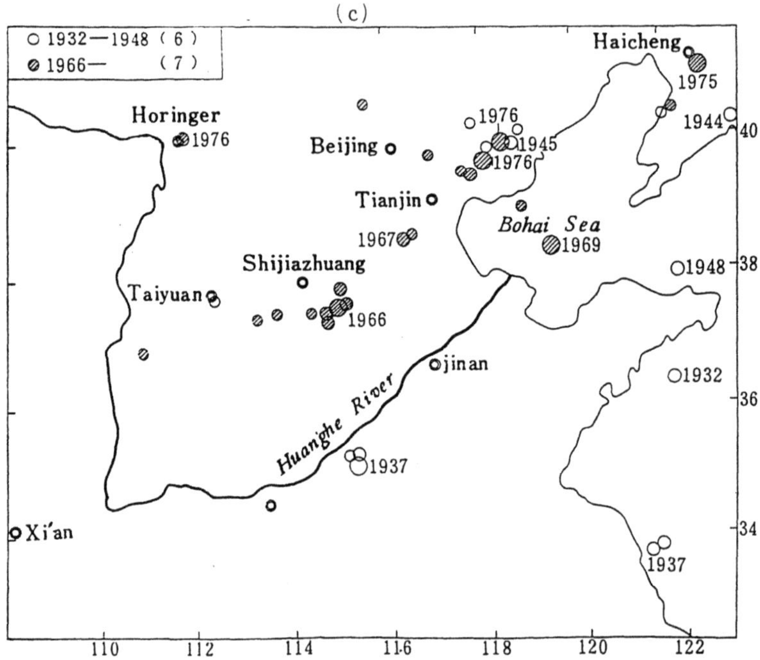

Fig. 13.4.*a–c*. Epicentral distribution of earthquakes in various episodes of the fourth active period in North China; *a* the first, second and third episodes; *b* the fourth and fifth episodes; *c* the sixth and seventh episodes

13.2.3 Comparison Between the Third and Fourth Seismically Active Periods

1) Similar general evolutions of the two periods.

There are some aspects comparable in the developmental history of earthquakes in the third and fourth periods: (1) the earthquake frequency was low, with only individual strong earthquakes in the fourth episode of the first half of the periods, while the frequency and magnitude were high in the second half of the periods; (2) the earthquakes in the first three episodes were mainly concentrated at the northern and southern boundaries, and in the south in particular; (3) the earthquakes in the following several episodes moved toward the central part from both the southern and northern boundaries, with one or two episodes showing obvious NE and NW orientated activities, and such episodes had a high earthquake frequency; (4) high energy releases occurred in a late phase of the seismically active periods.

This general evolution may indicate that each seismic episode was only a rhythmical evolutional step during a seismic period, and like the mechanical relation among the multiple stress concentration points in an earthquake field, the first occurrences of earthquakes altered the stress distribution in the earthquake field and the state of the medium of some areas, creating certain seismogenic conditions for forthcoming earthquakes. One episode in itself was not able to complete an event of crustal movement independently, but the development of several episodes might have accomplished gradually the first and late phases of a given seismic period, achieving thus a "circular–shaped" process, which is a process of overall activity from the boundaries to the interior of a fault block.

If the above ideas on the general evolution hold true, how should we look into the seventh episode of the fourth period which is of the highest concern at present? The first problem is whether the seventh episode is over or not, which implies the estimation of three elements of an earthquake episode, and will be treated in Chap. 14. The second problem is with which episode of the third period can this episode of the fourth period be compared, which means that this problem is essentially concerned with the study of the definite historical position of the seventh episode. There are two different opinions on this problem. The first holds that the seventh episode at present can be compared with the sixth in the third period, for the following reasons: (1) the earthquake frequency of both is very high; (2) the earthquakes of both are all concentrated in the northern part of North China; (3) there exist significant gap segments in the "circular" pattern of the fourth period; and (4) there are only 165 years up to now showing seismic activity for the fourth period, about 80 years less than the total time–span of the third period. The second opinion argues that the present seventh episode can be compared with the eighth episode of the third, the reasons being as follows: (1) the energy releases of the earthquakes in the seventh episode were very high, featuring intense releases of energy in later periods; (2) both the fifth and the sixth episode of the fourth period had a very high earthquake frequency, and were hardly distinguishable, being very similar to the sixth and seventh episode of the third period (Fig. 13.1). Obviously, the above two points of view hold a relatively different evaluation of the potential seismicity in future decades, however, it is very difficult to verify at present which argument is more realistic.

2) The relation between the oblique seismotectonic zone and the latitudinal seismotectonic zone.

There are altogether 17 seismically active episodes recognized in both the third and the fourth active periods, and usually the earthquakes in one episode were mainly distributed along a major tectonic direction, but sometimes along two tectonic directions. Of the 17 episodes, 9 were directly characterized by the seismicity lying in an east–west direction, and 6 in a north–east direction. However, if all the earthquakes in every episode were placed on a geographical map, the general arrangement of earthquakes would be dominated by the north–east orientation. Such latitudinally sequenced and obliquely lined distributions of earthquakes are very significant for forecasting and theoretically interpreting seismic activity.

The nodal plane solutions of focal ruptures indicated that a majority of strong earthquakes in North China are of NE dextral dislocation, or NW sinistral dislocation, and the major axes of the isoseismal lines of strong earthquakes are also mostly directed NE and NW (Yang Lihua 1980). These facts have demonstrated that these earthquakes were associated with the dislocations of locked segments at oblique faults. It is thus shown that the earthquakes in North China are controlled by an oblique structure, especially the NE one. Therefore, after a strong earthquake, the NE–SW or NW–SE migration of earthquakes was always first proposed. However, it is not the case actually, earthquakes having occurred instead in succession mostly along an EW direction, over a certain length of time in an earthquake episode.

Theoretically, if the controlling role of oblique faults isconsidered, it is only natural to arrive at a situation as shown in Figure 13.5a. In an oblique faults–intercrossed block, the right–lateral dislocations occurred at NE lying faults, and the left–lateral dislocations at NW lying faults, due to actions of regional dynamic forces in a nearly EW direction. Such extrapolation is simple and concise, in agreement with the basic facts of individual earthquakes, and so this has been generally accepted. Nevertheless, considering the tectonic activity due to the consecutive occurrence of earthquakes, the problems become much more complex, and the distribution of seismic activities in an east–west row and north–east line is one of the outstanding problems.

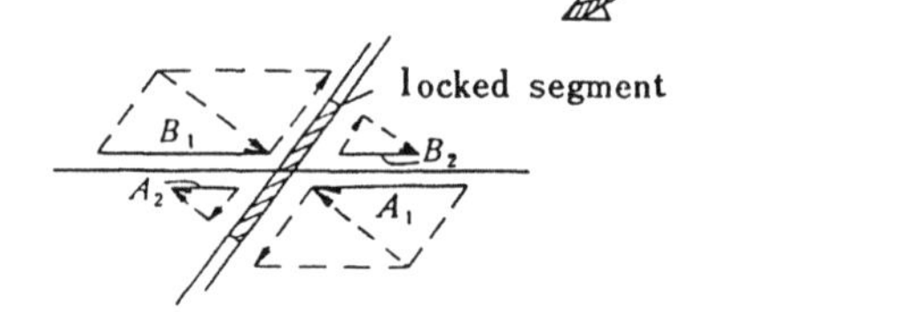

Fig. 13.5. Considerations on two earthquake–controled tectonic movements in the North China seismic region; *a* simple earthquake of a rhombic fault block; *b* control of movements of deep–seated latitudinal faults

Figure 3.7 shows the basic knowledge we gained on the seismotectonic framework in North China. From the figure it can be seen that the shallow crustal movements of the Earth are characterized by the relative motion of small rhombic blocks. But this is merely a portion of the crustal movement in North China. There exists also another latitudinal tectonic movement which propagates much farther than the extension of a shallow–seated fault block. If the latitudinal tectonic movement involves effective latitudinal motion ,as shown in Fig.13.5b, the stresses would be also concentrated near locked segments of the oblique fault zones, leading to dislocation earthquakes. Under the action of strike motion of the latitudinal zones, it is quite possible to have more than one earthquake occurring near the same latitudinal zone, causing thereby the earthquakes to be distributed in an east–west row along a north–easterly line.

Near the crossing points of the latitudinal and oblique structures, the latitudinal forces acting on the locked segments of the oblique fault can be simply expressed by the vector solution method (Fig. 13.5b). If a pair of approximately equivalent right–lateral couples act on both sides of a locked segment,accompanied with a NS extension, the dextral shear–extensional movement would be produced as a result, which coincides with the dislocation mode of most earthquakes in North China.

13.3 Generality of Seismic Episodes

Since the beginning of this century, the records of strong earthquakes have been well documented. By investigating national or global seismicities in regard of a seismic zone or seismic region, it has been also found that there have clearly existed fluctuations and alternations of sometimes strong and sometimes weak seismicities, very similar in character to the earthquake episodes. Before we are to study the possible relation between them,we shall describe some sets of successive occurrences of strong earthquakes or episode phenomena of earthquakes in this century.

13.3.1 Four Active Phases of Strong Earthquakes in Mainland China in This Century

The nine great earthquakes dealt with in this book occurred successively and correspondinqly in the North China and Sichuan–Yunnan regions, showing two and half times of to–and–fro distributions roughly along the same NE tectonic direction. Such phenomena implied a possible intrinsic relation among these events, which has been explained in the paper written by Xu Shaoxie et al. (1980) on the basis of their own viewpoint of crustal buckling .

Since 1970, when research on the overall situation of nation–wide seismicity began, many Chinese investigators have been engaged in studying the consecutive occurrence of strong earthquakes in continental China. Many of them inclined to divide seismic activities in and around Mainland China into four active phases (Fig. 13.6), each one lasting 13–16 years, and with a time interval of 6–9 years, as shown in Table 13.2. In fact, these are episodic phenomena of seismic activities in Mainland China and its neighbouring regions.

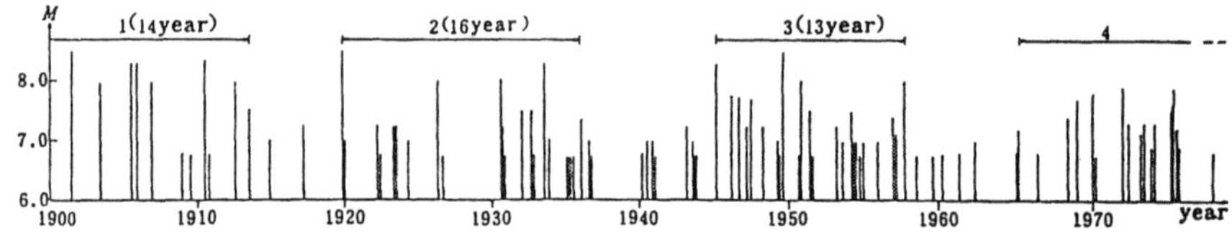

Fig. 13.6. Temporal diagram for shallow earthquakes with M > 6.7 in and around Mainland China from 1900 to 1980

By comparing these four active phases, the seismic activities in the first three phases seemed to be similar in magnitude and frequency, but the seismicity in the fourth phase was relatively lower, and shorter in duration. The first three active phases all began and ended with earthquakes of M7.5, while the fourth phase commenced with the 1966 Xingtai earthquake of M7.2 and after the 1976 Songpan earthquake, no earthquakes of M7 occurred ever again in Mainland China. If it is strictly assumed that this phase started with the 1970 Tonghai earthquake of M7.7, and concluded with the Songpan earthquake of M7.2 in 1976, this phase lasted only 7 years. In this case, the fourth phase may not be over, and possible earthquakes with M7.5 may follow. Still, it is also possible that the fourth active phase has been terminated. Why does this phase feature low magnitude and short duration? It can be presumed that such phenomena may be associated with the locations of the main seismic activities and the developmental stages of regional seismicities in every seismically active phase.

According to the catalogue of earthquakes from 1897 to 1957 given by Duda (1965), and the catalogue of epicenters determined by the Chinese Seismic Network thereafter, have the epicentral distributions of the four active phases in Mainland China been plotted (Fig. 13.7). From the figure it can be seen that the earthquakes in these four phases were concentrated in certain areas, and we called the area of earthquake concentration as a main seismicity area in an active phase.

The main seismicity area of the first active phase is located along the NE tectonic zone from Pamir to Baikal Lake; the main seismicity area of the second phase lies along the NWW tectonic zone from Haiyuan, east of the Hexi corridor to Fuyun, Xinjiang Autonomous Region; the main seismicity area of the third active phase lies along the nearly NS tectonic zone from Chayu, east of Tibet to central Mongolia; and the main seismicity area of the fourth active phase includes the Sichuan–Yunnan seismic region and the northern part of the North China seismic region. From the historical records of earthquakes generally, the northwest China region is the one with the most great earthquakes of M > 8. The first three main seismicity areas are located in the west part of China, so there are more high–magnitude earthquakes there. As for the fourth active phase, the Sichuan–Yunnan region is of frequent seismicity with rare occurrences of M8 earthquakes and, compared with the west of China, the North China seismic region is generally low in earthquake frequency. Although there were many historical earthquakes with M8 in the North China seismic region, most of them

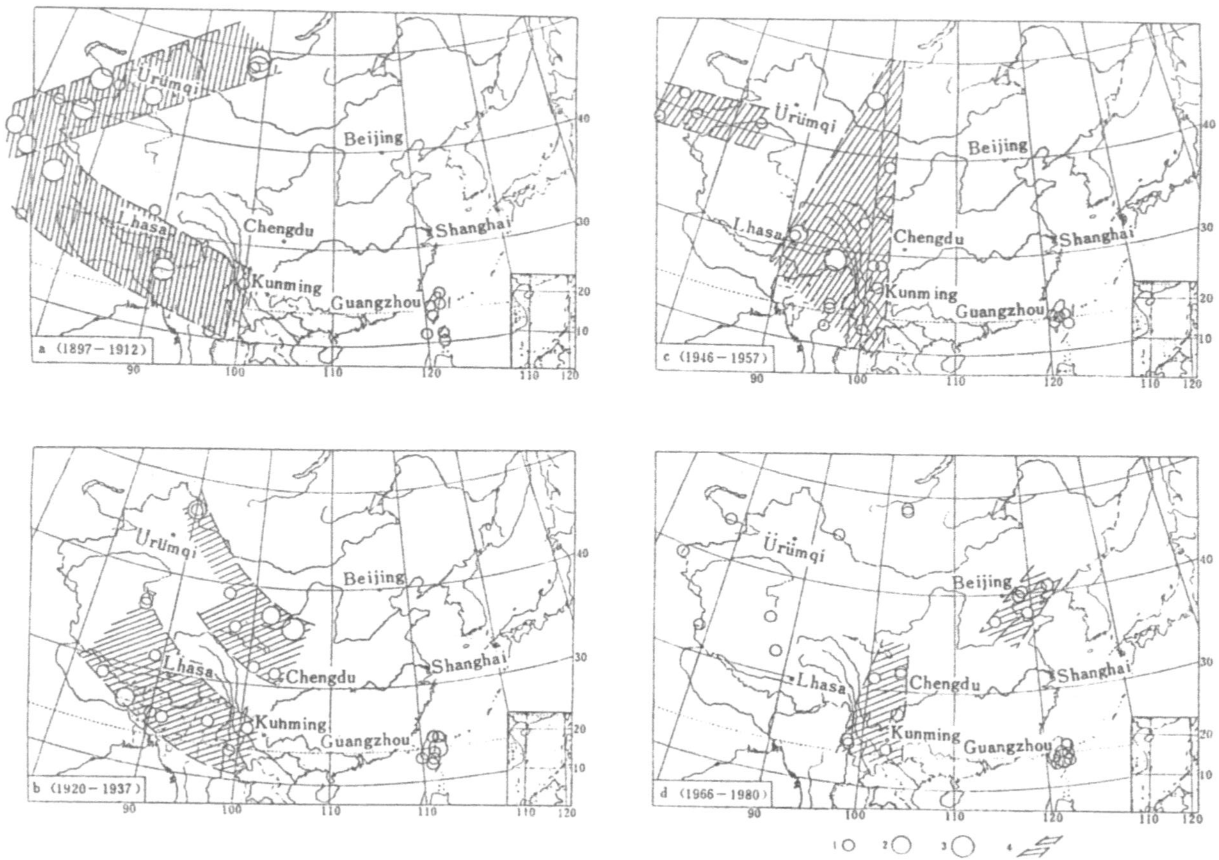

Fig. 13.7 *a–d*. The epicentral distributions of the four active seismic phases in Mainland China in this century;*a* 1897–1912; *b* 1920–1937; *c* 1946–1957; *d* 1966–1980. *1* M7.0–7.4; *2* M8.0–8.4; *3* M8.5; *4* main seismicity area

occurred in the several late earthquake episodes in the second half of each seismic period. But the total number of great earthquakes in this region is far smaller than in the west part. Therefore, this might be the reason why the magnitudes are fairly small for the earthquakes in the fourth phase, and also why possibly there will be earthquakes of M8 in the eighth or ninth episode of seismicities in North China.

What is the reason for the jump–like migration of the main seismicity areas of the four active phases in this century, and the movement roughly from west to east? This is a very interesting problem. Liu Baichi (1979) came up with the idea of the stress–adjusted earthquake field as a solution. In his opinion, seismic activity occurred in a seismotectonic region, then moved to a neighbouring region, showing the segmented occurrences and the migration of seismicity. He also suggested that due to the northward push of the Indian plate, seismicity was first induced at the Pamir zone which pertained to the NE apex of the Indian plate, then its eastward migration led to seismic activity in the north Qinghai–Xizang block as a whole, and was followed by the overall seismic activity in the nearly NS tectonic zone related to the NE apex of the Indian Plate, and finally the further eastward shift, brought the activity into the eastern China seismic region. The idea of the stress–adjusted field has been commonly adopted by a good number of researchers in recent years. However, such regional migration and adjustment may not be totally dependent upon the push of the Indian plate, and this problem will be discussed again below. Another intriguing problem is the question where will the main seismicity area of the next active seismic phase appear. Unfortunately, only insufficient data exist on seismicity in the western part of China before this century, hence we are unable to confirm whether the migration of earthquakes from west to east mentioned above is a kind of repeated phenomenon in history or a regular one, nor can we speculate about the main seismicity area of the next active phase. The spatial distribution of continental earthquake episodes is very complicated, and deserves further study in the future.

13.3.2 Mogi's periods

Mogi (1974) published his research results in his *Active Periods in the World's Chief Seismic Belts*. He suggested that the consecutive occurrences of large earthquakes from a general character of global seismicity, and the world's chief seismic belts exhibited fluctuating and alternating variations, with successive occurrences of large earthquakes in a certain period, and occurrences of minor and weak earthquakes in another period. Several typical diagrams given by Mogi are represented in Fig.13.8, which may show that:

1) In each belt, there are periods of 16–20 years with fluctuations and alternations of strong and weak earthquakes, which actually seem to be the rhythmical fluctuations of earthquake episodes.

2) In each seismically active phase, the earthquakes are basically distributed over an entire seismic belt, hence providing a dynamic regime of the whole seismic belt for earthquake alternation.

3) When the seismicity is high in the high–latitude Aleutian seismic belt, it is

relatively calm in the other three seismic belts of low latitude, and vice versa, thus showing the negative and positive rhythms of earthquake episodes.

Mogi also thought that there might exist common active periods of earthquakes in low–latitude seismic belts, and the correlations of large low–latitude ranges may represent the influences of the rotation of the Earth on plate movements.

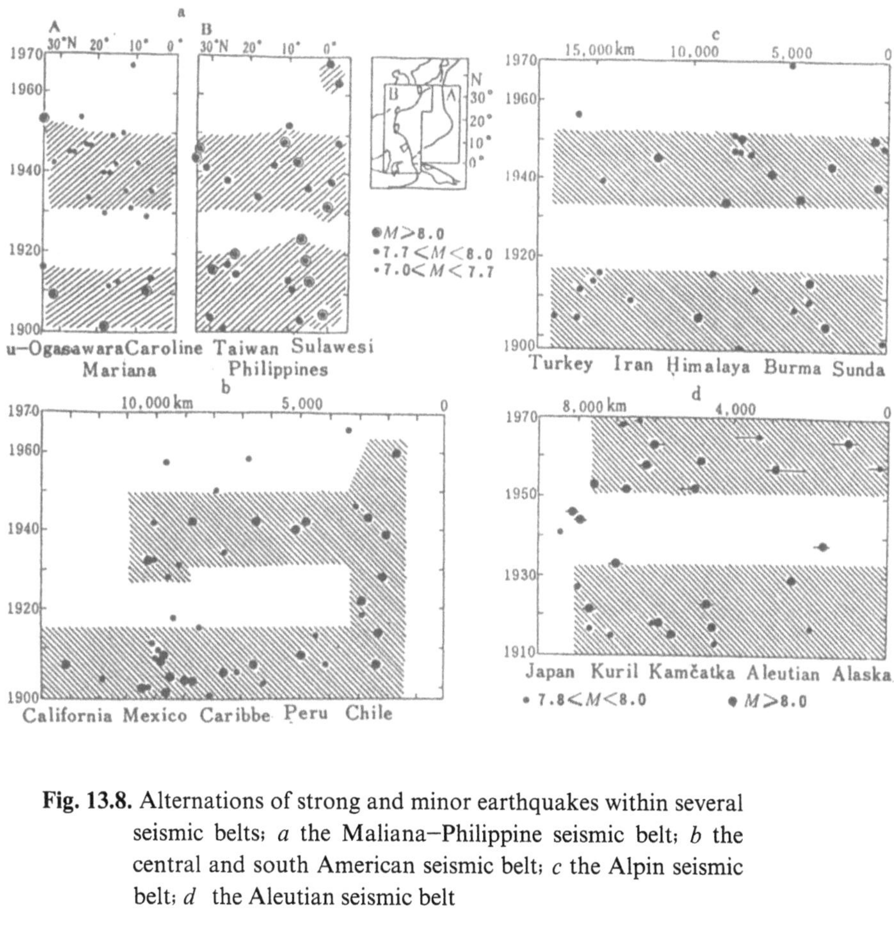

Fig. 13.8. Alternations of strong and minor earthquakes within several seismic belts; *a* the Maliana–Philippine seismic belt; *b* the central and south American seismic belt; *c* the Alpin seismic belt; *d* the Aleutian seismic belt

13.3.3 Relationship Between Regional Earthquake Episodes in Continental China and Its Surroundings

The relationship between the earthquakes in Mainland China and seismicity on the surrounding plate boundaries is a very interesting research subject. One of the fundamental tasks in this respect is to compare the time–space distribution of earthquakes. Xue Feng (1981) obtained some significant results from such a comparison. He divided Mainland China into three large regions, east (E) northwest (NW), and southwest (SW), and their surrounding seismic zones on the plate boundary into the West Pacific seismic zone (a), the Aleutian seismic zone (b), and the Himalayan seismic zone (c).

Then various seismic regions in Mainland China have been compared with their surrounding seismic zones on the plate boundary (Fig. 13.9). The results showed the fluctuating phenomena to be similar to earthquake episodes, and established the following relations between the regional seismic episodes in and around continental China:

1) Between a region and E region, b region and NW region, and c region and SW region, there existed approximately synchronous phenomena of earthquake activities, and there were no definite faster or delayed relations.

2) The episodic phenomena of earthquake activities in both b and c regions were very clear, and the rhythmical fluctuations of strong and weak earthquakes were just contrary in the two regions, as shown in Table 13.2 (4) and (10). The episodes of earthquakes in the SW and NW regions were similar to those in the b and c regions, respectively [see Table 13.2 (7) and (8)], but the time distributions were not as regular as in the b and c regions, thus indicating the relative complexity of continental earthquakes.

3) There was an apparent synchronous relation of earthquakes between a region and E region, but their episodic relation was rather complicated. The episodic bounds sometimes were equivalent in a and b regions or sometimes similar in a and c regions.

4) It is rather understandable that there existed a synchronous relation of earthquakes between a and E regions, c and SW regions, because the two regions border on each other and either the compression of the bordering plates on Mainland China or the interactions between the continent and the surrounding plates would lead to synchronous seismic activities. However, why did the synchronous episodic fluctuations of earthquakes also exist in the NW and c region, seeing that there is a great distance between them? It is really difficult to explain such a phenomenon by assuming that the northward push of the northern margin of the Pacific plate caused seismicity in the c region as well as in the NW region of Mainland China. The direction of regional dynamics for generating earthquakes in the c region was nearly NS, and so it was in the NW region too. Furthermore, both the c and the NW region are located at high–latitude zones of the northern hemisphere. In these similar aspects may the root cause of the synchronous seismicity in the two regions be found.

13.3.4 Episodic Phenomena of Global Seismicity

Gao Xu et al. (1980), using the catalogue of earthquakes with $M > 7$ in the world from 1897 to 1977 compiled by Duda (1965) and Bath (1979), revealed the episodic phenomena of global seismicity with the frequency variation curves. They suggested six episodes of seismicity, with three relatively calm seismic episodes, ?–1904, 1919–1934, and 1951–1967; and three relatively active episodes, 1905–1918, 1935–1950, and 1968– . Each episode lasted for about 14–17 years, and the active and quiescent episodes occurred alternately. If the sixth episode started in 1968, then it may not have ended yet.

The above–mentioned four sets of data may indicate that the global seismicity, the earthquakes at different segments of the Circum–Pacific seismic belt, and the earthquakes in various seismic regions of continental China are all characterized by phasic fluctuation and alternation of strong and weak earthquakes. For this reason, a comparison of global seismic episodes became possible.

13.4 Global Comparisons of Seismic Episodes and Preliminary Discussion on the Causes of Earthquake Episodes

13.4.1 Global Comparison of Seismic Episodes According to the Episodic Phenomena of Seismic Activities

We have compiled 13.2 which shows episodic phenomena in different segments of the Circum–pacific seismic belt, continental China with its various seismic regions, and Algeria as no detailed analyses of episodic phenomena of earthquakes in the Mediterranean region are available, we included only the data from Roussel 1973 on seismicity in Algeria, while in Fig. 13.10 the episodic times in various regions are compared in a columnar representation.

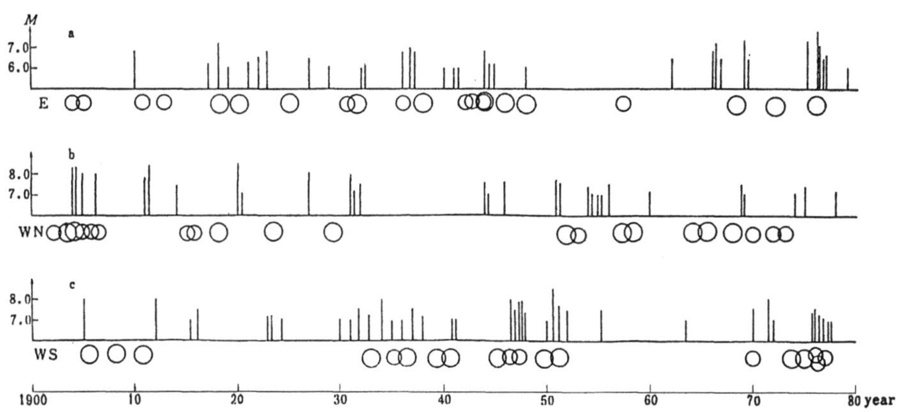

Fig. 13.9. Comparison of seismic belts in Mainland China and the surrounding plate boundaries;*a* the West Pacific seismic belt; *b* the Aleutian seismic belt *c* the Himalayan seismic belt; *E* the East China Mainland seismic region; *NW*the Northwest China seismic region; *SW* the Southwest China seismic region. The *lines* indicate earthquakes in Mainland China, and the *open circles* represent earthquakes on the plate boundaries (*great circles* mean M > 8, and *small ones*, M7–7.9)

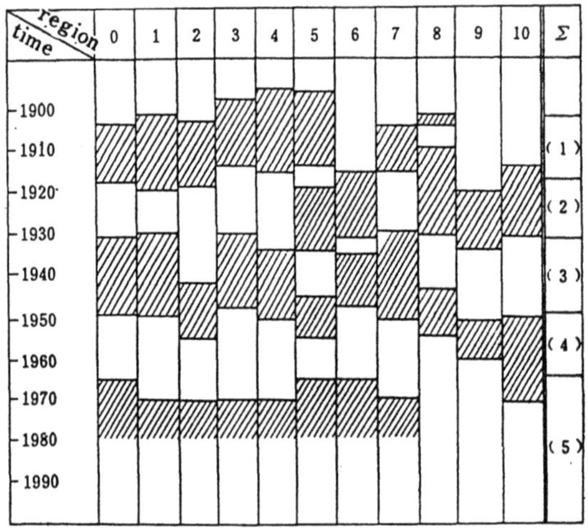

Fig. 13.10. Global columnar section of earthquake episodes. Area numbers are the same as those given in Table 13.2,Σ–global earthquake episodes

Table 13.2.

episode \ region	0	1	2	3	4	5	6	7	8	9	10
1	1905 -1918	1902 -1920	1904 -1920	1898 -1914	1897 -1916	1897 -1914		1905 -1915			
2						1920 -1934	1917 -1932		1911 -1932	1922 -1935	1915 -1932
3	1932 -1950	1931 -1950	1943 -1957	1931 -1948	1934 -1951		1936 -1948	1930 -1951			
4						1946 -1957			1944 -1955	1952 -1962	1951 -1969
5	1969–					1966–	1966–	1970–			
6											

Illustrations of regions: 0—globe; 1—Philippine-Mariana zone; 2—Tonga-Fiji zone; 3—west American coastal zone; 4—Indonesia-Alpes zone; 5— Mainland China and its vicinities; 6—east Mainlandi China region; 7—Southwest Mainland China region; 8—Northwest Mainland China region; 9—Algeria zone; 10—Aleutian zone

From Table 13.2 and Fig. 13.10, the following details can be appreciated:

1) The episodic phases are basically the same for seismic zones No.1,2,3 and 4 at low and middle latitude. Because in these four zones global seismicities made up two–thirds or more of the total, the fluctuations of earthquakes in these zones would decide the basic regime of global seismic activities. We suggested that the consistent rhythm of fluctuations of strong and weak global seismicities be

termed "the positive rhythm of global seismic activities."

2) For the North Pacific seismic belt, called also the Aleutian seismic zone, the rhythm of fluctuations of strong and weak earthquakes is exactly contrary to the positive rhythm, so it is termed negative rhythm.

3) The episodic phenomena of earthquakes in Mainland China are very complicated, with a nearly negative rhythm in the Northwest China region, a nearly positive rhythm in the Southwest China region, and positive and negative rhythm alternating in the eastern region, If continental China is considered as a whole, the time for the alternation of episodes in it would be quite close to that in the world, but the rhythm of the continental earthquakes is fairly complex, and the main seismicity areas are also very complicated in episodes and the directional variations of the earthquake migrations. All this fully demonstrates that continental China is a complex region, affected by multiple factors.

All in all, we can regard the rhythmical fluctuations of global seismic episodes as the basic rhythm of global seismicities, and then regional comparisons can be made based on the positive and negative rhythms. The seismic episodes in Mainland China can be classified with reference to the basic rhythm of global seismicities, but regional comparisons of positive and negative rhythms should be analysed concretely.

13.4.2 Causes of Earthquake Episodes

Why occur episodic phenomena of earthquakes in the whole world and in various major seismic regions? We made only a few studies in this field, and the following will furnish only some pointers regarding this problem.

Table 13.3.

earth rotation	atmosphere	ocean	geomagnetism	solar activity
>100 years				96 years
60 years			57.8 years	59 years
				48 years
30 years			(Z) (H)	28 years
22 years			22.2, 21.4years	21 years
18 years		18.6years		17 years
15 years				15 years
13–9 years	10.6 years	10.6years	10.7, 10.5years	12–9 years
		6.3years	<10 years	

1) The quasi–periodic rhythm of fluctuations of strong and weak earthquakes, with a time scale of about 10–20 years, is one of the basic characteristics of crustal movement, belonging to the domain of micro–dynamics. Such global micro–dynamic variations seemed very difficult to explain by partial vari-

ations of the thermal regime in the interior of the Earth, or partial variations of convections in it. By using various global periodic dynamic factors, including the variations of the Earth's rotation for shortly modulating and triggering the global stress field, such short–lived global dynamic variations can be interpreted.

Luo Shifang (1978) studied the maximal entropic spectra of the variations of the Earth's rotation and of other environmental factors (as shown in Table 13.3). These spectra with periods of about 20 or 10 years were very general and obvious. The environmental dynamic factors, as modulating and triggering, will provide a possible approach to explain the successive occurrence of earthquakes in localities where energy accumulations have reached a critical state during seismogenic processes in the global stress fields, within roughly the same time range.

2) The second characteristic of global earthquake episodes is the interchange of activity areas for alternations of positive and negative rhythms. For example, the interchange of negative and positive rhythms in both high and low latitudes may be either associated with the overall deformation of the earth, or with the alternative variations of global stress fields in high and low latitudes. Furthermore, there was a positive rhythm of seismic activity along the nearly NS lying seismic zone, but a negative rhythm along the nearly EW running one, which thus illustrates the orientation of the dynamic background for earthquake episodes, and is perhaps related with global crust movements as a whole.

3) The third basic characteristic of global earthquake episodes is that they did not develop in equal periods, and differences existed among them. Particularly, analysis of the episodes in Mainland China and the North China seismic region suggested that there were historical associations between the episodes, which like beats linked the development of seismically active periods. If these active periods may possibly affect the process of local crustal movement, then the environmental factors are variable ones superimposed on a regional basic stress field, and the common action of both leads to the development of one seismic episode after another.

These are our fundamental considerations concerning the causes of seismic episodes. We will use these considerations to discuss the prediction of three elements of an earthquake episode in Chap. 14.

CHAPTER 14

Fundamental Understanding About Earthquakes in Continental China and Ideas on Step–by–step Earthquake Prediction

In this chapter, we shall describe our fundamental understanding about earthquakes in continental China derived from the seismogenic environments and earthquake–producing mechanism, the relationship between crustal movement and precursory anomalies, the seismogenic processes, and the characteristic time–space distribution of earthquakes in Mainland China, and on this basis, we shall elucidate some viewpoints we considered for the prediction of earthquakes in continental China, and present our ideas on step–by–step earthquake prediction.

14.1 Basic Knowledge on Earthquakes in Mainland China

There are limitations in time and space for the nine great earthquakes, and so the basic features of earthquakes in continental China cannot be completely revealed only from the summing up of the nine earthquakes. However, their numerous similarities are typically representative, and their principal aspects are as follows:

14.1.1 Correlation of the Continental Earthquakes with Tectonic Environments of the Continental Crust

It is generally accepted that shock is an expression of the sudden dislocation of a rock body, but the origin and mechanisms of earthquakes may be significantly different, as the geological and geophysical environments for earthquakes differ. The earthquakes in continental China were very different from those in the plate boundary zones (Ma Zongjin 1987), and the preparation and generation of the continental earthquakes were directly associated with the large regional tectonic movements of the continental crust itself. In order to illustrate this, we have emphasized that the continental earthquakes are a kind of tectonic shocks in the continental crust. As mentioned previously, the nine great earthquakes and those with $M \geqslant 6$ occurring in the territory of China originated almost in the continental crust, and most of them occurred within the competent layers of the middle continental crust consisting of granitic and metamorphic rocks. Most of the earthquakes occurred due to repeated dislocation of the pre–existing fault zones, but a few took place due to newly dislocated faults. From a mechanical angle, the quakes can be regarded as a result of mutual constraint and development of stress, strain and the state of the medium in a fault zone, and this is a basic feature of the tectonic environment and the earthquake–producing mechanism for the direct preparation of continental earthquakes. With this in mind, researches on seismogenic regularities should be conducted on the basis of studies on tectonic motion of the continental crust, and in particular on the developmental regime of stress–and– strain–concentrated segments. Then, are there any possibilities for detecting the transitory stress and strain distribution in the continental crust as well as their variations? As the competent layer is fairly close to the surface, the overall motion and the faulting of the layer are proceeding extensively within the middle and upper continental crust. Therefore, we may directly or indirectly detect them as objective evidence in studying the regularities of earthquake preparation.

14.1.2. Non–uniqueness of the Relation Between Precursory Anomalies and Earthquakes

Earthquakes could be predicted on the basis of a variety of possible anomalous phenomena before quakes, and this is one kind of approach to the problem of earthquake prediction. In order to make a distinction between this kind of prediction and that made by means of the purely mathematical statistical method, we have termed it as the precursory earthquake prediction. Although it is presently difficult to be certain that precursors will be observed before every shock, it is still very well known that there are precursors before earthquakes, especially before larger ones. In the above–mentioned summarization of the nine great earthquakes, we have tentatively noted some precursory phenomena or such anomalous ones as were related to precursors, and some anomalous phenomena have been observed by Chinese and foreign scientists many times prior to earthquakes. These actual observation data have supported the idea on precursory earthquake prediction. That is why observational work of precursory phenomena has been rapidly developed in many seismically active countries. Furthermore, it is unavoidable that some anomalous phenomena should occur during tectonic movement of the continental crust, in the light of the above–mentioned seismogenic environments for continental tectonic earthquakes. Therefore, we think the idea of precursory earthquake prediction is of rather general and practical importance.

However, we encountered a fundamental problem in earthquake prediction practice, which is the non–uniqueness of the relation between "precursors" and earthquakes. As. said previously, the anomalous phenomena before shocks are called precursors because earthquakes always occurred in terrains where tectonic movements were most intensive, and which were precisely the places where the appearance of various kinds of anomalies was the most obvious. Hence, people naturally regarded these anomalies as earthquake precursors, and suggested that they were associated with a seismogenic source. However, practical findings showed that such "precursors" were sometimes followed by the occurrence of earthquakes, and sometime not. In our opinion, the key to this problem is how to interpret the relation between "precursors", anomalies, and earthquakes.

As regards continental tectonic earthquakes, these showed themselves as a kind of manifestation and outcome of crustal tectonic motions. It may be said that till the occurrence of earthquakes, the various kinds of anomalous phenomena are only a certain superficial expression of tectonic movements, including changes in the properties of the medium, and their mutual influences . Even for crustal tectonic movements under conditions of fairly high stresses or strains, there still existed multiple possibilities for their development. For instance, in a locking segment, as the media went through a stage of hardening and softening, the fissures would develop swiftly, resulting in an earthquake due to sudden dislocation; because of the high temperature or permeance of liquid, an aseismic slip would occur at a locking segment of a fracture zone in an anomalies–concentrated region. In case of no continous energy supply, or of decreasing regional tectonic stresses, the already strongly developed elastic strains can be restored gradually, thus causing the pre–existing anomalous phenomena to die away. Since there are various prospects for the development of anomalies, direct identification of a single anomalous phenomenon as "precursor" is bound to be in non–unique contradiction with the occurrence of earthquakes, just as the relation between the process of gathering clouds and precipitations is non–unique.

In view of the above–mentioned contradiction, we suggested that the concept of "precursors" cannot be used arbitrarily, but in principle the word "anomaly" should be utilized to mirror a certain crustal movement. Accepting this as a prerequisite, not only a single anomaly must be analyzed, but also the time–space development of multiple anomalies should be analyzed mechanically in a comprehensive way so as to judge where stress and strain will become stronger, and when will the possibility for an abrupt dislocation in the course of their development. Through the summarization of the nine great earthquakes, we have only deduced from the analyses of each kind of anomalous phenomena, certain clues which implied the possible occurrence of earthquakes, and have gained some experience in applying the anomalous phenomena for speculation about forthcoming earthquakes, but have failed to discover "precursors" which must be followed by earthquakes. The application of such an understanding to monitoring and analyzing various anomalies is different, in principle, from that of precursors which necessarily lead to earthquakes. Then why do we still regard the non–unique anomalies as the expression of a seismogenic process? This is because, for the purpose of prediction, it is necessary to describe these developmental processes of anomalies as potential precursors of forthcoming earthquakes as soon as possible, so that the study of earthquake prediction and prevention can be developed correspondingly. This procedure is quite similar to the description of the developmental process of clouds before rain as a process of rainfall. However, this is only one possible process of earthquake generation.

14.1.3 Stages of the Seismogenic Process and Variety in the Behaviour of Earthquakes

Since continental tectonic earthquakes originated from the tectonic movements of the continental crust, generally speaking there should be long– or short–term seismogenic processes before the occurrence of earthquakes. This has been supported by the examples furnished by the nine great earthquakes. But are there any common stages for seismogenic processes? And do the large or small magnitudes of earthquakes stand in relation to the long– or short–term seismogenic processes? In line with the concept of earthquake field, earthquakes are directly associated with developments of the variable stress field with reference to the basic stress field of the crust, while the variable stress field is connected with the variations of regional tectonic forces and the self–adjustment of stresses within the field. During an earthquake episode, each event also had a very complex relation with the earthquake field. Therefore, in the seismogenic process, there did not exist a simple linear dependency between earthquake magnitude and the time length of the seismogenic process. We used to compare the relation between the magnitudes of the Chinese continental earthquakes (not limited to the nine great earthquakes) and the lengths of seismogenic processes ($\log T = 0.38M - 3.4$) with the relation ($\log T = 0.76M - 1.83$) given by Rikitake (1976) based on certain events all over the world. The results showed that there was a remarkable difference in slopes for both relations. However, the original data used for both relations indicated that it was common for both that their data points were largely discrete, with the earthquakes of M5—7 displaying the anomalies on time scales of equivalent dimensions. It is thus demonstrated that the relation between the magnitudes and the lengths of seismogenic processes is rather complicated.

The data on one aspect from a single station show that in many cases no complete process with several stages seemed to occur. But why do we still emphasize stage divisions as consisting of long, intermediate, short, and imminent periods during a seismogenic process in this book, especially in the comprehensive analyses in Chap. 10? There are three reasons for this:

1) The actual observations of anomalies before the nine great earthquakes have indeed presented the stagewise development of anomalies, namely (1) the intensification of regional tectonic movements; (2) the relative concentrations of anomalous phenomena in some local areas or zones; (3) the intense changes of continuous anomalies; and (4) the sudden occurrence of anomalies in large quantity. This is an overall manifestation, but the single–point and single–item anomaly is only a local expression in the earthquake field, which is not able to represent the whole process before an earthquake.

2. The time lengths of these stages before earthquakes are not constant, but the long–, intermediate–, short–term, and imminent anomalies all have their own characters different from each other, and they also appear generally in succession. Statistically speaking, long–period anomalies occurred several years before the earthquakes; the intermediate–period ones generally appeared several years to months before the earthquakes; the short–period ones months to days before the earthquakes; and the imminent ones, which were the most striking, several days prior to the earthquakes.

3) The above general description of seismogenic processes are basically in conformity with the rock rupture experiments and stage division.

This objectively general process may be taken as a subjective general working procedure for earthquake prediction.

It should be stated emphatically that the general knowledge on seismogenic processes has been gained on the basis of the most salient phenomena exhibited in various stages (either defined, or obtained in experiments), and even on typified phenomena. As a matter of fact, the manifestations of most seismogenic processes were not typical, but considerably different, with a conspicuous variety of phenomena. The analysis of the structural characters of seismic sources and the variety of earthquake phenomena tend to be associated with the regionalization and classification of earthquakes and the influences of historical factors. Therefore, our work on earthquake prediction cannot be based on the occurrence of the typical anomalous phenomena during each corresponding stage before every strong earthquake, but instead we should analyze the actual observations, and take into account the influence of regionalization, classification and historical factors, and so make a comprehensive evaluation guided by such general knowledge of the seismogenic process.

14.1.4 Inhomogeneity of Time–Space Distribution of Earthquakes

There are two cases here to consider: One concerning a single earthquake, and the other a group or an episode of earthquakes, whose time–space distributions are all inhomogeneous. This fundamental character is very important for the prediction of continental earthquakes, and it has not been adequately recognized previously.

One of the elementary features of the continental earthquake field is the inhomogeneity of stress–strain distribution within the crust. The multiple stress–concentrated points within the field interact mutually under the unified development of the field. An earthquake is always prone to occur near an anomalies–concentrated point. However, during an entire seismogenic process multiple anomalies–concentrated points may develop at the same time, thus inevitably causing great difficulty in predicting the sites of the earthquakes. This is an extremely grave problem in the prediction of continental earthquakes. The concept of "sensitive points" to anomalies in a large–scale region may be, in turn, conducive to the monitoring and prediction of earthquakes.

Although the seismogenic process of an event is gradually approaching the time of occurrence of an earthquake with the development of anomalous phenomena, the developments are not gentle, very often progressing by jump, and particularly during the short–term and imminent stages of the earthquake the modulating and triggering factors may possibly play a certain role in generating the quake. Therefore, the probability of earthquake occurrence is also heterogeneous in time distribution, and this is not only a complicated factor in predicting the time of an earthquake, but also one which can be made use of.

The heterogeneity of time–space distribution of an earthquake episode was dealt with in Chap. 13. It is essentially similar to that for a single earthquake. The estimation of the three elements of an earthquake episode constitutes a subject which should be studied for long–term prediction, and it will be further discussed in the following section.

The above–described four aspects are the most basic characteristics we have summarized from earthquake prediction practice, and represent the fundamental understanding we have gained from shallow–depth tectonic earthquakes in continental China with a view to earthquake prediction.

14.2 Thoughts About a Program for Step–by–step Earthquake Prediction

The four points mentioned in the previous section constitute a working basis for considering how to do prediction work, and also involve some practical problems which must be solved. In accordance with these insights, and by summing up the experience gained and lessons learned during earthquake prediction work in the past decade or earlier, we suggested the program for step–by–step earthquake prediction in the light of the concept and mechanism of the phasic seismogenic process (Ma Zongjin et al. 1981).

The step–by–step earthquake prediction basically implies that there are five procedural stages for prediction, of which four include the long–term, intermediate–term, short–term, and imminent phases of a seismogenic process, and the fifth is the post–earthquake stage after the first strong shock. During each phase, different problems for prediction must be solved. By means of rough estimation and also careful consideration, by using both distant location and near site, and taking into account multiple and single points, prediction of the three elements will be gradually made possible. We have called such a working sequence step–by–step earthquake prediction. The most prominent features for comprehensive research and the tasks to be accomplished during each prediction stage are as follows:

Long–term prediction stage:

1) Comparative studies on earthquake episodes.

2) Regional estimations of three elements of an earthquake episode. The three elements include the starting and terminating time, the main activity area, and the total intensity and frequency of an episode.

Medium–term prediction stage:

1) Discrimination and determination of continuous anomalies.

2) Contour of anomalies–concentrated areas, and selection of critical working areas.

3) Identification of the features of crustal movements and of possible types of seismogenic tectonics in these areas.

Short–term prediction stage:

1) Discrimination and determination of short–term anomalous phenomena.

2) Comparison of the respective developmental tendencies of anomalies–concentrated areas for choosing the possible earthquake–generating area which is to be intensively monitored.

3) Forecast of the magnitude of a forthcoming earthquake in accordance with the distribution area and the duration of anomalies in the potential earthquake–generating area.

Imminent prediction stage:

1) Discrimination and determination of abrupt anomalies.

2) Study of the modulating and triggering factors and investigations of related phenomena for predicting risky intervals of earthquake occurrences.

3) Further delimitation of the place and time of an earthquake occurrence based on time–space distributions of sudden anomalies and short–term anomalous phenomena and by considering phenomena of population anomalies migrating toward the seismic source.

Prediction stage after the first strong shock:

1) Research on types of earthquake sequence and re–determination of seismo–structural types.

2) Estimation and prediction of the potentiality of successive strong earthquakes and of latest strong shocks.

3) Studies and evaluation of the possibility of succession of regional strong earthquakes.

The following will concisely illustrate the reasons and practices for the main points defined for the various stages.

14.2.1 Long–term Prediction Stage

We have divided earthquake prediction into two major types, distant–period and near–period predictions. The goal is to make a clear distinction between the estimation of seismic risk in the future 100 years or so and ability to predict the imminence of the three elements. This is what we are striving to achive in the next 10 years or so. Studies on seismicity showed that there indeed exist rhythmical fluctuation of earthquake episodes in global or regional extention, hence, the long–term prediction stage for near–period prediction was assigned to the study of earthquake episodes. For instance, currently seismic activities in continental China may be part of the fifth episode in the world since the beginning of this century, but when will this episode come to an end? The northern part of the North China region and the Sichuan–Yunan region during the fifth episode are the main activity areas in continental China. Have strong seismic activities terminated in these two areas or, if not, when will either of them separately end ? Where will the main activity areas be in the next earthquake episode in Mainland China? What kind of important early manifestations will take place in these areas if they happen in Northwest China or the coastal region of Southeast China?

Presently, continuous observation data are scarce and they cover also a short time–span, so studies of three elements in an earthquake episode should be comprehensive, taking as starting points mainly three aspects, such as the appropriate delineation and correlation of seismic regions and zones, the relation between earthquake occurrences, and also various environmental factors. Research should first deal with the division and comparison of episodes, then with the studies and forecasts of the regularity of migration in the main activity areas. Once a main activity area is clearly defined, the magnitude of an episode can be readily assessed. In addition to forecasting a main activity area, secondary ones should be also considered. Therefore, it is necessary to study the general behaviour of various seismic regions in Mainland China in accordance with the positive and negative rhythms of global seismicity, and to determine in which stage of a seismically active period the individual seismic areas find themselves.

If a large earthquake could be predicted several years or even more than 10 years in advance, the wish for and attempts at prediction would be certainly encouraging, e.g. in the past seismic gap phenomena used to indicate earthquake risk areas. However, predictions of this kind are characteristic rather of background research, and are always very general, like studies and evaluations of earthquake episodes, and they need to be further elaborated during the medium–term prediction stage.

14.2.2 Medium–term Prediction Stage

The data on the nine great earthquakes have shown that there might be remarkable anomalies in the band–like distribution of earthquakes, strain release, earthquake frequency, b value, ratio of seismic wave velocities, ground deformation, apparent resistivity, ground–water regime, gravity, geomagnetism, earth stress (using the electro–induction method) and so on, 1–3 years before large earthquakes. Therefore, the occurrence of relatively remarkable successive anomalies would mark the entry into the stage of medium–term prediction research.

The basis for medium–term predictions involves various continuous anomalies, hence it is necessary first to discriminate and judge the reliability of the anomalies themselves. Mainly the influence of some factors, such as man–made disturbances, zero drift of instruments and seasonal variations, should be identified, and the consistence in the time–space distribution of multiple anomalous phenomena can serve as a criterion for determining anomalies.

In order to make a medium–term prediction of the three elements of an earthquake, the anomalies–concentrated area should be first delimited. Empirical and theoretical analyses have all revealed that the anomalies–concentrated areas might reflect a rather intense stress–strain state, which might develop and become a possible future source region. However, not all anomalies–concentrated areas would be locations for future strong earthquakes. Some anomalies–concentrated areas, delimited on the basis of the regional earthquake field, should be simply sites for intensified research and monitoring, but should not serve as prediction areas. The contour of anomalies–concentrated areas and their shapes are very often influenced by the actual distribution of stations. This is a practical problem which deserves consideration in choosing areas for intensified work.

As the medium–term anomalies are developing, it is difficult to anticipate when the short–term prediction stage will be reached. That is why the estimation of the occurrence time for a potential earthquake (unit in years) is rather tentative. Similarly, since the duration of medium–term anomalies is uncertain, the areal distribution of these anomalies may not be wide enough, causing thus difficulties in estimating the magnitude of the forthcoming earthquake.

With the gradual development of medium–term anomalies, a comprehensive study can be made of them in a selected work–intensified area. In this way, certain speculations can be made on the mode and features of crustal movements, the basic framework of seismogenic structures, and the possible type of the earthquakes in the given area. Knowledge in this aspect will be helpful in planning short–term prediction work and analyzing certain phenomena.

14.2.3 Short–term Prediction Stage

The major characteristics of short–term anomalies consist in significant variation in the shape of continuous curves, such as an obvious increase in the variational rates of the anomalies, returning to normal level or reversing development in several tens of days or several days before the mainshocks.

The chief task in this stage is still to discriminate and define short–term anomalies. The disturbances may originate mainly from precipitation and artificial factors.

Once the short–term anomalies are confirmed, the duration time of medium–term anomalies can be also determined. Meanwhile, in several anomalies–concentrated areas, the behaviour of short–term anomalies may be different, and these anomalies would be more clear–cut with their areal development. Under these conditions, the magnitude of the future earthquake can be calculated, and the place of occurrence of the earthquake can be nearer estimated. This should be made into an area of intensified monitoring. Furthermore, the occurrence time of an earthquake can be also estimated in monthly units.

The short–term anomalies are very often centered along some bands in spatial distribution, and seemingly start to migrate toward the epicentral areas from the outside. Such phenomena may imply that the medium of a locked segment in a source area is evidently becoming softened and beginning to turn unstable, thus leading to creeping of nearby relevant fault zones, which generally extend toward the hypocentral area or its surroundings. An analysis of these phenomena will be useful for estimating time and place of the earthquake's occurrence.

The short–term prediction stage is a very important one in the whole prediction system. Nevertheless, there are many uncertainties and difficulties in actually judging anomalies. One of the solutions is to have some items recorded continuously in regions with intensified monitoring, make comparisons and observations of disturbance factors, and carry out fast automatic processing of the observational data.

14.2.4 Imminent Prediction Stage

Two major kinds of macro– and microscopic abrupt anomalies seem to occur in the imminent stage. The imminent stage is a continuation and development of the short–term stage (unstability started in the source region), hence many abrupt anomalies and continuous anomalies alternate one after another, or even mixed together. For example, abrupt anomalies may sometimes occur at an earlier time before a large earthquake, as they did 2 months before the Haicheng and Songpan earthquakes, during which the anomalies featured migrations three times up and three times down to the sources. Similarly, the short–term variations of continuous anomalies also appeared very late. In fact, these two kinds of anomalies may be partly anomalous expressions of phenomena of the same nature but of different form. The following three points may be considered as clues for further researching imminent earthquake information.

1) In addition to scattering in large–scale space range, the imminent abrupt anomalies appeared to be mostly concentrated in the future epicentral area, with an intense variation and a variety of types. However, these empirical understanding is qualitative and relative. There must be large–area and multiple–point observations made for recognising these variations of the imminent anomalies before earthquakes. Besides, it is also necessary to study the comprehensive behaviour of these anomalies. If the imminent anomalies in source areas may really imply the rapid development, and the closing and opening movements of underground fissures, and the intense flow of ground fluid, as well as marked variations in the thermal regime and eletromagnetic field at the earth surface, which result in the appearance of synchronous anomalies shown by animals, earth light and local climate, then they can be used to make imminent earthquake prediction with more confidence. At present, there are few studies made on the origin mechanisms of multiple imminent anomalous phenomena, so definite conclusions can not be drawn. According to previous experience, the warning time for imminent prediction could be considered as several days after these grouped anomalous phenomena appeared.

2) Migration of short–term and imminent anomalies. It is impossible currently to confirm the generality of this migration phenomenon. Considering that the apparent velocity of migration toward source epicenters tends to quicken gradually, when the velocity exceeds a critical speed, surpassing, say, 3–5 km / day, this tendency may imply the beginning of macroruptures in the source area. Such phenomena can be used for predicting either occurrence time, or location of an impending earthquake.

3) Effect of modulating and triggering factors on earthquake occurrences. This is a research subject which has already attracted much attention. An earthquake is merely a local transient phenomenon in the process of crustal tectonic movement, so it should be necessarily affected by some accidental and local factors. The idea of time–span prediction, represented by modulating phenomena of "times two" and "times nine" of variable geomagnetic fields, was practised in the predictions of the nine great earthquakes. Since few studies have been made on mechanisms of the modulating phenomena, and disturbance factors are also very perplexing, many problems need to be discussed. Nevertheless, the time–leapt points were in good agreement with the occurrence times of some earthquakes. This fact is indeed encouraging, and still remains at present a worthwhile reference for predicting the time of an imminent earthquake. If there is a background for medium–term and short–term precursory phenomena, the triggering effects of intense variation in the tidal forces, rainfall and atmospheric pressure, are also very valuable for predicting an imminent earthquake.

14.2.5 Prediction Stage A fter a First Strong Quake

Are there any successive strong quakes, any strong aftershocks, any late strong aftershocks, or any corresponding earthquakes or distant related ones, within the surrouding areas of hundreds of kilometers after the first strong shock? These problems actually arose in the nine great earthquake sequences, which should be studied and solved.

In accordance with previous work experience, forecasts of successive strong earthquakes and strong aftershocks can be partly made mainly with regard to the following two aspects: (1) studies on variational regularities and cycles of activity criteria like b and P values of earthquake sequences, and identification of types of earthquake sequences; and (2) assessment of seismotectonic types through coseismic isoseismal lines, surface ruptures, earthquake mechanism and aftershock distribution, so as to estimate the features of earthquake sequences, and predict locations of successive strong shocks.

The phenomena of tractional earthquakes and distant related ones seem to be associated with stress adjustments of earthquake fields, but any concrete diagnosis must be made on the basis of analysis of local anomalous phenomena. Since we have not gained any definite guideline for discriminating phenomenally between " precursors" and " aftermath" , it is extremely difficult to make diagnostic decisions. The phenomenon of earthquake " aftermath" which extended outwardly from epicenters deserves attention, being a possible reference for forecasting tractional earthquakes and distant related ones.

In general, for the continental earthquakes in China, the imminent prediction of the three elements of earthquakes can be finally achieved through a long–, medium–, short–, and imminent–term prediction process and by means of a gradual approach connecting one link after the other. This working design is a primary summarization of experiences in earthquake prediction obtained in the past decade or earlier, but there are still some problems waiting to be tackled, such as studies of earthquake episodes, relation of the development of successive anomalies to crustal stress–strain variations, mechanism of multiple abrupt anomalies, types of earthquake sequences and seismotectonic categories, interaction between single point and multiple points in an earthquake field, and the mechanism of the effects of modulating and triggering factors. For practical prediction purposes, we have linked these research subjects together as a schedule for work, while for prediction research we regard them as individual research domains and subjects, which deal with basic scientific problems of many disciplines.

We do not think the above working designs are suitable to the prediction of all earthquakes, as the main basis for them only came from studies of the nine earthquakes. However, by comparing the present prediction designs with the knowledge on earthquake prediction before the 1966 Xingtai earthquake, we may say that our understanding in this field has been greatly improved. We are confident these designs can be adapted to the prediction of some future earthquakes.

Since the August 1976 Songpan earthquake of M7.2 about 5 years have elapsed, during which no earthquakes with M reaching 7 occurred any more in continental China. This is a relatively calm period, which provides us with a good opportunity to understand and study various observational data collected during it. We expect to use the results of these observation to reevaluate and revise the conclusions and information included in this book.

At present, it is still difficult to accurately predict earthquakes, but under favourable circumstances it is possible to make a fairly good prediction of certain types of earthquakes.

One earthquake after another has affected the humanity, and finally they will force us to find a better way to accurately predict, control, and make use of them. The level of earthquake prediction will be gradually improved through constant practice.

References in Chinese

Bullen (1965), An Introduction to Seismology, Scientific Press.

Cai Zuhuang et al. (1980), Geological Outline of Seismic Fluid,Seismological Press.

Chen Lide et al. (1979), The 1976 Longling Earthquake, Seismological Press.

Chen Peishan et al.(1977), Research of Earthquake Rupture Process and Earthquake Prediction from View Point of Fracture Mechanics, Acta Geophysica Sinica, V.20, No.3.

Chen Tianchang et al. (1981), Studies of Seismicity Before the M7.2 Songpan Earthquake, Selected Papers of the 1979 International Symposium on Earthquake Prediction, Seismological Press.

Chen Yong (1977), Application of Acoustic Emission Technique in Rock Mechanics Research, Acta Geophysica Sinica, V.20, No.4.

Chen Yuntai et al. (1975), Studies of Seismogenic Process of the 1966 Xingtai Earthquake Based on Deformation Observations, Acta Geophysica Sinica, V.18, No.3.

Chen Yuntai (1979), The Dislocation Model for the 1976 Tangshan Earthquake Derived from the Inversion of Geodetic Data, Acta Geophysica Sinica V.22, No.3.

Chen Yuntai (1980), Gravity Variations Before and After the 1975 Haicheng and 1976 Tangshan Earthquakes, Acta Seismologica Sinica, V.2, No.1.

Compiling Group of the Tangshan Earthquake Precursors (1977), Precursors of the Tangshan Earhtquake, Seismological Press.

Deng Qidong et al. (1976), Discussion on Hypocentral Model of the Haicheng Earthquake, Geological Science, No.3.

Deng Qidong et al. (1979), Features of Tectonic Stress Fields in China and Their Relations with Plate Motion, Seismology and Geology, V.1, No.1.

Ding Guoyu (1979), Aspects of Earthquake Prediction, Research of Earthquake Science, No. 1 (trial).

Feng Deyi et al. (1975), Anomalous Variations of Wave Velocity Ratio Before the May 1975 Yongshan–Daguan Earthquake of M7.1 in Yunnan Province, Acta Geophysica Sinica, V.18, No.4.

Feng Deyi et al. (1976), Preliminary Studies on Wave Velocity Anomalies Before and After Some Strong and Moderate Earthquakes in West China (Part one) —Wave Velocity Ratio Anomalies, Acta Geophysica Sinica, V.19, No.3.

Feng Rui et al. (1976), Variations of Seismic Wave Velocity Ratio Before and After the Haicheng Earthquake, Acta Geophysica Sinica, V.19, No.4.

Fu Chengyi (1976), Ten Lectures on the Earth, Scientific Press.

Fu Zhengxiang (1979), Tentative Discussions on Spatial Distribution of Volumetric Deformation and Earthquakes, Northwest Earthquake Bulletin, V.1, No.4.

Fu Zhengxiang (1983), Discussion on Concentration and Diffusion of the Regional Seismicity before and after Some Great Earthquakes, Seismology and Geology, V. 5, No 2.

Fu Zhengxiang (1981), Shallow–depth Strong Earthquake and Variations of the Earth Rotation, Journal of Seismological Research, No.3.

Geodetic Brigade, State Seismological Bureau (1977), Crustal Deformation for the Haicheng Earthquake of M7.3, Acta Geophysica Sinica, V.20, No.4.

Geodetic Brigade, State Seismological Bureau (1975a), Crustal Deformation for the 1966 Xingtai Earthquake, Acta Geophysica Sinica, V.18, No.3.

Geodetic Brigade, State Seismological Bureau (1975b), Features of Crustal Deformation for the 1970 Tonghai Earthquake in Yunnan, Acta Geophysica Sinica, V.18, No.4.

Gravity Group of Comprehensive Research Team of Seismological Bureau of Hebei Province (1979), Gravity Changes Before and After the Tangshan Earthquake, Selected Papers of the 1979 International Symposium on Earthquake Prediction, Seismological Press.

Gu Haoding et al. (1976), Focal Mechanisms of the Haicheng Earthquake, Liaoning Province, of 4 February 1975, Acta Geophysica Sinica, V.19. No.4.

Gu Haoding et al. (1980), Precursory Swarm and S Wave Polarization, Acta Seismologica Sinica, V.2, No.4.

Gu Jicheng et al. (1979), The Time–space Distribution Regularities of Strong Aftershocks and Theoretical Interpretations, Acta Geophysica Sinica, V.22, No.1.

Guan Xingguo (1980), Features and Analysis Method of Tilt Anomalies for the Tonghai Earthquake of M7.3, Acta Geophysica Sinica, V.23, No.3.

Guo Shifeng et al. (1980), Photoelastic Simulation of Modern Tectonic Stress Field of North China, Journal of Seismological Research, No.2.

Guo Shunmin (1977), Discussions of Regional Tectonic Background and Occurrence Model of the Tangshan Earthquake, Geological Science, No.4.

Guo Shunmin et al. (1979), Analyses of Regional Tectonic Background of Occurrence Conditions of the Zhuwo Earthquake, Luhuo, Sichuan, Compilation of Seismogeological Investigations in Macroseismic Regions of Sichuan and Yunnan, Seismological Press.

Guo Zengjian et al. (1977), "Times–Nine Method" for Earthquake Prediction, Earthquake Front, No.5.

Guo Zengjian et al. (1979), Physics of Seismic Sources, Seismological Press.

He Zhitong et al. (1977), Discussions on Spatial Distribution, Tectonic Stress Field and Occurrence Process of the Xingtai Earthquake Sequence, Acta Geophysica Sinica, V.20, No.2.

Historical Group of the Earthquake Working Committee, Academia Sinica (1956), A Chronological Table of Historical Earthquakes in China, Scientific Press.

Huang Deyu et al. (1981), Variations of Duration Ratio τ_H / τ_V of Seismic Wave Vibrations Before the Tangshan Earthquake, Journal of Seismological Research. No.2.

Huang Fuming (1981), Fault Creeping and Earthquake Prediction (An

Overview), Seismology Abroad, No.5.
Huang Xiqing (1960), Preliminary Summary of Basic Characteristics of Geological Structures in China, Acta Geologica Sinica, V.40, No.1.
Hydrochemical Group of Beijing Seismological Crew (1977), Laboratory Studies of Rock Fracture and Radon Content Variation, Acta Geophysica Sinica, V.20, No.4.
Institute of Biological Physics, Academia Sinica (1977), Animal and Earthquake, Seismological Press.
Institute of Geology, Academia Sinica (1974), A General Introduction to Seismogeology in China, Scientific Press.
Jiang Jinchang (1980), Peculiar Capabilities, Seismological Press.
Jiang Pu et al. (1980), Evolutions and Tectonic Mechanical Conditions of the Anomaly Field During the Seismogenic Processes of the Haicheng and Tangshan Earthquake Sequences, Seismology and Geology, V.2, No.4.
Jiang Xiue et al. (1981), Studies of Wave Velocity Anomalies of the Tangshan Earthqauke with Four–seismic Phase Method of Single Station, Acta Geophysica Sinica, V.24, No.1
Jin Anshu et al.(1979), Variations of Seismic Wave Velocity Ratio in Beijing Area, Acta Seismologica Sinica, V.2, No.3.
Jin Yan et al. (1976), One Feature for Dislocation Mode of Hypocenter Before the Haicheng Earthquake in Liaoning, Acta Geophysica Sinica, V.19, No. 3.
Kan Rongju et al. (1977), Studies on Recent Tectonic Stress Field and Recent Tectonic Activities in Southwest China, Acta Geophysica Sinica, V.20, No.2.
Li Cundi (1980), Magnetic Effects of the Tangshan Earthquake, Kexue Tongbao, V.25, No.15.
Li Fangquan et al. (1979), Stress Measurements in North China, Acta Geophysica Sinica, V.22, No.1.
Li Ping et al. (1979), Compiled Reports of Seismogeological Studies in the Sichuan–Yunnan Macroseismic Region, Seismological Press.
Li Qinzu (1980a), Basic Features of Crustal Stress Field of North China, Acta Geophysica Sinica, V.23, No.4.
Li Qinzu et al. (1980b),Characteristics of Grouped Seismic Activity in North China, Studies of Earthquake Science, No.1.
Li Quanlin et al. (1978), Time–Space Scanning of b Values, Acta Geophysica Sinica, V.21, No.2.
Li Xianggen (1977), Discussions on Neotectonic Features, Motion Types and Geological Criteria for Macroseismic Region in the Sichuan–Yunnan Plateau, Studies on Seismogeology and Seismic Zoning in Southwest China, Seismological Press.
Li Zhiqiang et al. (1980), Discussion on Seismogenic Process of Recent Clustered Strong Earthquakes in North China, Acta Seismologica Sinica, V.2, No.4.
Lin Banghui et al. (1979), Studies of Asymmetric Bilateral Rupture Process and Its Application to the Haicheng Earthquake, Acta Seismologica Sinica, V.1, No.2.
Lin Changyou et al. (1979), Preliminary Study on Seismogenic Process of Large Earthquakes, Northwest Seismological Bulletin, V.1, No.3.

Liu Baichi (1979), The Dynamic Model for Stess Adjusted Fields of Continental Earthquakes in China, Seismology and Geology, V.1, No.3.

Liu Wangqin et al. (1979), Study of Seismogenic Process of the 11 May 1974 Zhaotong Earthquake in Yunnan with Directivity Function of Rayleigh Waves, Acta Seismologica Sinica, V.1, No.1.

Liu Zhengrong et al. (1977), The Yongshan–Daguan Earthquake of 11 May 1974 in Yunnan, Acta Geophysica Sinica, V.20, No.2.

Liu Zhenrong et al. (1978), The Features of the Longling Conjugate Earthquake Swarm, Journal of Seismological Research, V.1, No.2.

Liu Zuyin (1979), Discussions on Time–space Distribution and Seismogenic Process of the Tonghai Earthquake Sequence, Journal of Seismological Research, V.2, No.3.

Luo Guangwei (1984), Gravity variation and Earthquake Occurrence, Acollection of Papers of International Symposium on Continental Seismicity and Earthquake Prediction, Seismological Press.

Luo Huanyan et al. (1980), 3–D Finite Element Analysis of the Relation Between the Stress State and the Driving Forces of Structural Rock Bodies, Seismology and Geology, V.2, No.1.

Luo Zhuoli (1980), Hypocentral Stress Field, Deformation Field and Tilt Field. Acta Seismologica Sinica, V.2, No.2.

Ma Zongjin (1974), Imminent Earthquake Phenomena and Earthquake Luminescence, Earthquake Front, No.1.

Ma Zongjin (1980a), Multiple (Stress Concentration) Stress Fields and Earthquakes in North China, Seismology and Geology, V.2, No.1.

Ma Zongjin et al. (1980b), Tectonic Classification of Recent Nine Strong Earthquakes in China, Northwest Earthquake Bulletin, V.2, No.1.

Ma Zongjin, Chen Zhanli et al. (1984a), The Basic Characteristics of the Continental Earthquakes, A Collection of Papers of (ISCSEP), seismological Press,

Ma Zongjing, Fu Zhengxiang, xie Feng (1984b), The Continetal Seismotectonic System in 20 ° –50 ° N,Acta, Geol. Sin. 58 (2).

Ma Zongjin, (1987a), Study on Intermediate–term Earthquake Prediction, Seismic Research V.10, No 4.

Ma Zongjin, (1987b), Comparison of Continental Seismotectonics between Eastern China and Eastern U.S., Earthquake Research in China, V.1, No 2, 241–251.

Magnetotelluric Sounding Group of Lanzhou Seismological Brigade, State Seismological Bureau (1976), Electric Properties of the Crust and Upper Mantle of the North Segment of the North–South Seismic Zone in China, Acta Geophysica Sinica, V.19, No.1.

Map Compiling Group for National Seismic Intensity Zoning, State Seismological Bureau (1979), Album of Isoseismal Lines in China, Seismological Press.

Matsuzawa Takeo (1979), Theory of Seismology and Its Applications, Seismological Press.

Meteorology and Earthquake Group of the Lanzhou Seismological Brigade

(1976), Meteorological Heat Anomlalies Before Large Earthquakes in China, Earthquake Front, No.1.
Mirzoeb. K.M. et al. (1980), The Search for Spatial and Temporal Regularities of Changes in Some Parameters Before Large Earthquakes, Seismological Press.
Niu Zhiren et al. (1976), Pursuance Model for Seismogenic Preparation, Acta Geophysica Sinica, V.19, No.3.
No.3 Research Section of Institute of Geophysics, Academia Sinica (1970), An Approach for Seismic Risk Zoning from Regularities of Strong Seismic Activities in North China, Earthquake Front, No.1.
Pan Qiuye et al. (1979) Photoelastic Experiments of Typical Tectonics, Earthquake Front, No.3.
Qian Fuye et al. (1980), Ten Examples for Resistivity Variations Before Earthquakes, Acta Seismologica Sinica, V.2, No.2.
Qian Jiadong et al. (1985), Application of Apparent Resistivity Method in Earthquake Prediction, Seismological Press.
Qiu Qun (1976), Seismic Background and Seismicity for the M7.8 Tangshan Earthquake of July 28 1976 in Hebei Province, Acta Geophysica Sinica, V.19, No.4.
Ren Jishun et al. (1980), Geotectonics of China and Its Evolution, Scientific Press.
Rikitaki (1978), Earthquake Prediction, Seismological Press.
Seismological Bureau of Hebei Province (1976), Catalogue of Xingtai Earthquakes, Seismological Press.
Seismological Bureau of Sichuan Province (1979), The 1976 Songpan Earthquake, Seismological Press.
Shao Xuezhong et al. (1980), Deep Sounding Results Obtained from Converted Waves of Earthquakes in the Beijing–Tianjin–Tangshan Region, Seismology and Geology, V.2, No.2.
Shi Zhenliang et al. (1973), Strong Earthquake Activities in China and Plate Tectonics, Geological Science, No.4.
Shu Shui (1974), Features of the Luhuo Earthquake of M7.9 and Seismicity in Luhuo, Acta Geophysica Sinica, V.17, No2.
Shu Shui (1976), Rock Dilatancy and Water Diffusion of Source Stress Field the Yongshan, Yunnan, Earthquake of M7.1 and Its Precursory Development, Acta Geophysica Sinica, V.19. No.2.
State Seismological Bureau (1977), Concise Catalogue of Strong Earthquakes in China, Seismological Press.
Tang Rongchang et al. (1976), Preliminary Studies on Ground Cracks and Origin of the 1973 Luhuo Earthquake of M7.9, Acta Geophysica Sinica, V.19, No.1.
Teng Dezhen (1978), Basic Seismogeological Characters of the Central Xiaojiang Fault Segment in Yunnan, Earthquake Research V.1, No.2.
Teng Jiwen et al. (1974), The Xingtai Earthquake and Deep Tectonic Background of the Central North China Plain (Part One), Acta Geophysica Sinica, V.17, No.4.

Teng Jiwen et al. (1975), The Xingtai Earthquake and Deep Tectonic Background of the Central North China Plain (Part Two), Acta Geophysica Sinica, V.18, No.4.
Wang Biquan et al. (1979), Quasi–Periodicity of Strong Aftershocks, Acata Seismologica Sinica, V.1, No.2.
Wang Changling et al. (1980), Features and Causes of Water Radon Content Anomalies in Springs to the North of the Hypocenter Before the Songpan–Pingwu Earthquake, Acta Seismologica Sinica, V.2, No.1.
Wang Chengmin et al. (1981), Characteristic Variations of Water Level in Deep Wells Before and After the Tangshan Earthquake, Selected Papers of the 1979 International Symposium on Earthquake Prediction Seismological Press.
Wang Chengmin et al. (1988), Study of Microbehavior of Groundwater, Seismological Press.
Weng Chunyong et al. (1978), Determination of Seismic Fault Parameters for the Tonghai Earthquake with Deformation Data, Acta Geophysica Sinica, V.21, No.3.
Wang Qiming (1980), Preseismic Anomalies Indicated by the String–type Borehole Strainmeter, Earthquake Studies, V.3, No.4.
Wang Ren et al. (1979), Elementary Solid Mechanics, Geological Press.
Wang Ren et al. (1980), Mathematical Modeling of Regularities of Earthquake Migrations in North China, Acta Seismologica Sinica, V.2, No.1.
Wang Suyun et al. (1987), Seismotectonic Stress Field in Eastern Sea Area of China, Earthquake Research in China, V. 3, No 3.
Wang Yipeng (1979), Intraplate Earthquakes and Mid–Cenozoic Stress Fields in China, Seismology and Geology, V.1, No.3.
Wang Zegao (1979), Increased Frequency of the Xingtai Aftershocks and Following Strong Earthquakes in North China, Acta Seismologica Sinica, V.1, No.2.
Wei Guangxing et al. (1978), Seismic Gap Characteristics Before Large Earthquakes in North China, Acta Geophysica Sinica, V.21, No.3.
Working Team of the Tangshan Earthquake (1977), Short and Imminent Precursory Data for the Tangshan Earthquake, Seismological Press.
Wu Jiayi et al. (1979), The Correlations Between Earthquakes in North China and That in Japan, Acta Geophysica Sinica, V.22, No.4.
Wu Kaitong et al. (1976), Features of the Haicheng Earthquake Sequence, Acta Geophysica Sinica, V.19, No.2.
Wu Kaitong et al. (1981), On the Detectability of Seismic Networks in China, Journal of Seismological Research, V.4, No.1.
Xu Ji et al. (1982), Tectonic Classification of Shallow–depth Earthquakes in Mainland China, Proceedings of Seismogeology, Tianjin Science and Technology Publishing House.
Xu Shaoxi et al. (1980), Distribution Characteristics of Earthquakes Around Beijing and Crustal Buckling, Acta Seismologica Sinica, V.2, No.2.
Xu Shaoxi et al. (1981), Foreshock Series and Swarm for the Haicheng Earthquake, Acta Seismologica Sinica, V.3, No.3.

Xu Zuzhe (1978), Pre–earthquake Electromagnetic Wave Anomalies, Earthquake Front, No.6.

Xue Feng (1981), Relations Between Earthquakes in China and That Around the World, Journal of Seismological Research, No.3.

Yan Fengtong et al. (1978), Focal Mechanisms of the Longling Earthquake, Earthquake Research, No.1.

Yan Jiaquan et al. (1979), Regional Features of Recent Tectonic Stress Fields of China and Its Vicinity, Acta Seismologica Sinica, V.1, No.1.

Yang Lihua et al. (1980), Crustal Stress Field of North China, Seismological Press.

Ye Hong et al. (1980), Preliminary Discussion on Features of Intraplate Seismic Faults in North China and Their Relations to the Crust and Upper Mantle Dynamic Process, Seismology and Geology, V.2, No.1.

Zhang Guomin et al. (1981), Characteristics of Some Precursors Before Damaging Earthquakes in the Yanshan Zone, Journal of Seismological Research, No.1.

Zhang Shurong (1980), Features of the 5 January 1970 Tonghai Earthquake, Earthquake Research, V.3, No.2.

Zhang Sichang et al. (1978), Discussion on Geologic Features of the 1976 Longling Earthquake, Journal of Seismological Research, V.1, No.3.

Zhang Wei et al. (1981), Discussion on Predicting Earthquakes with Hydrochemistry, Acta Seismologica Sinica, V.3, No.1.

Zhang Yingzhen (1981), Anomalous Crustal Uplift and Aseismic Creep Before the Tangshan Earthquake, Acta Seismologica Sinica, V.3, No.1.

Zhang YngZhen (1985), Relationship between the Regional Stress Fields and Sources in the Preparation Process and Occurrence of Large Earthquakes, North China Earthquake Sciences, V.3, No 3.

Zhang Zhili et al. (1980), The Rupture Process and Its Mechanical Analysis of the Tangshan Earthquake. Acta Seismologica Sinica, V.2, No.2.

Zhang Zusheng (1984), Some Problems in Inersion of Seismic Source Parameters from Geodetic Data, Acta Seismological Sinica, V.6, No 2.

Zheng Zhizhen et al. (1975), Variations of Average Dislocations in the Xingtai Area, Acta Geophysica Sinica, V.18, No.2

Zheng Zhizhen et al. (1977), Spectral Variations of Foreshocks and Aftershocks for the Haicheng Earthquake, Acta Geophysica Sinica, V.20, No.2.

Zheng Zhizhen et al. (1978), Time–space Distributions of Average Dislocations Before and After Large Earthquakes, Acta Geophysica Sinica, V.21, No.1.

Zheng Zhizhen (1979), Discussion on the Seismogenic Process of the Tangshan Earthquake with Numerical Filtering of Ground water Data, Acta Geophysica Sinica, V.22, No.3.

Zhu Chengnan et al. (1976), Rupture Mechanism for the 1974 Zhaotong Earthquake in Yunnan, Acta Geophysica Sinica, V.19. No.4.

Zhu Chennan et al. (1978), Cenozoic Tectonic Development of South Yunnan and Discussions on Geological Background of the Tonghai Earhtquake, Journal of Seismological Research, V.1, No.3.

Zhu Ying (1979), Geotectonics of North China Block and Distribution Regularity of Anshan–type Iron Ore, Geophysical Prospecting and Chemical Exploration, No.1.

References in English

Anderson D.L. (1974), Earthquake and the Rotation of the Earth. Science, V.186, No.4158.

Barnes D. G. (1966), Gravity Changes During the Alaska Earthquake. J.G.R, V.71, No.2.

Benioff H. (1951a), Earthquakes and Rock Creep, T. 1: Creep Characteristics of Aftershock, B.S.S.A., V.41, No.1.

Benioff H. (1951b), Crustal Strain Characteristics Derived from Earthquake Sequences. Trans. Amer. Geophy. Union, V.32, No.4.

Brace W.F. et al. (1966), Stick–slip as a Mechanism for Earthquake. Science, V.153, No.3739.

Brace W.F. et al. (1968), Electrical Resistivity Changes in Saturated Rocks during Fracture and Frictional Sliding. J.G.R., V.73, No.4.

Brace W.F. (1975), Dilatance–related Electrical Resistivity in Rock. Pageoph., V.113, P.207.

Bullen K.E. (1965), An Introduction to the Theory of Seismology. Cambridge University Press.

Byerlee J. (1978), A Review of Rock Mechanics Studies in the United States Pertinent to Earthquake Prediction. Pageoph., V.116, No. 4–5.

Chen Zhangli et al. (1984), Characteristics of Regional Seismicity before Major Earthquake, Earthquake Prediction, Proceedings of the International Symposium on Earthquake Prediction, Terra Scientific Publishing Company (TERRAPUB), Tokyo, UNESCO, Paris.

Derr J.S. (1973), Earthquake Light: A Review of Observations and Present Theories. B.S.S.A., V.63, No.6.

Dieterich J.H. (1978), Time–dependent Friction and the Mechanics of Stick–slip. Pageoph., V. 116, No.4–5.

Duda S.J. (1965), Secular Seismic Energy in Circum–Pacific Belt. Tectono– physics, V.2, No.5.

Evison F.F. (1977), Precursory Seismic Sequences in New Zealand. New Zealand Journal of Geology and Geophysics, V.20, No.1.

Fedotov S.A. (1965), Regularities of the Distribution of Strong Earthquakes in Kamchatka, the Kurile Islands, and Northeastern Japan. Trudy Inst. Earth Physics Acad. Sci. U.S.S.R., No.36.

Frank F.C. (1966), Deduction of Earth Strain from Survey Data. B.S.S.A., V.56, No.1.

Fujii Y. (1966), Gravity Changes in the Shock Area of the Niigata Earthquake 16 June, 1964. Zisin, V.19. No.3.

Gordon S.S. et al. (1977), The 27 July 1976 Tangshan Earthquake——A complex Sequence of Intraplate Events. IASPEL and IAVCEL Joint General Assemblies Durham, 1977.

Gutenberg B., and C.F. Richter (1954), Seismicity of the Earth and Associated Phenomena. Princeton University Press.

Jaeger J.C. (1956), Elasticity Fracture and Flow with Engineering and Geological Applications. Methuen London.

Johnson A.G., et al. (1973), Water Level Fluctuations on the San Andreas Fault South of Hollister, California. Stanford Univ. Publ., Geol. Sci., V.13.

Kanamori H. (1972), Mode of Strain Release Associated with Major Earthquakes in Japan. Annual Review of Earth and Planetary Sciences. Palo Alto, Calif., V.1.

Knopoff L. (1958), Energy Release in Earthquake. Gophys. J. Roy. Astron. soc., V.1, No.1.

Lagan J.M. (1978), Creep, Stable Sliding, and Premonitory Slip. Pageoph., V.116, No.4–5.

Leonard R.S. et al.(1965), Observation of Ionospheric Disturbances Following the Alaska Earthquake. J.G.R., V.70, No.5.

Luo Huanyan et al. (1983) Finite Element Modeling of Crust Temperature Distribution in the North. China Bay Area, Tectonophysics, 92 T_1–T_9.

Ma Zongjin et al.(1987a), Study on Intermediate–term Earthquake Prediction, Physical and Observational basis for Intermediate–term Earthquake Prediction, Open–file Report 87–591, Menlo Park, California, 1987.

Ma Zongjin, Fu Zhengxiang (1987b), Study on Intermediate–term Earthquake Prediction, U.S.G.S.Proceedings of Workshop XXXVII, 1986, 13–17 Nov., Menlo Park CA, U.S.A.

Mescherikov J.A. (1968), Recent Crustal Movements in Seismic Region, Geodetic and Geomorphic Data, Tectonophysics, V.6, No.1.

Mogi K. (1962), Study of Elastic Shocks Caused by the Fracture of Heterogeneous Materials and Its Relations to Earthquake Phenomena, Bull. Earthq. Res. Inst., V.40., P.125.

Mogi K. (1968), Source Locations of Elastic Shocks in the Fracturing Process in Rocks. Bull. Earthq. Res. Inst., V.46, P.53.

Mogi K. (1974), Active Periods in the World's Chief Seismic Belts. Tectonophysics, V.22, No.3–4.

Mogi K. (1979), Two Kinds of Seismic Gap. Pageoph., V.117, No.6.

Nishimura E. (1953), On Some Destructive Earthquakes Observed with the Tiltmeter at a Great Distance. Disaster Prev. Res. Inst. Kyoto Univ. Bull., No.6.

Papazachos B.C. (1966), studies of Foreshock and Aftershock series of large Earthquakes, Scientific Report, No.5.

Reid H.F. (1910), The California Earthquake of April 18, 1906, V.2: The Mechanics of the Earthquake. The Carnegie Inst. Washington.

Rikitaki Tsuneji (1976), Earthquake Prediction. Developments in Solid Earth Geophysics 9, Elsevier Scientific Publishing Company.

Roussel J. (1973), Les Zones Actives et la Frequence des Seises en Algerie. Bull. Soc. Hist. Nat. Afr. Nord Alger, V.64, Fasc. 3et 4.

Sadovsky M.A. et al. (1979), Short–term Hydrogeodynamic Precursors of Earthquakes. International Symposium on Earthquake Prediction, Paris.

Savage J.C. et al. (1970), Accumulation of Tectonic Strain in California. B.S.S.A., V.60, No.6.

Scholz C.H. (1968), The Frequency–magnitude Relation of Microfracturing in Rock and Its Relation to Earthquake. B.S.S.A., V.58., No.1.

Scholz C.H. et al. (1973), Earthquake Prediction: A Physical Basis. Science, V.181, No.4102.

Soboreb G.A. (1979), Physical Processes Responsible for Earthquakes. International Symposium on Earthquake Prediction. Paris.

Steketee J.A. (1958), Some Geophysical applications of the Elasticity Theory of Dislocations. Can. J. Phys., V.36, No.9.

Stesky R.M. et al.(1974), Friction in Faulted Rock at High Temperature and Pressure. Tectonophysics, V.23, P.177.

Stesky R.M. (1978), Rock Friction–effects of Confining Pressure, Temperature, and Pore Pressure. Pageoph, V. 116, No.4–5.

Sundaran R.N. et al. (1976), Precursory and Coseismic Water Pressure Variations in Stick–slip Experiments, Geology, P.108–110.

Terada T. (1931), On Luminous Phenomena Accompanying Earthquake. Bull Earthq. Res. Inst., Tokyo Univ, V.9.

Utsu T. (1961), A Statistical Study on the Occurrence of Aftershock. Geophysical Magazine, V.30, No.4.

Wunan Lin, Chi–yuen (1981), P–wave Velocities in Rocks at High Pressure and Temperature and the Constitution of the Central California Crust. J.G.R., V.61, P.379–400.

Wyss M. et al. (1973), Earthquake Prediction Based on Station Residual. Nature, V.245, No.5421.

Zhang Guomin et al. (1981), Some Features of the Medium– and Short–term Anomalies Before Great Earthquakes. Third M. Ewing Symposium on Earthquake Prediction, New York, U.S.A.

References in Russian

Kondratenko A.M. (1962), Some Results of Studying of Varying of Velocity of Longitudinal Wave and Ratios Between Velocity of Longitudinal Wave and Transversal Wave in Zone of Seismic Source. Works of the Institute of Physics of the Earth of the Academy of Sciences of U.S.S.R.

Kostrov. B.F. (1975), Mechanics of Seismic Source of Tectonic Earthquake. Science. Moscow.

Kovalevski V.S. (1976), Change of Many Years Fluctuation Level of Underground Water and Underground Flow–off. The Institute of Underground water Problem of the Academy of Sciences of U.S.S.R.

Nikolaev A.B., Kissin I.C. (1984), Hydrogeodynamic Earthquakes Precursors, Akad. Nauk, SSSR, Moscow.

MIX
Papier aus verantwortungsvollen Quellen
Paper from responsible sources
FSC® C105338

If you have any concerns about our products,
you can contact us on
ProductSafety@springernature.com

In case Publisher is established outside the EU,
the EU authorized representative is:
Springer Nature Customer Service Center GmbH
Europaplatz 3, 69115 Heidelberg, Germany

Printed by Libri Plureos GmbH
in Hamburg, Germany